An Introduction to Oekologie

Useful for

B.Sc., M.Sc., Environment NET, SLET, PMT, CPMT, AFMC, AIIMS, BCECE, JCECE, EAMCET, MP-PMT, etc.

Author

Dr. Prasanna Kumar Pankaj (M.Sc., Ph.D., PGDCA)
Lecturer
Department of Botany
S.B. P.G. College, Baragaon, Varanasi, U.P., INDIA.

Co-author

Dr. A.K. Gupta (M.Sc., Ph.D.)
Reader & Head
Department of Botany
S.B. P.G. College, Baragaon, Varanasi, U.P., INDIA.

PILGRIMS PUBLISHING
◆ Varanasi ◆

An Introduction to Oekologie
Dr. Prasanna Kumar Pankaj

Published by:
PILGRIMS PUBLISHING

An imprint of:
PILGRIMS BOOK HOUSE
(Distributors in India)
B 27/98 A-8, Nawabganj Road
Durga Kund, Varanasi-221010, India
Tel: 91-542- 2314060,
Fax: 91-542- 2312456
E-mail: pilgrims@satyam.net.in
Website: www.pilgrimsbooks.com

ISBN: 978-81-7769-716-2

Printed in India at Pilgrim Press Pvt. Ltd. Lalpur Varanasi

CONTENTS

Chapter 1: Ecology and Ecological Factors 1-62

- Introduction 3
- Ecological organisation 3
- Basic concept of ecology 4-5
- Branches of ecology 5-6
- Environmental factors (Biotic and Abiotic) 7
  - Abiotic factors (Climatic, Edaphic & Topographic Factors) 8-9
    - Climatic factors (Light, Temperature, Water and Rain, Humidity, Wind and Atmospheric gases) 10-18
    - Edaphic factors (Mineral matter, organic matter, soil water, soil air, soil organism and soil reaction) 19-35
    - Topographic factors (Altitude, Slope, Steepness, Direction of Slope and Exposure) 36-39
  - Biotic factors (Microorganisms, Plants and Animals) 39-42
- Master Strokes and Entrance corner 42-62

Chapter 2: Ecosystem 63-128

- Introduction 65
- Characteristics 65-66
- Components of ecosystem 66-68
- Trophic level 68-69
- Food chain and food webs 69-73
- Ecological pyramid 73-76
- Energy flow in ecosystem 76-78
- Ecological efficiency 78-79
- Productivity of ecosystem 79
- Productivity levels 79-81
- Measurements of productivity 81-83
- Biogeochemical cycles (Carbon cycle, Oxygen cycle, Water cycle, Nitrogen cycle, Phosphorus cycle, Sulphur cycle) 83-89

- Types of ecosystem (Desert, Grasslands, Taiga, Tundra, Forest, Aquatic, Ocean, Lake and Pond ecosystem) 90-101
- Master Strokes and Entrance corner 101-128

Chapter 3: Plant Communities 129-160

- Characteristics of community 131-133
- Study of community (Analytical characters and Synthetic characters) 133-134
  - Analytical method 134-142
    - (i) Quantitative characters i.e. Frequency, Density, Abundance, Cover and Basal Cover;
    - (ii) Qualitative Characters i.e. Life form, Physiognomy, Stratification, Phenology and Periodicity, Sociability and Vitality)
  - Synthetic method (Presence of Constance, Fidelity, Dominance, IVI and other methods) 142-144
- Master Strokes and Entrance corner 144-160

Chapter 4: Ecological Succession 161-174

- Definition and Types of succession 163-164
- Process of Succession (on Hydrosere and Xerosere) 164-171
- Master Strokes and Entrance Corner 171-174

Chapter 5: Energy Resources 175-198

- Inexhaustible resources (Solar energy, Nuclear power, Geothermal energy) 177-179
- Exhaustible resources – 179-187
  - (i) Renewable resources (Forest resources, Animal power, Ocean energy)
  - (ii) Non-renewable resources (Coal, Natural gas, Oil, Biogas, Hydrocarbon, Hydrogen, Biuomass and Alcohol)
- Master Strokes and Entrance Corner 188-198

Chapter 6: Plant Adaptations 199-220

- Definition and their classification 201
- Hydrophytes (Types, Morphological, Anatomical and Physiological Adaptations) 202-207

- Xerophytes (Types, Morphological, Anatomical and Physiological Adaptations) 208-211
- Halophytes (Classification, Morphological, Anatomical and Physiological Adaptations) 212-215
- Master Strokes and Entrance Corner 215-220

Chapter 7: Soil Erosion and Its Conservation 221-234

- Factors of soil erosion 223
- Types of erosion 223-225
- Methods of soil conservation (Biological, Mechanical and By dry forming practices) 225-231
- Master Strokes and Entrance Corner 231-234

Chapter 8: Plant Indicators 235-242

- Introduction 237
- Types of plant indicators 237-241
- Master Strokes and Entrance Corner 241-242

Chapter 9: Biodiversity and Its Conservation 243-278

- Definition 245
- Magnitude of diversity 245-247
- Necessity of conservation 247-248
- Causes of depletion of biodiversity 248-250
- Red data book and Green data book 250
- Category of threat (Extinct, Endangered, Vulnerable, Rare, Threatened etc.) 251-252
- Conservation of biodiversity (By Environmental legislation, Ex-situ and In-situ conservation) 252-258
- Natioal parks, Sanctuary and Biosphere reserves 258-264
- Hot spots and Types of diversity 264
- Master Stroke and Entrance Corner 265-278

Chapter 10: Environmental Pollution 279-328

- Universe and Solar system 281-282
- Concept of environment (About Abiotic and Biotic component) 283

➢ Lithosphere, Atmosphere and Hydrosphere 283-292
➢ Kinds of Pollution 292-318
- Solid waste pollution – Causes, Effect and Control measurement 292-295
- Atmospheric pollution – Causes, Effect and Control measurement 296-308
- Water pollution – Causes, Effect and Control measurement 309-315
- Radio active pollution – Causes, Effect and Control measurement 315-316
- Noise pollution – Causes, Effect and Control measurement 317

➢ Master Stroke and Entrance Corner 318-328

PREFACE

As per latest U.G.C. guidelines for model curriculum for Indian universities, the present book on "**An Introduction to Oekologie**" has been written in simple, lucid and easy to understand language. The book has been profusely illustrated with self-explanatory labelled diagrams and aims to catering the needs of undergraduate and postgraduate students and those who are appearing for competitive examinations like- NET, SLET, PMT, CPMT, AFMC, AIIMS etc. The special features of the book are- Master strokes, Entrance corner and recent information about each topic.

We are extremely thankful to Dr. Shatrughana Singh, Reader, Department of Botany, Udai Pratap Autonomous P.G. College, Varanasi; Dr. M.P. Singh, Reader, Department of Botany, Udai Pratap Autonomous P.G. College, Varanasi; Dr. Ajit Kumar Srivastava, Sr. Environmental Engineer, Allahabad; Dr. Kamalesh Kumar Srivastava, Reader, Department of Botany, Government P.G. College, Chunar, Mirzapur; Dr. Vinita Singh, Reader, Department of Botany, S.B. P.G. College, Baragaon, Varanasi; Dr. Sudha Rani Govil, Lecturer, Department of Botany, S.B. P.G. College, Baragaon, Varanasi; Dr. Kanak Manjari, Lecturer, Department of Botany, S.B. P.G. College, Baragaon, Varanasi; Mr. Piyush Mishra, Lecturer, Department of Botany, S.B. P.G. College, Baragaon, Varanasi and Dr Nandlal Singh for their cooperative and eager efforts for the timely release of this book.

Suggestions for further improvement of the book are expected and solicited.

Authors

Chapter 1
Ecology and Ecological Factors

ECOLOGY AND ECOLOGICAL FACTORS

- In 1859, French zoologist **Geoffroy St. Hillare**, had proposed the term **ethology**, According to him "ecology is the study of relations which exist between the organisms and their environment".
- In 1894, English naturalist **St. George Jackson Mivart**, coined the term **hexicology** and his definition is similar to "Hilare".
- The term **ecology** was **proposed** for the first time in 1866 by a German biologist "**Earnst Haeckel**" in the book "**Generalle morphologie der organismen**".
- The word **ecology (Greek Oekologie)**, first "**coined**" by "**Hanns Reiter**" in 1885, is derived from Greek words, Oikos = meaning house or dwelling place and Logos = the study, hence the study of living beings in their natural habitats (house) is called ecology.
- In 1893, "The Madison Botanical Congress" removed the word "O" from "**Oekologie**" and it was englishized as "Ecology", so the main credit goes to Reiter. So, "ecology is the study of interrelationship between living organisms and their environment".

Ecology is the branch of Botany, but it differs in three aspects from other branches-

i) It also includes the abiotic component
ii) Ecological studies are made above the organismic system.
iii) There are two factors in nature, they are extrinsic and intrinsic. In ecology, only extrinsic factors are studied.

ECOLOGICAL LEVEL OF BIOLOGICAL ORGANISATION

The various levels of biological organisation are called **biological spectrum,** while due to the interaction between abiotic and biotic systems, a system develops which is called **Biosystem**. Entire living components of the planet earth are called **Biota** or **Biome**.

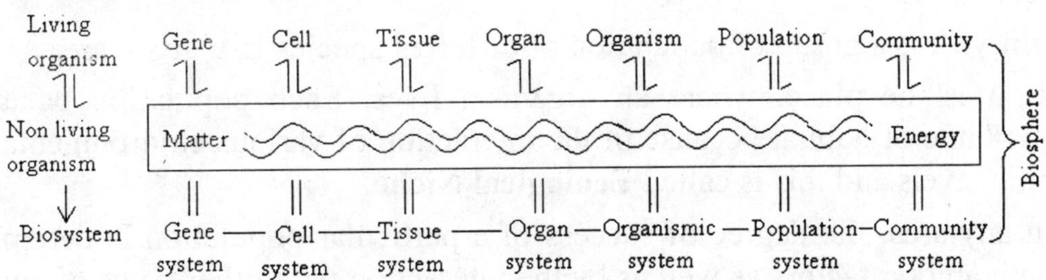

Fig. showing level of organisation (After E.P. Odum)

Basic Concepts of Ecology

A. Structural concept (Descriptive Ecology):

- All the living organisms and their surrounding environments mutually react with each other and affect each other in different ways.
- Any constituent or condition of environment which affects the organisms in any ways is called as environmental factor.
- Environment is much dynamic and works as a sieve for the selection of organisms for the growth, from so many forms.
- All the species of organisms make effort to maintain its proper structure, function, reproduction, growth and development by preservation of its **genetic pool**, but most of organisms have a sufficient **plasticity** to modify their morphology as well as genetic characters according to the changing environment, and are called as **Ecads** and **Ecotypes** respectively. So, we can say that the species can increase the capacity of tolerance by developing Ecads and Ecotypes.
- By the activities, organisms can modify their environment to make it suitable for their healthy growth. Modified environment becomes less suitable for the existence of plant communities, as a result, other plant communities are invited which also change the environment.

"The development of communities over a period of time at the same site is called **succession**". The process of succession and change in environment continues till equilibrium is established between the changed environment and a community and that community is called **climax community**.

B) Functional concept (Functional ecology):

Ecosystem: Ecosystem is a segment of nature, consisting of living organisms and their physical environment, both interacting with each other and exchanging materials between them. It is the structural and functional unit of nature.

Population: Population is a group of individual organisms of the same species in a given area.

Community: It is a group of population of different species in a given area.

Habitat: It is the place, where an organism lives. Each population occupies the specific volume of habitat because of the interaction of various environmental factors and trophic levels and this is called **Ecological Niche**.

- In any area, the degree of success of a particular population is determined by both abiotic factors as well as by the interaction with other types of population. This interaction may be positive, negative or neutral.

- The energy provided by Sun is a driving force for the ecosystem. The flow of energy in an ecosystem is **unidirectional** or non cyclic in nature.
- The inorganic substances required for the synthesis of organic substances are called inorganic nutrients or **biogenetic** substances. These nutrients circulate between biotic and abiotic components of the ecosystem. This phenomenon of circulation of nutrients is called **Bio-geo-chemical cycle**.
- Healthy and successful growth of an organism is regulated by **limiting factors**. For example, tolerance of plants to different temperature range has been shown in the following figure-

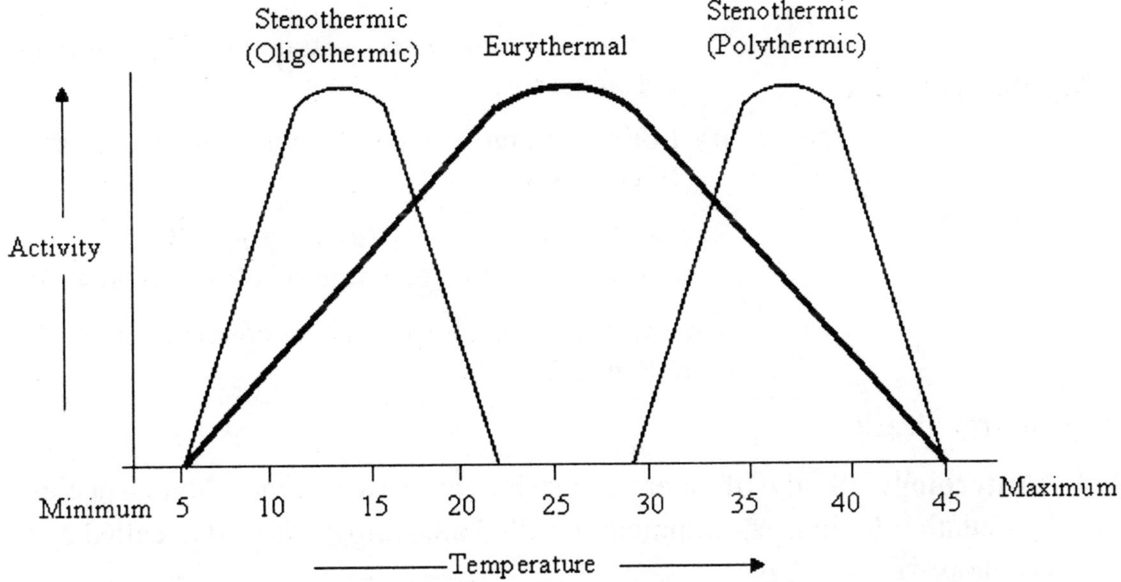

Fig. showing range of different temperature tolerance of plants

C) Evolutionary concept (Evolutionary ecology):

Now-a-days, organisms present on this earth are the product of evolution of the past organisms. By knowing evolutionary aspects, we can guess the future shape of different environmental conditions.

Branches of Ecology

A) On the basis of taxonomy-

i) Plant ecology

ii) Microbial ecology

iii) Animal ecology

B) On the basis of habitat-

i) Terrestrial ecology
- Grassland ecology
- Forest ecology
- Desert ecology

ii) Aquatic ecology
- **Fresh water ecology**- When salt concentration is less than 0.5%, the water is called as fresh water and the study of fresh water ecology is called as limnology. It is of two types-
  i) Lentic – When water is stagnant and the ecosystem is **Lacustrine** type
  ii) Lotic – When water is flowing and the ecosystem is **Riverine** type
- **Brackish water ecology (estuary)**- Salt concentration is in between 0.5-3.5%, ecosystem is **Palustrine** type
- **Marine water ecology**- When concentration of salt is more than 3.5%

Level of organisation

1. **Autecology**: Study of interrelationship in between individual species or its population in their environment is called autecology. It is also called as species ecology.
2. **Synecology**: Study of interrelationship between community and their environment is called synecology.
3. **Population ecology**: Member of a species occurring in an area form population, and its study called population ecology. It deals with growth, density, size, multiplication, natality (Birth), mortality (Death) and tolerance etc.

   There are two types of growth rate shown by population i.e.-

1) **J-shaped**: In this, growth is in inexponential fashion (2, 4, 8) i.e. the growth is density independent and such populations are unstable.
2) **S-shaped**: It depends upon density i.e. population is stable.

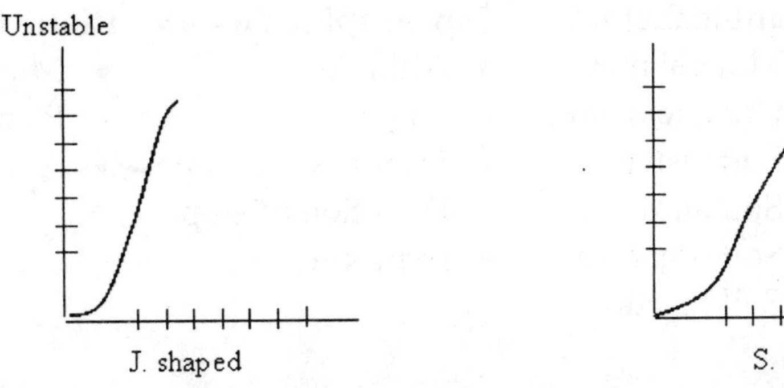

Fig. showing pattern of growth rate

3) **Gene ecology or Genetic ecology**: Study of genetics at genetic level is called gene ecology.

Ecads or Ecophene: In this, variations are not genetically fixed, they are only due to environmental conditions; and when species of different environment are grown in uniform condition, they become similar.

Ecotypes: In this, variations are genetically fixed and plants maintain their characters even when they are grown in uniform condition.

Ecotone: The transition zone (meeting point) between two ecosystems is called as ecotone e.g. the meeting point of fresh water ecosystem and marine water ecosystem is called as **Estuary** containing brackish water. As a rule, number of species in any ecotone is greater than either of the ecosystem and this phenomenon is called as **edge effect**.

Ecocline: The change takes place from one ecotype to other, is gradual and intermediate forms of ecotypes are called Ecocline.

ENVIRONMENTAL FACTORS

Everything which surrounds any organisms is called environment. The environment of an organism may be subdivided into two components-

1) Physical or abiotic or non-living component
2) Biotic component

Any constituent or condition of the environment which affects directly or indirectly to the form and functioning of the organism in any specific way is called environmental or ecological factor. It is of following types-

1) Climatic factors
2) Edaphic factors
3) Topographic factors
4) Biotic factors

} Related to Abiotic component

Climatic factors	Edaphic factors	Topographic factors	Biotic factors
• Light	• Mineral matter	• Altitude	• Micro-organisms
• Temperature	• Organic matter	• Slope	• Plants
• Water and Rain	• Soil water	• Steepness of slope	• Animals
• Humidity	• Soil air	• Direction of slope	
• Wind	• Soil organism	• Exposure	
• Atmospheric gases	• Soil reaction		

Climatic Factors

Climate is the one of the most important natural factor which controls the growth and development of plants. Its' study is called climatology. The climate includes the following main factors-

A. Light

Light is one of the most important environmental factors influencing various activities of both plants and animals. The main sources of natural light are sunlight, moonlight and light producing organisms. Of these, sun light has the greatest ecological significance. Light may be defined as the "visible part of spectrum of electromagnetic radiation". The radiation travels in the form of waves, which may be longer as well shorter. Shorter wavelengths have greater energy in their photon and they are also refracted more than longer wavelengths.

Visible radiation = 390-700 nm of wavelength

[Higher wavelength] IR-radiation = More than 700 nm

[Shorter wavelength] UV-radiation = 100-390 nm of wavelength

UV-rays have the greatest killing effect on the protoplasm. It helps in the synthesis of anthocyanin in leaves and inactivates the growth hormones. Microbes injured by UV-rays, X-rays, γ-rays, show high degree of ionising effect and cause mutation in living system.

Earth is a very small object in the solar system and it receives only 1-50 millionth of the total solar radiation. Out of total solar energy, only 2% is used in photosynthesis.

Depending upon the penetration of light, oceans are divided into the following three zones-

1) **Euphotic zones (Upto 80 metres depth)**: The rays of longer wavelengths are absorbed near the surface, while rays of shorter wavelength penetrate deepest, IR rays penetrate up to 4 metres, red and orange colour light penetrates upto 20

metres and yellow rays upto 80 metres, hence plants are unable to photosynthesize below 80 metre depth.

2) **Disphotic zone (80 to 200 metres depth)**: Green and blue light penetrate up to 100 metres depth and UV up to 200 metres depth. Both euphotic and disphotic together known as **Photic zone.**

3) **Aphotic zone (below 200 metres of depth)**: No light penetrate beyond 200 metres depth.

Instruments for measuring light

Photometer or Luxmeter: By this instrument we can measure the intensity of light.

Radiometer: Used for measuring intensity of light.

Pyranometer: Used for measuring total quantity of light.

Sun-shine recorder: Duration of light is measured by this instrument.

Secchi disc: Used for measuring turbidity of water.

Effect of light

1) **Chlorophyll synthesis**: Light is essential for chlorophyll synthesis except in some plants, like – algae, ferns and few conifers.

2) **Photomorphogenesis**: Organ differentiation in plants is called morphogenesis. If there is role of light in organ differentiation then it is called photomorphogenesis. In presence of light, proplastid is converted into plastid but in absence of light, protoplast is converted into "**etioplast**" and the plant is called "**Etiolated plant**" (Because in absence of light, seedlings develop a non green (yellowish), hooked appearance, long weak internodes, unexpanded leaves and poorly developed roots).

3) **In seed germination**: In majority of plants, seed germination is independent of light condition, such seeds are called as **non photoblastic seeds**. In some plants, seed germination is dependent on light conditions, such seeds are called as **photoblastic seeds**. The photoblastic seeds are of two types. If light is must for seed germination, it is called as **positive (+ve) photoblastic** e.g. *Viscum, Lactuca* etc. and if seeds fail to germinate in light condition are called **negative (-ve) photoblastic** e.g. Onion, Tomato.

4) **In closing and opening of stomata**: Light plays an important role in opening and closing of stomata.

5) **Bending movement**: At the tip of the plant, growth hormones- Auxins are found, so when the light reaches, the plant moves at that side where light reaches directly; this movement of plant in presence of light is called Bending movement. Generally, the yellow and green colours affect the plant movement.

6) **Arrangement of chloroplast in Mesophyll cells**: When the high intensity of light falls on mesophyll cells, the chloroplast is arranged towards the radial wall of mesophyll cells, but at low intensity, the chloroplast is arranged on upperwall and in diffused light, the chloroplast is distributed uniformly in mesophyll cells. This type of arrangement of chloroplast is called **apostrophe, epistrophe** and **parastrophe**, respectively.

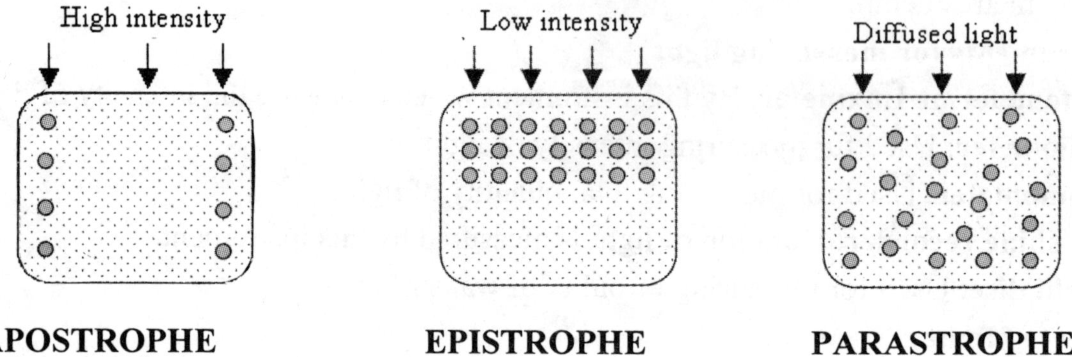

Fig. showing role of light in arrangement of chloroplast in mesophyll cells

7) **Distribution of plants**: On the basis of their relative light requirements, the plants can be classified into two categories:

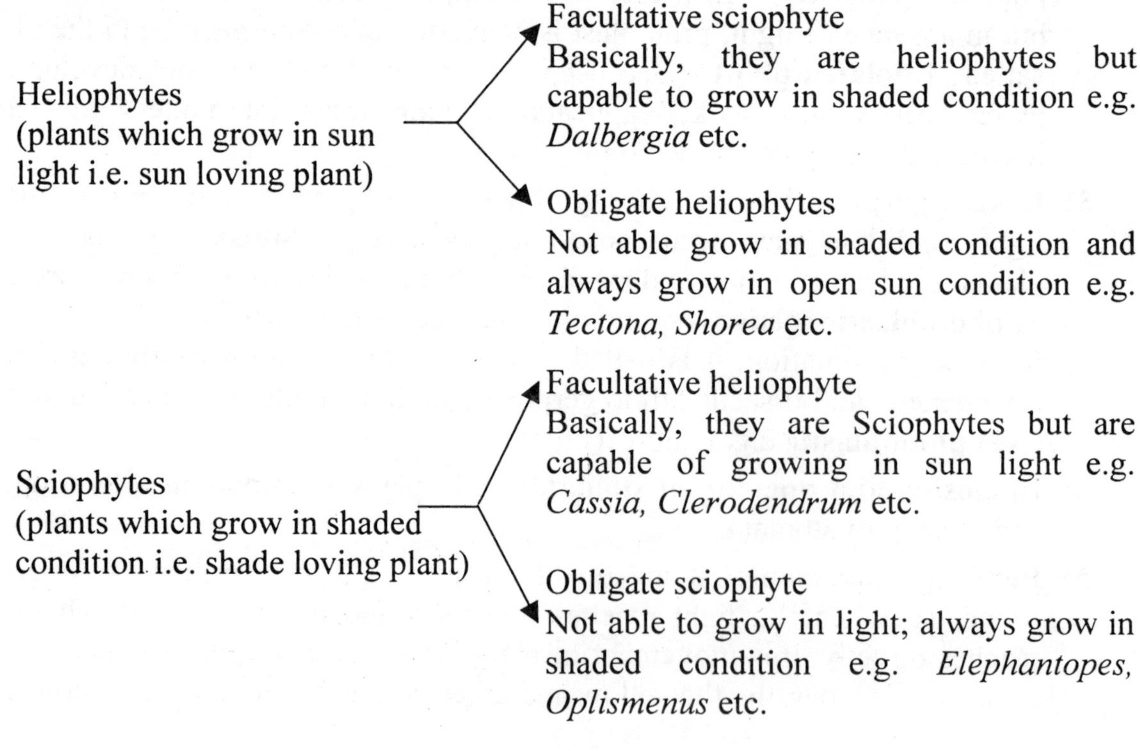

Photophilous or Heliophilous or Sun loving plants	Sciophilous or Heliophobous or Shade loving plants
• Internodes are short	• Internodes are long
• Profusely branched	• Sparsely branched
• Less chloroplasts	• Chloroplasts are more
• Having more dry weight	• Having less dry weight
• Plants are resistant to temperature and draught	• Plants are less resistant to temperature and drought

8) **Photoperiodism**: Duration of light responsible for flowering, fruiting and their geographical distribution are called as their "**Critical photoperiod**" and the process is known as photoperiodism. On the basis of photoperiodism, the plants can be classified into 5 categories-

i) **Short day plants**- In this, flowering occurs when day length is below the critical period, eg. *Xanthium* sp., *Saccharum officinarum*.

ii) **Long day plants**: In this, phoroperiod required is more than the critical photoperiod, eg. *Brassica* sp., *Raphanus sativus*.

iii) **Short-long day plants**: They require short photoperiod for floral initiation and long photoperiod for blossoming, eg. *Tropiolum* sp.

iv) **Long-short day plants**: They require long photoperiod for floral initiation and short photoperiod for blossoming, eg. *Bryophyllum* sp.

v) **Day neutral plants**: These plants are not sensitive to photoperiod for flowering, eg. *Lycopersicum esculentum*, *Gossypium herbaceum,* etc.

B) Temperature

All the physiological and biological reactions in animals are affected very much by the variation in the temperature of the surroundings. Temperature also affects the morphology and phenology of plants like- seed germination, vegetative growth, flowering, fruiting, seed dispersal, transpiration, respiration, photosynthesis and distribution of plants.

According to **Vant Hoffs law,** all plants grow within a limited range of temperature. There are three types of temperature limits-

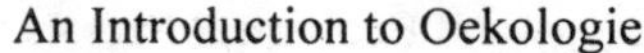

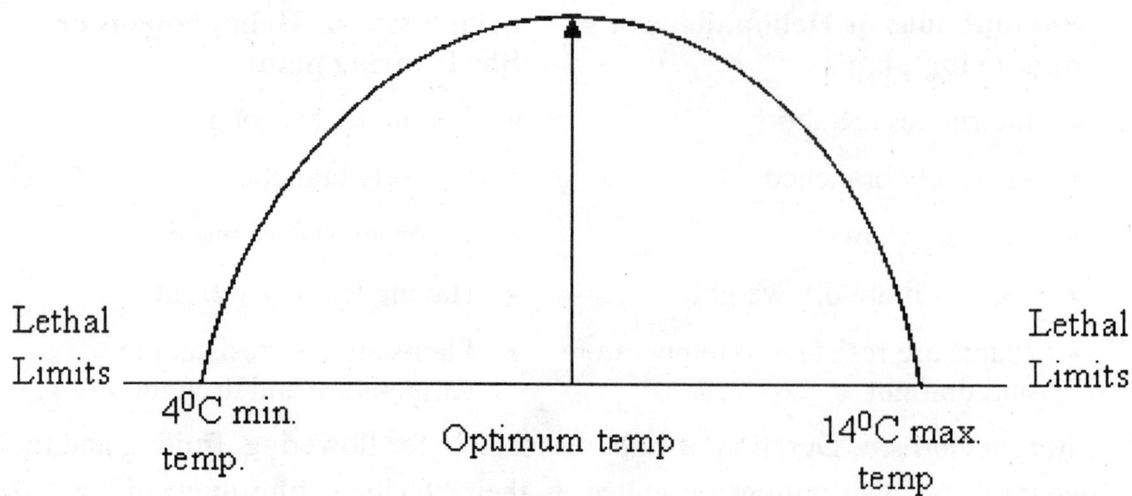

Fig. showing different temperature limits

i) Minimum temperature
ii) Maximum temperature
iii) Optimum temperature

The limit of temperature between minimum to optimum is called as '**lethal limits**', if the temperature is constant for particular species, that temperature limit is called as "**Cardinal temperature**".

According to temperature limits, two types of plant species are found-

i) **Urythermic species**: Those species, which have broad range of temperature tolerance are called Urythermic species e.g. *Artimitia* sp.

ii) **Stenothermic species**: Those species, which grow in narrow temperature tolerance range, are called as Stenothermic species. Stenothermic species are of two types-

(a) **Oligothermic species**: Those species which grow at narrow temperature tolerance range towards low temperature zone are called as Oligothermic species e.g. *Chlamydomonas nivalis*.

(b) **Polythermic species**: Those stenothermic species having narrow temperature tolerance range towards high temperature zone are termed as polythermic species e.g. species of Palm.

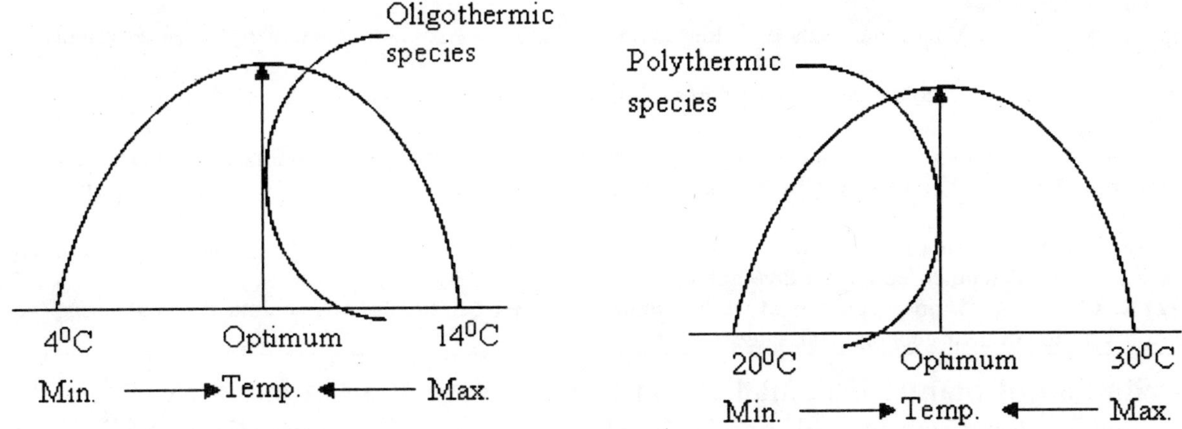

Fig. showing temperature tolerance range for oligothermic and polythermic species

2) Thermal stratification: Thermal stratification is found in aquatic bodies. The upper surface of water is called "**epilimnion**", in this zone temperature is variable, and this variation may be seasonal, diurnal or it may be spatial. The lower zone is called "**Hypolimnion**", where temperature is almost constant. In between upper and lower zones, a transition zone takes place, which is called **Metalimnion**; it is equivalent to compensation point.

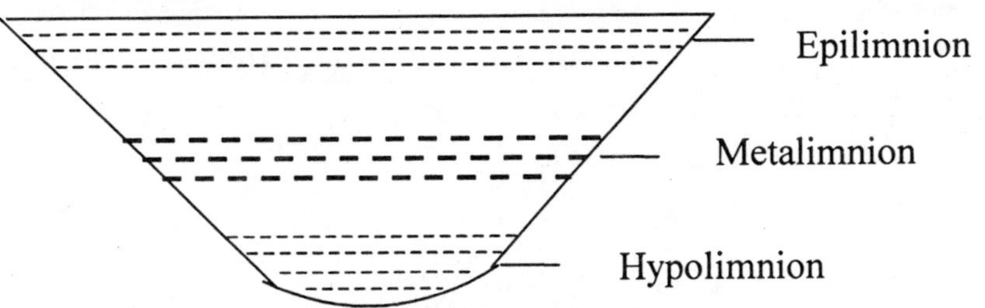

Fig. showing thermal stratification

3) Thermoperiodism: The effect of change in temperature on the phenological behaviour of plant is called thermoperiodism.

Vernalization: In annual plants, light is important for flowering and temperature is secondary factor, but for biennial species, prolonged exposure to cold temperature is required for flowering which is called as vernalization eg. wheat. **Lysenko** in **1938** used the term vernalization for the first time.

Very low temperature causes injuries to the plant. When injury is caused by low temperature (0°C), then it is called "**chilling injury**", but when injury is caused by the temperature below than 0°C, is called as "**Frost injury**".

Some important rules-

Bergaman's rule: Birds and mammals of colder areas are larger in size as compared to their equivalents in warmer areas.

Rensch's rule: Birds of colder areas have narrow and acuminate wings while those of warmer areas have broader wings.

Jordan rule: Fish size as well as number of vertebrate increase in colder areas as compared to warmer areas.

Allen's rule: Animals of colder areas have shorter extremities (e.g., tail, ears, head) as compared to animals to warmer areas.

Vant Hoff's law: The rate of respiration is doubled (like in animals) at the increase of 10°C above the optimum temperature, provided other factors are favourable.

Gloger's rule: The birds and mammals of warm humid regions tend to be darker in colour than inhabiting the cold or dry region of their geographical range.

Distribution of plants: If rainfall is constant, the vegetation of any area is directly governed by temperature, this condition is called as "**Schimper's first law**". Four climatic and vegetation zones have been recognized on the basis of mean temperature along the latitude

Sl.No.	Zone	Latitude	Mean Annual Temperature	Winter	Vegetation
1	Tropical climate	0°-20°	Above 24°C	No winter	Tropical forest
2	Sub –tropical	20°-40°	17-24°C	Mild winter	Sub-tropical deciduous forests
3	Temperate	40°-60°	7-17°C	Winter with snow	Coniferous forest
4	Arctic and Antarctic	60°-80°	Below 7°C	Winter with heavy snow	Arctic vegetation

Above forest types are found, due to the effect of temperature and governed by Schimper's first law.

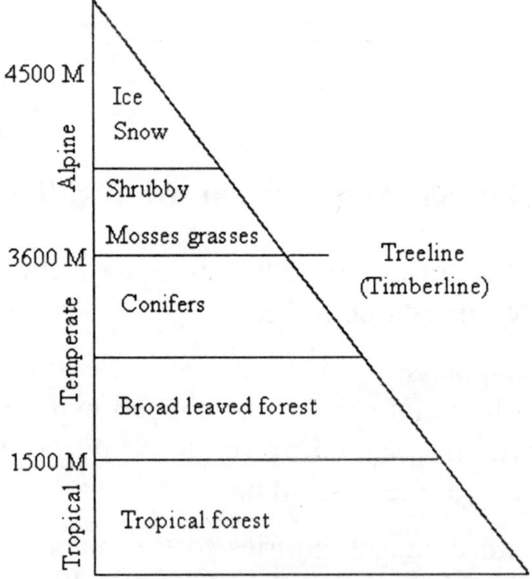

Fig. showing types of vegetation at different heights

If we move towards the higher altitude, above 10 thousand feet, generally woody plants are not found and this height is called "**Tree line**". Above 13 thousand feet almost all the vegetation is absent, only some specialized **algae** and **lichens** may be present, and this height is called "**Snow line**".

Elfine scrub: It has been observed that, if trees are present above the 10 thousand feet of height, they are very dwarf and succulent and show shrubby appearance due to snow fall and high wind velocity, this type of vegetation is called "**Elfine scrub**".

On the basis of temperature condition, world's vegetation is classified into following categories-

i) **Megatherms**: The vegetation growing in the condition in which high temperature prevails throughout the year (30-45°C). The dominant vegetation is **tropical rain forest**.

ii) **Mesotherms**: The vegetation growing in the condition in which high temperature (30-45°C) alternates with low temperature (20-30°C). The dominant vegetation is **tropical deciduous forest**.

iii) **Microtherms**: The vegetation growing in the low temperature (10-20°C) conditions. The vegetation is "**mixed coniferous**" type.

iv) **Hekisotherms**: The vegetation growing in the very low temperature (0-10°C) conditions. In this condition, the dominant vegetation is "Alpine" type. The ratio of rate of respiration at the difference of 10°C is called **temperature coefficient** (Q_{10}). Rate of respiration doubles with the increase of 10°C above the optimum temperature, if other factor is favourable. It is "**Vant Hoff's**" law.

Thermal constant: Thermal constant is the minimum amount of heat unit which is required before flowering. The minimum effective temperature at which the seed germinate is called **plant zero**, it is 37°F (2.8°C) for spring wheat, 55°F (12.8°C) for maize and 62°F (16.7°C) for cotton.

The difference between the plant zero and the subsequent mean of daily temperature is counted in Fahrenheit degrees and called "**Heat units**".

C) Precipitation

Precipitation includes all moisture that come to the earth in the form of rain, snow, hail, dew etc. In nature, there is constant interchange of water in between earth surfaces and atmosphere. This interchanged cycle is called **water cycle** or **Hydrologic cycle**. In hydrologic cycle, two temperature events are involved-

i) Precipitation
ii) Evapo-transpiration

Precipitation: Precipitation takes place in different forms e.g. Drizzle, rain, snow, dew, frost and hail.

Drizzle: Minute water drops, floating in the air are called drizzle. If the water droplets are larger and heavier than drizzle, it is called Rain.

Snow: Moisture in solid state is called as snow. It is angular in shape.

Dew: Dew is formed due to condensation of moisture, directly on the surface of the object. When the temperature of air it below 32°F, it causes freezing and called as frost or frozen dew.

Sleet: When the moisture in solid state falls in the form of small grains or pellets of ice, it is called sleet.

Hail (Ola): When the moisture in solid state falls in the form of balls or lumps of ice, it is called Hail.

Mist (Kuhra): It is less dense and drops are smaller than rain.

Fog (Dhundh): It occurs due to cooling of highly humid air near the earth surface, forming cloud like appearance and disappears at high temperature. Water determines the distribution, structure and functions of organisms in any ecosystem. One very important characteristics of water is, that it is **Universal solvent** i.e. why it is important for cycling of matter in the ecosystem and it is also much important for cycling of matter in the ecosystem and it is also much important for photosynthesis.

According to **Hutchinson 1957** total quantity of water present on earth is 2,86,000 geogram [1 geogram = 10^{20} gm] in which 97% is present in vast ocean, 2% in form of snow and only 1% is fresh water.

Rainfall: Water is essential for life. In India about 80% water is used for irrigation, 50% are used as drinking water, and for thermal power and 1.2% of water is used for domestic purposes, while 12% are used for other purposes.

Rain shadow: Rainfall is determined largely by geograph and large air movement pattern of weather system. When the moisture laden winds blow from the ocean towards the high mountain, they deposit most of the moisture on the ocean facing side of mountain and are called seaward or **windward side** or slope and its opposite side is called **Lee side**; this phenomenon is called "**Rain shadow effect**".

According to **Schimper's** third law, the types of vegetation in temperate and tropical zone are determined by amount of rain fall and humidity of air. If the heavy rain fall takes place throughout year in warm tropical region, it produces "**Evergreen rain forest**". Heavy rainfall, but confined to a few months, results **deciduous forest**. The area with little precipitation in summer and sufficient in winter produces **Scrub**

forest bearing **sclerophyllous trees**. The area with heavy rain fall in summer and little in winter produces "**grassland**". Heavy rainfall and high temperature in equatorial and tropical zones cause **Tropical rain forest**. In case of forest, most of the rainfall is retained by the tree canopy and this phenomenon is called as "**Canopy interception**" or if it is in presence of vegetation, direct rainfall on soil surfaces is not possible. The rain drops fall at the surface of leaf is called as "**Canopy interception**". In this condition, water comes down very slowly along with surface of the stem and called as "**stem flow**", as a result the chances of erosion are very low.

Shola forest: Shola forest is common in India around Nilgiri hills. The plants have dense canopy due to the presence of leathery leaves, so the rain water comes in the form of small droplets called "**forest rain**". It also determines the line of vegetation known as "**Schimper's second law**".

Annual rain fall	**Vegetation**
1. 0 to 13.24 cm/annum	Desert
2. 13.25 to 35.1 cm/annum	Semi arid grassland
3. 35.2 to 63.5 cm/annum	Dry subtropical grassland, savanna (community between grassland and forest i.e. a grassland with scattered trees)
4. 63.6 to 114.3 cm/annum	Humid subtropical forest
5. 114.4 to 203.2 cm/annum	Tropical rain forest

The ratio of $\frac{\text{Rainfall (R)}}{\text{Evaporation ratio (E)}}$

i) If R/E is 0 = Desert
ii) If R/E is 0.2 to 1.0 = Grassland
iii) If R/E is 1.0 to 1.6 = Subtropical forest
iv) If R/E is more than 1.6 = Tropical forest

D) Wind

The moving air is called as wind. Different terms are given for different wind velocity-

High air ⇒ 1.5 – 4.5 km/hr
Breeze ⇒ 6 – 50 km/hr
Gale ⇒ 51 – 102 km/hr
Storm ⇒ 103 – 115 km/hr
Hurricane ⇒ More than 115 km/hr.

The dust particle carrying wind is called "**Dust storm**" and snow carrying wind is called "**Squall**". Wind has two types of effect on plant life-

1) **Direct effect**: It has mostly mechanical effect. Some of them are as follows-

i) **Lodging effect (flattering of plants e.g. cereals)**: Lodging is a type of wind injury, common in grasses, in which plant falls flat on the ground.

ii) **Breaking and Uprooting**: The plants, which have soft tissues, wind of high velocity breaks the branches and sometimes uproot the complete plant. However, such effects are uncommon in forests where the canopies of individual tree at different heights reduce the velocity of wind to about 80%. Such vegetation in forests serves as natural "**wind breaks**".

iii) **Deformation**: Strong wind froms a constant direction, causing permanent alteration in the form and position of the shoots. This alteration is called "**deformation**". It is common in trees, growing on ridges and along the coasts; such type of tree is called the "**Flag tree**".

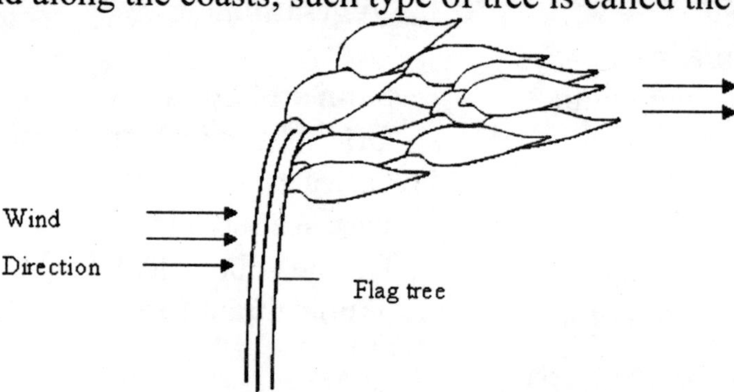

Fig. showing effect of wind on plant

iv) **Abrassion**: It is caused by ice **or** sand carrying winds. It removes the buds and barks of the trees; the injured bark causes desiccation and infection of different kinds of pathogens.

v) **Dwarfing**: At high altitude **or** at sea-shores, the plant body becomes dwarf due to adverse mechanical and physiological condition.

vi) **Rainfall**: Wind is very essential for movement of cloud and rain fall in different parts of world.

vii) **Absence of zoophilous plants**: Those plant species in which pollination takes place by insects **or** birds are not found in those areas having high wind speed.

viii) **Anemophily**: The wind is helpful in pollination and dispersal of fruits and seeds.

ix) **Soil erosion and deposition etc.**: In bare deserts, strong wind causes constant **wind erosion,** which makes soil unfit for the growth of plants. It can be reduced by plantation of "**wind break**" or "**shelter belts**".

2) Physiological effect:

i) **Desiccation**: High evaporation and transpiration due to high wind velocity result in the failure of plants in maintaining the internal water balance and thus they suffer from desiccation.

ii) **Dwarfing**: Dehydration and consequent loss of turgidity result in the dwarfing of plant organs generally in trees on sea coasts, arctic and alpine timberline.

iii) **Salt spray**: Salts of the water are carried by strong winds in the vicinity of the ocean. Such salts have an injurious effect on plants growing there.

iv) **Compression wood**: As a result of wind deformation, trees develop a dense and reddish colour xylem called "**compression wood**".

E) Atmospheric gases

Heavy gases (Amount in volume) = N_2 = 79.03%, O_2 = 20.99%, Ar = 0.93%, CO_2 = 0.03% and H_2 = 0.01%

Light gases (Amount in volume) = Ne = 0.0018%, He = 0.0005%, Cr = 0.0001%, O_3 = 0.000004%

EDAPHIC FACTOR

Factors related with soil formation, physical and chemical properties of soil are known as "**edaphic factors**". The process of soil formation is known as "**Pedogenesis**", and its study called "**Pedology**".

Earth is a spherical planet of the solar system. It is differentiated into three main layers-

i) **Crust**: Outermost solid zone; about 9-40 km thick. Surface is covered by soil.

ii) **Mantle**: Middle layer; about 2900 km above core.

iii) **Core**: Inner layer; about 2500 km. The magnetic character of soil is due to presence of Fe and Ni in core.

Warming (1909) classified the different kinds of vegetation on the basis of soil types-

Halophytes – on saline soil
Psychrophytes – on cold soil
Lithophytes – on rock
Chasmophytes – on rock crevices
Chersophytes – on waste lands
Eremophytes – on desert
Oxylophytes – on acidic soil
Calciphytes – on basic soil
Psilophytes – in Savanna
Psammophytes – on sand

Soil properties

- Soil separates
- Textural composition
- Soil structure
- Density of soil
- Components of soil
- Inorganic constituents
- Organic matter
- Soil water
- Soil air
- Soil organisms
- Soil moisture constants
- Soil reaction and soil pH

1) **Soil separates**: To study the mineral particles of soil, scientists separate them into groups according to the size and these groups of the particle are referred to as "**soil separates**". Gravel, sand, silt and clay are the chief soil separates.

2) **Soil texture**: Soil texture denotes, relative proportion of different types of soil particles (soil separates). Growth of plant is directly influenced by the size of particles. Various relationships that exist between plant and soil are mainly controlled by soil texture. The absorption and retention of water, the circulation of water and air in the soil etc. are determined by soil texture.

There are three broad textural classes recognized, i.e. sands, loams and clays, the size of different kinds of soil's particles are as follows-

i) Gravel = >5.000 mm in diameter

ii) Fine gravel = 2.000 mm to 5.000 mm in diameter

iii) Coarse sand = 0.200 mm to 2.000 mm in diameter

iv) Fine sand = 0.020 mm to 0.200 mm in diameter
(Having minimum water holding capacity)

v) Silt = 0.002 mm to 0.020 mm in diameter

vi) Clay = <0.002 mm in diameter
(Having maximum amount of water holding capacity)

Types of soil

In soil, the soil particles are present in different ratio. On the basis of percentatge of different particles, soil can be classified into following categories:

a) **Sandy soil**: Having 85% sand and 15% clay and silt. Except this, having the particles of **Mica**, **Felspar** and **Quartz**. Having low water holding capacity and nutrient. Due to intercellular spaces, water percolates down. Generally xerophytic plants can grow in such soil.

b) Silt soil: Having 90% silt + 10% sand. It is not good for agricultural purposes and it is poor in aeration.

- **i) Silt loam**: 30 to 50% sand + 50% silt + less than 2% clay
- **ii) Silt clay**: Less than 20% sand + 59-70% silt + 30-50% clay
- **iii) Silt clay loam**: 30% sand + 50-80% silt and clay.

c) Clay soil: 50% clay + 50% silt + some % of sand. Having maximum water-logging capacity, less CO_2 and maximum O_2, it is better for plants.

- **i) Clay loam**: 20-50% sand + 20-50% silt
- **ii) Sandy clay loam**: 50-70% sand + 10-15% silt.

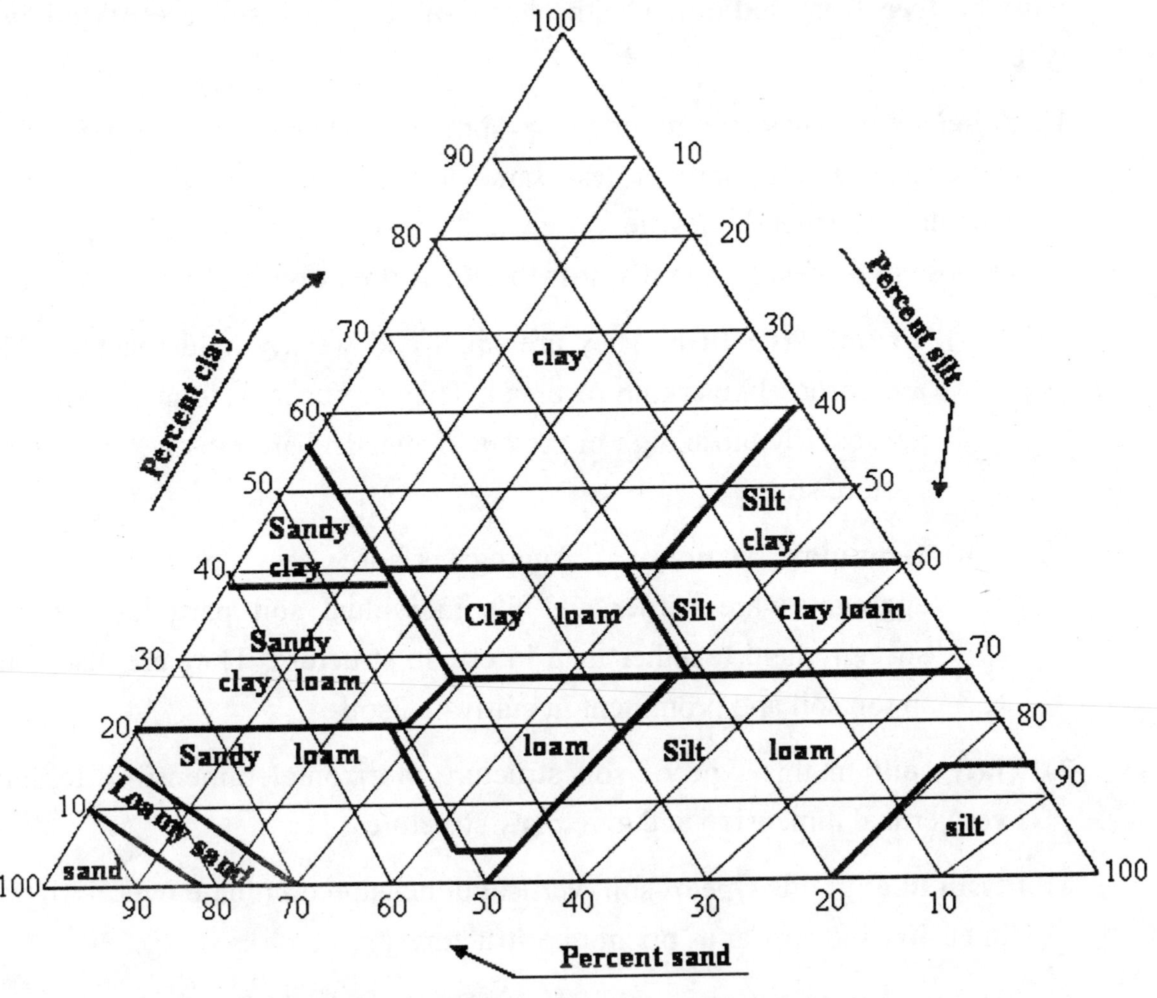

Fig. Textural process of soil

d) **Loam soil**: 30-50% sand + 30-50% silt + 20% clay and 10-15% humus, having moderate fertility and water holding capacity.

   i) **Loamy sand**: 80 to 85% sand + clay
   ii) **Sandy loam**: 50 to 80% sand + clay
   iii) **Sandy clay loam**: 50 to 80% sand + 20-30% clay.

3) **Structure of soil**: Different mineral particles are found in the form of aggregates and these aggregates form a definite soil structure for proper development of soil structure. Soil must contain enough organic matter and it must be **free** from **sodium**. On the basis of structure, soil are of following types-

   1) **Spheroidal**: They are rounded **or** sphere in shape. All the axis of the aggregates are of more or less same length with their faces curved or irregular. In general, they lie loosely and are separated from each other. It is suppose to be best for plant's growth. It is further divided into two types-

      a) **Crumb structure**: They are small and weakly held together. They are porous like crumb of bread. They are usually found in top soils, particularly those high in organic matter and are mostly prominent in pasture soils.

      b) **Granular structure**: Aggregates are less porous. Granular aggregates are harder and the individual soil particles are more strongly held together than in crumb structure. They are also found in top soil and prominent in cultivated soil.

   2) **Platy soil**: In this type of soil structure, horizontal dimension dominate over vertical dimension and give platy structure.

   3) **Prism like**: In this type of soil, vertical dimension dominate over horizontal dimension and appear as prism like structure.

   4) **Blocky soil**: In this type, soil is isodiametric with flat faces to appear cubical structure.

1. SPHEROIDAL

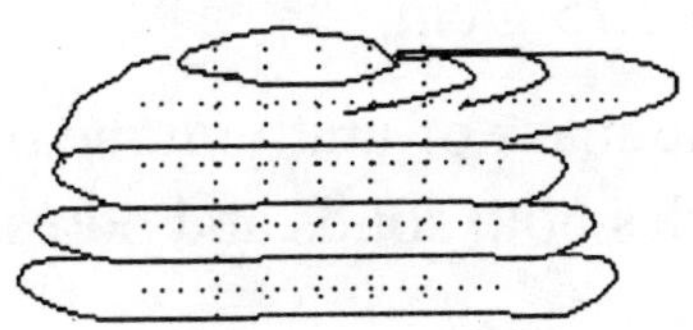

2. PLATY

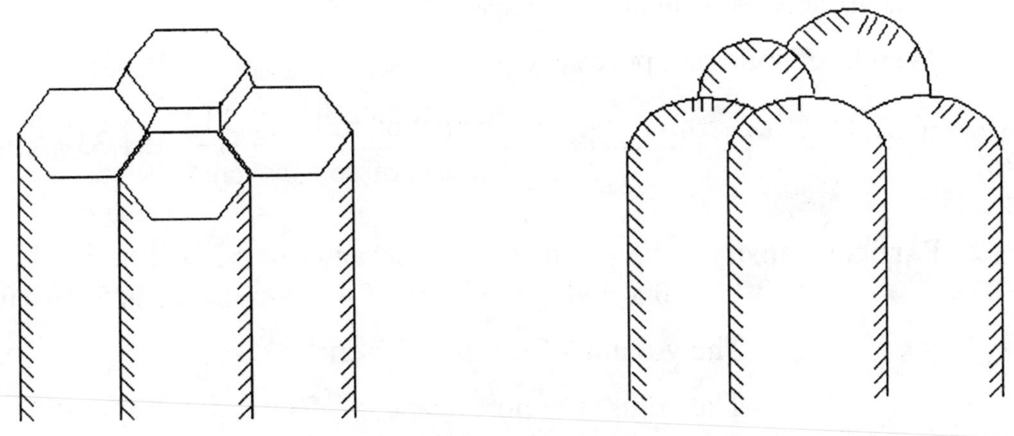

3. PRISM LIKE

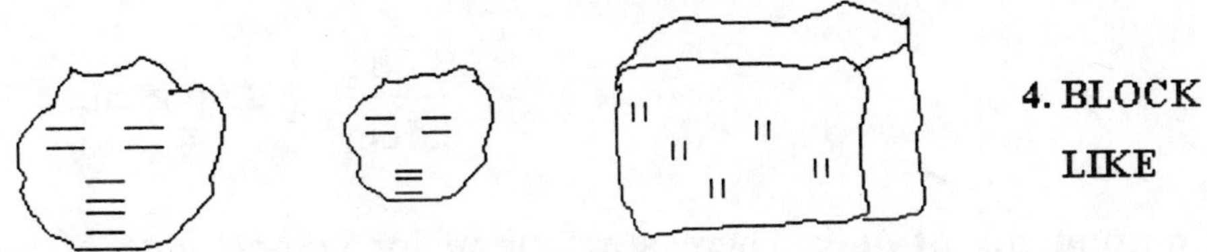

4. BLOCK LIKE

Fig. Structure of soil particles

4) **Density of soil**: "The mass of unit volume of soil" is called density of soil. It is of two types-

1) **Particle density of soil**: Density of solid soil particles excluding fluid space or pore space is termed as "**Particle density**" or in other word, the mass of a unit volume of soil solids is termed as "**Particle density**". It is usually expressed in g/cm^3. Thus, if one cubic centimeter of soil solids weighs 2.6 grams, the particle density is 2.6 g/cm^3. Particle density of important minerals existing in soil, like- Quartz, Felspar, Mica etc. may range from 2.60 to 2.75 g/cm^3.

2) **Bulk density**: "The mass of unit volume of dry soil" is called bulk density. This volume includes both solids and pores.

Calculation of soil density

1. Bulk density:

Suppose:

The volume of solid and pore space = 1 cm^3

Weight of solid and pore space = 1.33 g

$$\text{Therefore, the bulk density} = \frac{\text{Weight of soil}}{\text{Volume of soil}} = \frac{1.33\,\text{g}}{1\,\text{cm}^3} = 1.33\ \text{g/cm}^3$$

2. Particle density: The volume of same amount of soil solids is about 50% and the volume of the pore space is 50%. Hence,

The volume of solids = 0.5 cm^3

The volume of pore space = 0.5 cm^3

$$\text{Therefore, the particle density} = \frac{\text{Weight of solids}}{\text{Volume of solids}}$$

$$= \frac{1.33\,\text{g}}{0.5\,\text{cm}^3} = 2.66\ \text{g/cm}^3$$

5) **Components of soil**: There are four major components of soil- mineral or inorganic matter, organic matter, water and air. These components are present in following ratio-

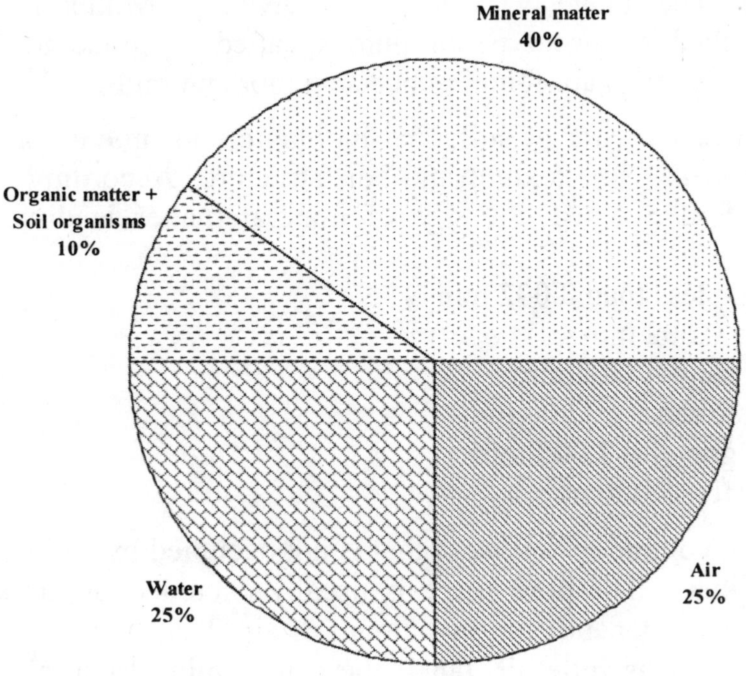

Fig. Components of soil

6) **Organic matter**: It is derived from plant refuse (fallen leaves, twigs, roots etc.), dead bodies of organisms and their excreta. The plant debris fallen on the ground is called as "**litter**". The organic matter has two components-

   i) **Major constituents**: It has longer residence in the soil. The major constituents include resins, lignin, cellulose etc.

   ii) **Minor constituents**: It is with short residence in the soil. It includes carbohydrates (Except cellulose and hemicellulose), proteins, organic acids, lipids, amino acids etc., they get easily decomposed. The major constituents undergo partial decomposition to produce a dark amorphous substance called humus. Humus are of two types-

      (a) **Mor humus**: It develops over well drained acidic soils with moderate rain. It's pH is less (3-4). The population of earthworm is poor or absent, microbial population is also very less, as a result organic matter get stored. This type of humus is called as **mor humus.**

      (b) **Mull humus**: This soil is less acidic (pH above 5). The soil is rich in earthworms and other decomposers. As a result, the process of decomposition is very fast, it is called mull humus.

The above classification is given by Muller in 1800. He also described a third type of humus called as moderate humus, whose characteristics are found in between mor and mull.

7) **Inorganic constituents of soil**: The inorganic constituents or mineral matter depends upon the soil texture and soil structure. According to International Society of Soil Science, the size of mineral matter is divided in form of textural classes-

   Clay = less than 0.002 mm
   Silt = 0.002 mm
   Find sand = 0.02 mm
   Coarse sand = 0.200 mm
   Fine gravel = 2.000 mm
   Gravel = 5.000 mm

8) **Soil air**: The volume of soil mass that is not occupied by soil particles is called as "**pore space**". The pore space is usually occupied by air and water. The amount of water depends on the amount of air. That is if, the soil is saturated with water, there is little air, but if the soil is fully dried, air fills the empty space. Plant roots grow and exist in pore space. It controls the amount of water and air in the soil and thus indirectly controls plant growth and crop production. In aerobic soil the percentage of O_2 is about 21%, whereas in water logged soil, there is deficiency of O_2. In a normal soil, the percentage of CO_2 is 0.5 to 1%. It has been found that, if the percentage of O_2 is 0.1 to 3%, it is responsible for killing of roots and if the percentage of O_2 is upto 5-10%, then development of root and absorption of water takes place.

9) **Soil water**: Soil water is more important than any other ecological factor, in the description of plant communities because it is not only required for the metabolic and transpirational needs of the plant, but also acts as a medium in which soil salts are present in dissolved form.

   Soil water is derived from rain; all the rain water falling on a soil is not retained by it, some of it is lost as run away water and gravitational water, while rest of the water is retained in the soil in various forms. According to Briggs (1897), following types of soil water is found-

   i) **Run away water**: It is that part of water which does not enter in soil, but is drained away from the soil surface along the slope. It is also called **run-off** water. The amount of run away water depends upon the permeability of the soil, moisture content of the soil, degree of slope and number of ditches.

ii) **Gravitational water**: It is that water which is not retained by soil but is drained downwardly under the influence of gravity. The phenomenon is called "**Percolation**". It occurs through macropore, larger than the size of 0.05 mm, within 1-2 days the gravitational water percolates down to reach the area of the parent rock, it is called **groundwater**. The upper layer of ground water is known as **water table**, but sometime due to development of pan water gets accumulated above the pan (impervious to water) and it is termed as "**hanging water tables**".

iii) **Capillary water**: It is the water present inside the micropores (less than 20 μ in diameter). It is the only water which is absorbed by the plant roots.

iv) **Chemically combined water**: This water does not found in free form. It remains combined with chemicals for e.g. $CuSO_4.5H_2O$.

v) **Hygroscopic water**: It is also known as imbibitional water. It is also not available for plant.

pF-scale: It determines the force by which water is attached by soil particles. The tension in between soil and water is measured by tensiometer.

10) **Soil organisms**: The organisms present in the soil are known as soil organisms. Total number of species in per unit area varies from each other.

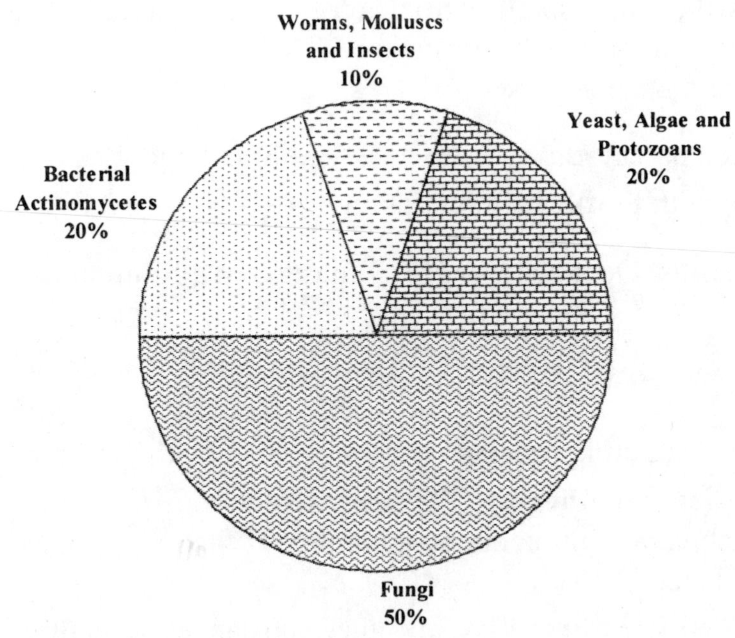

Fig. showing relative biomass of different soil organisms

Some of the important roles of soil micro-organisms are as follows-

- They decompose the dead organic matter and increase the available form of plant nutrients.
- They produce toxins.
- They produce several growth stimulating substances.
- They help in mixing of soil.
- Nitrogen fixation is brought about by micro-organisms.
- They improve the soil aeration which helps in better growth of roots.
- Some organisms cause injury to the plants.

Water holding capacity and field capacity

Water holding capacity is also known as storage capacity of soil. Water holding capacity of a soil is the maximum amount of total water that a soil can hold against the pull of gravity.

Field capacity are also called as capillary capacity. It is the maximum amount of water retained by a soil per unit of its dry weight. The sandy soil, sandy loam and clay have 5%, 15-25% and 45% field capacity respectively.

Formation of soil

Soils are derived from parent rocks by the process called **weathering**. In weathering process, rocks are disintegrated and decomposed into smaller pieces by physical, chemical and biological agencies. The products of weathering are called regoliths which are small particles of rock materials. These particles are basic materials which under the influence of pedogenic or soil forming process finally develop into mature soil.

Consolidated parent rocks —Weathering→ Small particles of parent matter or regolith
Mature soil with profile differentiation ←Pedogenesis—

Russian scientist Dockuchayer (1889) has given a formula for the soil formation.

$$S = g.e.b\ \Delta t$$

where s = Soil, g = Geology, e = Environment, b = Biological influences, Δt = Time interval.

The weathering process is accomplished by the following three processes-

1) **Physical weathering**: Physical weathering of rocks, is a mechanical process which is brought about by a number of factors, such as- temperature, water and wind etc.

   a) **Temperature**: Like all other substances in nature, rocky minerals expand on heating during day time and contract on cooling. All the parts of rock do not expand or contract at an equal rate. The coefficient contraction and expansion is highly variable in a complex type of rock and due to this rock breaks into finer and finer particles.

Rock is the poor conductor of heat. The temperature difference between different layers of rocks may result in formation of a uniform layer and they may separate from the rest, the process is called "**exfoliation**" and is quite common in hot areas.

Water present inside the rock crevices and joints freezes when the temperature falls below 0°C. The frozen water expands by 9% of its previous volume, which exerts a tremendous force (150 ton/ft^2) in the narrow spaces, enough to break large rocks.

b) **Water**: Water is one of the most important weathering agent and along with temperature, it is a main cause of physical weathering. The carrying capacity of water is enormous. The transporting power of a stream varies at the 6th power of its velocity for eg., a water current moving at 6 inches/sec, can carry a fine sand, and of 12 inches/sec can carry gravel.

c) **Wind**: After temperature and water, wind is also important agent of physical weathering. Abrassing power of wind is well known, just like water. Clay, silt, sand and even gravel are carried by wind, causing weathering. It also removes materials from talus.

d) **Glaciers**: At mountain tops, ice formation takes place in winter season. When the summer approaches, ice starts melting and glaciers (huge sliding masses of ice) move downward on the slopes. In the glacier's movement, the rocks are corroded and finally broken into sand particles.

2) **Chemical weathering**: The agents of chemical weathering are oxidation, reduction, carbonation, hydration and solution.

a) **Oxidation**: It means, addition of oxygen to mineral compounds. The reaction produces oxides, which when dissolve in water, weaken the rock and bring about weathering iron, aluminium, manganese oxides and sulphides which are easily oxidized-

For e.g. $4FeO + O_2 \rightarrow 2Fe_2O_3$
(Ferrous oxide) (Ferric oxide)

b) **Reduction**: It is reverse of oxidation and it is very rare in nature.

c) **Hydration**: The property of solids to take on water and enter into combination is known as **hydration** and substances thus formed are called **hydrates**, which are easily washed out and cause weathering. For e.g. Blue stone is the name of hydrate of cupric sulphate ($CuSO_4.5H_2O$). Upon heating, combined water splits and $CuSO_4$ becomes colourless. Like cupric sulphate, there are so many rocks containing hydrated compounds which take water and increase in volume, which cause disintegration of rocks.

d) **Carbonation**: In this process, CO_2 unites with water and form carbonic acids.

$$CO_2 + H_2O \rightarrow H_2CO_3 \text{ (Carbonic acid)}$$

Carbonic acid again reacts with carbonates forming bicarbonates-

$$H_2CO_3 + CaCO_3 \rightarrow Ca(HCO_3)_2$$

These bicarbonates are soluble and they are easily carried by water, causing weathering. Similarly CO_2 present in solution having free Ca and Mg ions may cause precipitate of carbonates.

e) **Solution**: The property of substance dissolved in another substance, is known as solution. Water is most abundant solvent found in nature. Pure water is dissolved slowly but in presence of salt and acid, the solubility gets increased. Practically, no any substance is soluble in absolutely pure water.

3) **Biological weathering**: In true sense, there is no biological weathering; actually it is physical and chemical weathering which take place due to biological agents. Plant roots are important agent of physical weathering, for eg. Roots of tree penetrate into the crevices and cracks of the rock, by expansion rock splits into the smaller particles.

Plant roots and lichens secrete H_2CO_3 and other acids which act chemically on rocks, resulting into decomposition and disintegration. Similarly micro-organisms are also important agents for biological weathering, they produce a large number of acids like- citric acid, mallic acid and oxalic acid etc. causing decomposition and disintegration. These are microbes which decompose silicate, phosphate, and carbonates; the action is direct as well as indirect, for example, the oxidant directly oxidizes sulphur into sulphuric acids.

Soil Profile

After weathering; pedogenesis takes place which is very complex and time taking process, resulting in the formation of mature soil, which is differentiated into different horizontal layers called "Horizons". These horizons are found one over the other called soil profile or a vertical section of soil in which all horizons (A, B and C), are present and it is called "Soil profile".

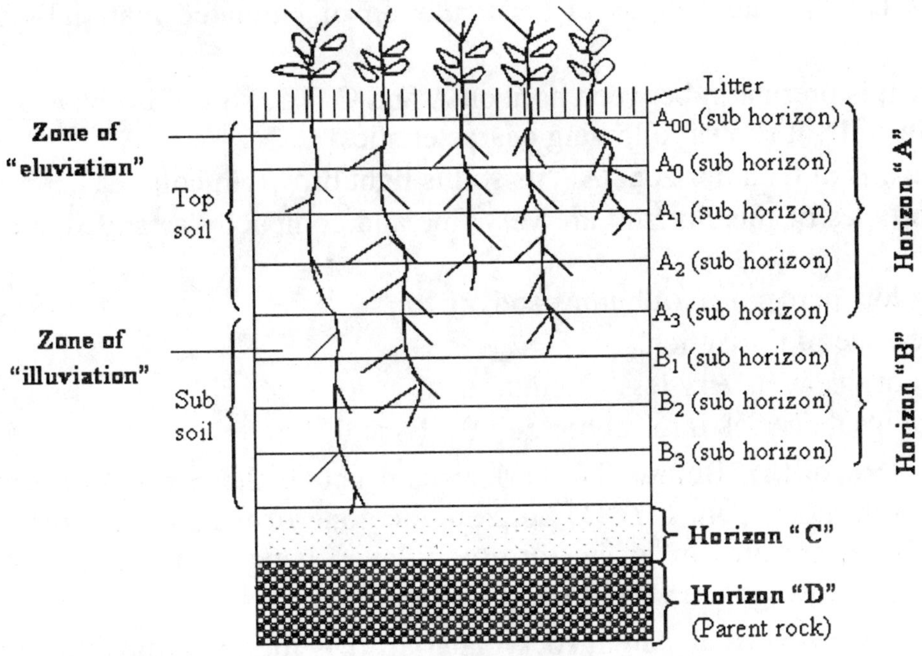

Fig. Soil Profile

Horizon "A": It is the top most layer of soil called "Top soil". It shows following characteristics-

- It is the fertile horizon
- Plant roots are widely distributed
- It is rich in humus
- Soil are dark brown in colour and loosely arranged.
- The thickness varies from 1-10 feet.
- Soils of this horizon is mainly "sandy".

Horizon "A" is again differentiated into the several subhorizons i.e.- A_{00}, A_0, A_1, A_2 and A_3.

Litter: The uppermost layer of soil contains organic matter in the form of dry leaves, twigs and debries etc. which is totally undecomposed and called "Litter". Litter undergoes partial decomposition and form "hums".

Eluviation and Illuviation: Water is important agent which helps in the development of soil profile. While percolating downwardly through mineral and organic substances of "A" horizon or top soil, it removes a number of soluble chemicals from the top soil. This process of washing away of soil constituents is termed as "eluviation" (meaning wash out) and the surface layer from which component are lost is called "zone of eluviation" or A horizon. The eluviated substances move downwardly and are deposited in the lower zone or "B" horizon, which is called "illuvial layer" (meaning "wash in"). The process of accumulation of eluviated materials is called "illuviation".

Horizon "B": It is present in between horizon A and C. The soil of horizon B are also known as "**sub soil**". It shows following characteristics-

- In comparison with horizon "A", its soil is light brown in colour.
- Soil particles of this horizon are very fine and compactly arranged, and soil is clayey.
- Having low percentage of humus and air.
- It is the zone of illuviation.
- Plant roots grow in very less amount.
- It is again subdivided into subhorizon B_1, B_2 and B_3.

Horizon "C" (Regolith): Below "B" horizon and above the surface of weathered rock, there is a third horizon, the "C" horizon, the zone of regolith. In this horizon, $CaCO_3$ and $CaSO_4$ are present in the form of a layer, hence only those plant roots reach in this horizon which are very strong in nature.

Horizon "D": Also known as parent rock material. It is the lowermost horizon. The soil formation starts from this horizon via pedogenesis, hence it is considered as parent rack material.

MODE OF ORIGIN OF SOIL

Depending upon the mode of origin, soils can be classified into two main groups-

1) **Sedimentary or Residual Soils (*in situ* development of soil)**: Soil developed at the same-place where parent rock is present. for eg.:
   i) **Residual soil of Vindhyan range**: These soils are deficient in N, P, Mg and Ca, so insectivorus plants are commonly found here.
   ii) **Black cotton soil of South India**: These soils are rich in clay, N, P, Mg and Ca, so it is beneficial for cotton plants.
2) **Transported soil (*ex situ* development of soil)**: These are shifted soils which are brought from one place to another by various agencies. It is of four types-
   i) **Colluvial soils**: The rock particles **move along the slope by force of gravity**. Rock fragments get disloged from time to time and roll down to the slope to accumulate at the base of tallus. Tallus is coarse and is also unstable. These types of soil are found in foot hills of Himalayan region.

ii) **Alluvial soil**: The rock particles are **transported by running water** The particles are smooth and rounded. They are deposited in strata or layers depending on the particle size and speed of running water, at different times. In India, alluvial soils is found in Indo-Gangetic plain.

Deposits, which are flooded at intervals and receive additional sediments are called **flood plains**. When a flood plain lies above the reach of running water is called **terrace delta**. It is the alluvial soil deposited near the mouth of river, when it passes into lake or sea. Alluvial deposits also occur near the foot hills where the speed of running water decreases, they are called "**alluvial pans**".

iii) **Glacial soils**: It is the soil **transported by the movement of glaciers**. The glacial deposits are also called **moraines** e.g. soil of Dehradun.

iv) **Eolian soil**: It is **wind transported soil**. It is unstable and poor in nutrients. They are of two types- **Sand Dunes** and **Loess**. Dune is made up of coarse of sandy particles. Loess is a fine textured deposit derived from Volcanic ash or rock flour. Eolian soil is found in Rahasthan area.

Pedogenesis is a complex process and depend upon climatic conditions. On the basis of climatic conditions, four types of soil is found-

a) **Laterite (Laterization)**: In tropical and subtropical regions, when rainfall occurs, the organic matter and minerals are leached away and hydroxides of aluminium and iron are precipitated in the form of residue which is called "**Laterite**". This process is termed as laterization. In this area, **temperature is high and soil is Red**.

In this process, silica is completely removed. Laterites usually do not show well differentiated horizons.

b) **Podzol (Podozolisation)**: In temperate zones, where climate is cold and moist, the rate of decomposition is slow and in the decomposition process, acids are produced which make the soil acidic. The acidic unproductive soil is called "**Podzol**" and the process is called "**podzolisation**". In the process of podzolisation, humus and some minerals including dissolved Si, Fe and Al salts from "A" horizon move downwardly with percolating water and accumulate in the lower horizons. This process is more effective in sandy base poor parent materials under intense leaching and thick vegetational cover. The colour of soil is ashy.

c) **Calcifies (Calcification)**: In subhumid and dry regions, due to lack of excessive moisture, the soil accumulates considerable amount of soluble materials and carbonates of calcium and Mg, which are deposited in "B" horizon. The soil having such features is called **Pedocol** (Ca accumulating soil). Pedocols are not acidic and Ca content in them is very high. This process is common in grassland.

Glei (Gleization): This process takes palce in wet and cold Tundra regions, where saline conditions do not exist. In this process, a compact structureless and sticky surface layer is developed. This layer is blue green in colour, poorly aerated and has reduced content of iron compounds. It favours surface accumulation of peat materials and undergoes a series of chemical, physical and biological changes which produce a characteristic soil.

CHIEF SOIL TYPES OF INDIA

Geographically soils of India are divided into three main groups-

1. Mature soils of Peninsular India (Include Red soil, Black soil and Lateritic soil).
2. Alluvial soils of Indogangetic plain
3. Scanty soil of Himalayas.

Soils of India are classified into the following main groups-

1. Laterite or Lateritic soil
2. Black cotton soils or "regur"
3. Alluvial soils, including deltaic coastal and inland alluvium.
4. Alkali soils and saline soils
5. Peaty and other organic soils of forests
6. Red soils including red loam, yellow earths etc.
7. Desert soils or arid soils
8. Scanty soils of mountains of hills.

1. **Laterite or Lateritic soil**: It is generally reddish or yellowish red in colour which turns black on exposure to Sun. It occurs on the plateau of Malwa, Madhya Pradesh, Central India, Bihar, Orissa, Tamil Nadu, Eastern and Western ghats, Assam, Bengal and Hyderabad.
2. **Black cotton solid or Regur (Regada = Black)**: Best suited for cotton cultivation. It is locally, known as **regur** in some provinces. **The soil is poor in "P" and rich in "Nitrogen"** and organic matter. It is very productive and its water holding capacity is much high. Soil requires proper irrigational facilities. It mainly occurs in Tamil Nadu, South East Bombay, Eastern Hyderabad, Maharashtra, Andhra Pradesh, Gujrat, Orissa, Tonk and Bundelkhand region of Uttar Pradesh.

3. **Red soil including Red loam and Yellow earth**: These soils are distributed in India, mainly in M.P., S.E. Bombay, Mysore, Andhra Pradesh, West Bengal, Assam, Orissa, Chhota Nagpur area. It develops from metamorphic sediments of granite, gnesses and schists. The soil is generally rich in potash but nitrogen content is low. The water absorbing capacity is very low.

4. **Alluvial soils including deltaic coastal and inland alluvium**: Soil is distributed in parts of Rajasthan, Punjab, U.P., Bihar, West Bengal and Orissa. The soil is best suited for agriculture, provided rains occur regularly. It is suitable for the cultivation of wheat, rice, cotton, maize, millet, oils, sugarcane etc. The various kinds of alluvial soils have same origin. They are formed from the materials transported by rivers over long period of time.

   The soils are immature and can be divided into two geological groups viz. Old and New. The Old alluvial soils are also called "**Bangar**" while the New alluvial soil group are called as "**Khadar**". Bangar is having particle of clay or silt, while khaddar is having sand with kankars.

5. **Alkali and Saline soil**: In these soils, percentage of salts is very high. These soils are distributed in Kanpur, Lucknow, Hardoi, Unnao, Raibareli, Azamgarh and in many other districts of U.P., Punjab, West Bengal, Bihar, Orissa, M.P., A.P., Gujrat, Tamilnadu, Maharashtra and Delhi. Alkali soils are barren and are also called Usars in Hindi.

6. **Peaty and other organic soils of forests**: Peaty soils are also known as peaty saline soils. At some places, it is called "Kari". It is blue coloured soil. The water logged peaty soils appear in the low lying submerged areas. They are also termed as "**marshy soils**". The peculiar colouration is due to presence of free aluminimum ferrous compounds. Organic matters are accumulated in poorly decomposed condition. Mainly occurs in Uttar Pradesh, South East Tamilnadu and Kerala.

7. **Desert soils or Arid soil**: It is commonly found in arid and semi arid regions of India eg. North West Rajasthan and South Punjab. In arid area, rainfall is very poor and temperature goes very high during the summers. Wind moves sand and sand dunes. The water holding capacity is very low.

8. **Scanty soils of Hills and Mountains**: It is distributed from north west to eastern part of the country. At higher altitudes, entire soil surface is covered with snow. It is of two types-

   1. Acidic soil with acidic humus.
   2. Basic soil having high base status.

TOPOGRAPHIC FACTORS OR PHYSIOGRAPHIC FACTORS

Factors related with the physical geography of the earth are called "**topographic factors**". The chief topographic factors are as follows-

1) **Altitude or Height of Mountain**: The altitude affects the influence of light, temperature, rainfall, humidity and wind velocity of region and thus influence the growth and distribution of vegetation. Temperature decrease **with an increase in altitude of the place. On mountain for every rise of 1,000 metres a lowering of 6-7°C has been observed (1°C for every 165 meters above sea level)**. There is more rain on high mountains. Temperature and rainfall both affect the vegetation at a given place.

   High mountain not only act as a climatic barriers to air masses between neighbouring areas or zones but also cause **local air circulation** giving rise to mountain winds. The most common wind in such areas are "**slope wind**" which **blow downhill during night** and **uphill during day**. It is frequent over glaciers in Antarctica.

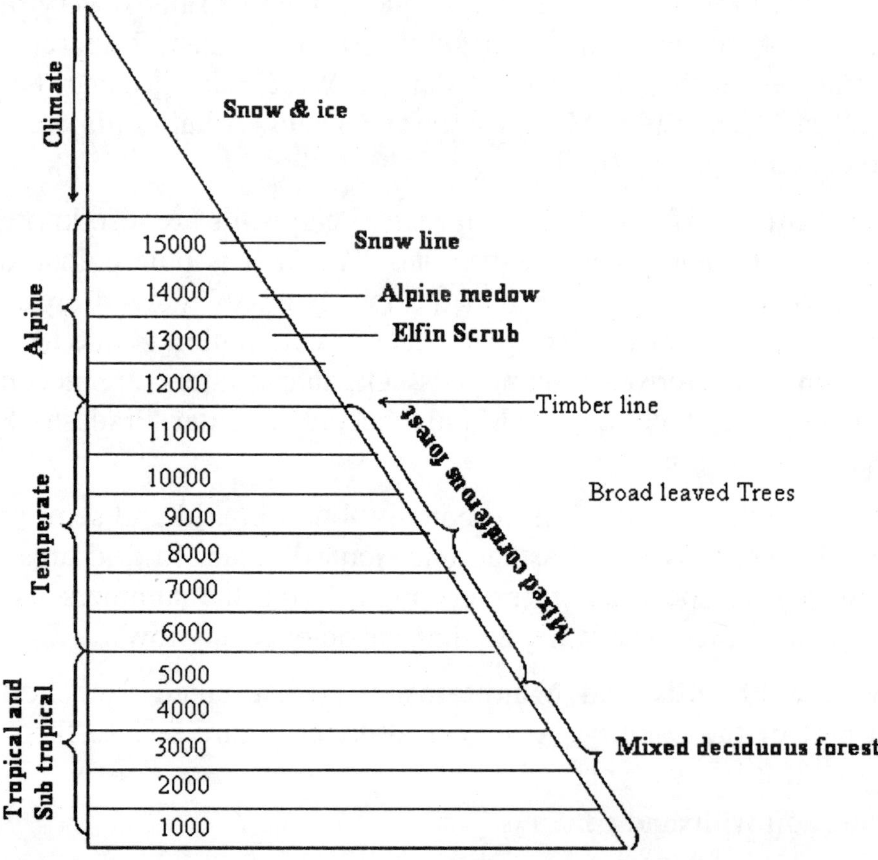

Fig. : Showing different kinds of vegetation at different altitude

- The wind facing direction of mountain are known as **windward side** and its opposite side is called as **Leeward side**.

2) **Steepness of slope**: The steepness of slope is directly related to amount of water erosion. It has been calculated that, when the steepness of slope is doubled, the carrying capacity water gets increased four times and the chances of erosion gets increased and the chances of percolation becomes reduced. In such type of mountains, soils are highly unstable and generally without any vegetation; such area is called as **Topographic desert area**, as a result there is difference in vegetation in the same area depending on the steepness.

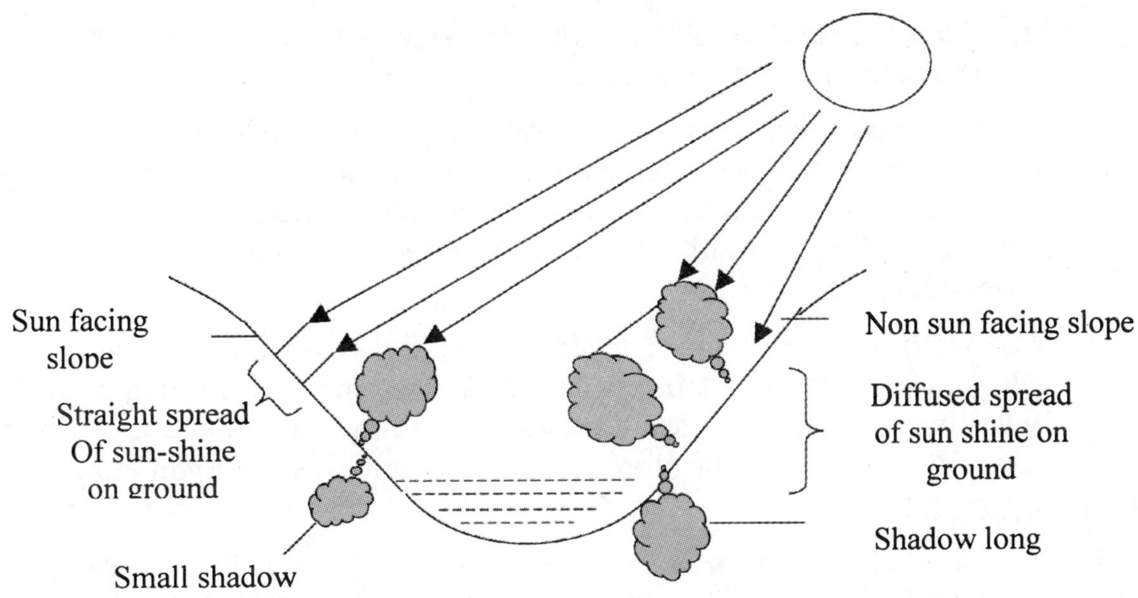

Fig. showing effect of steepness of slope on vegetation

3) **Exposure of mountain**

The sunward slope receives sun rays, almost perpendicular during mid day, while the non-sunfacing slope receives oblique sun rays which is shown in above figure. This slope also produces long shadows therefore, irradiation is less on the non facing slope as compared to the sun facing slope. Ashbel (1942) noted that at 30°C inclination, the sun facing slope receives 212 k. cal/cm^2 of solar radiation as compared to 102 k.cal/cm^2 on non sun facing slope a the same inclination

4) Direction of mountain chain

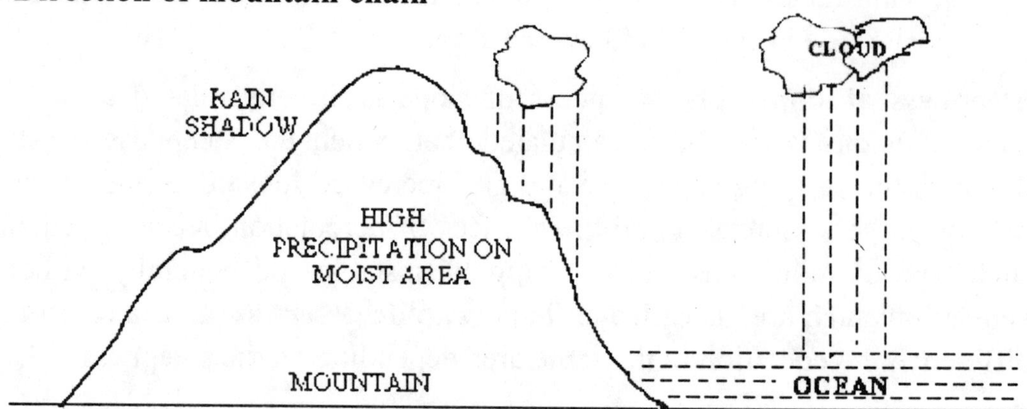

Fig. Occurrence of precipitation on the side of the mountain which lies in the path of cloud formation

The mountains condense the water vapour to produce cloud and hence rain. The mountain also intercept the moisture laden winds coming from the sea, such an interception some times produces very heavy rain fall near the foot hills before the altitude begins to increase appreciably, this is called **approach effect**.

Valleys: In between two mountains, depressions are found which are called valleys. Generally, they are provided with river and they are rich in moisture. Water is evaporated and gets condensed above the valleys and it comes down in the form of rain, just like forest rain, and is called as "**Canyon effect**".

E1 Nino Phenomenon

Under normal circumstances, the water of the Eastern pacific off Ecuador, Peru and Northern Chile are surprisingly cold, as much as 10°C cooler than the waters of the Western Pacific. This part of the Eastern Pacific is teeming with fish, since here cold waters, rich in nutrients, well up from the deep ocean. But once every five to ten years from December to march, the waters of the Eastern Pacific warm up a little (28°C i.e. 4°C higher than normal) which disrupts the upwelling of the rich, cold water. This in turn disrupts the anchovy fishery, key to Peruvian economy. This phenomenon is called E1 Nino (Spanish term for 'The Christ Child') since it starts in December.

E1 Nino is not a regular event as it takes place once every five years, on an average. But when it occurs, the local environment suffers from enormous disruption. The anchovy fishes die due to shortage of food, followed by birds that normally feed on the anchovy. This has disastrous effect on Peruvian economy since the birds deposit guano (dung) around their nesting is harvested by Peruvians and sold as a very rich fertilizer.

Local winds

Mountains and valleys are characterized by local winds. During night they become cool, dense and flows down with support of mountains called **Katabatic winds** or **gravity winds** or **fall winds**. During day they become warm and light and rises up and thus valleys become hot called **Anabatic winds**. Air flow becomes strong and gusty due to **funnel effect**.

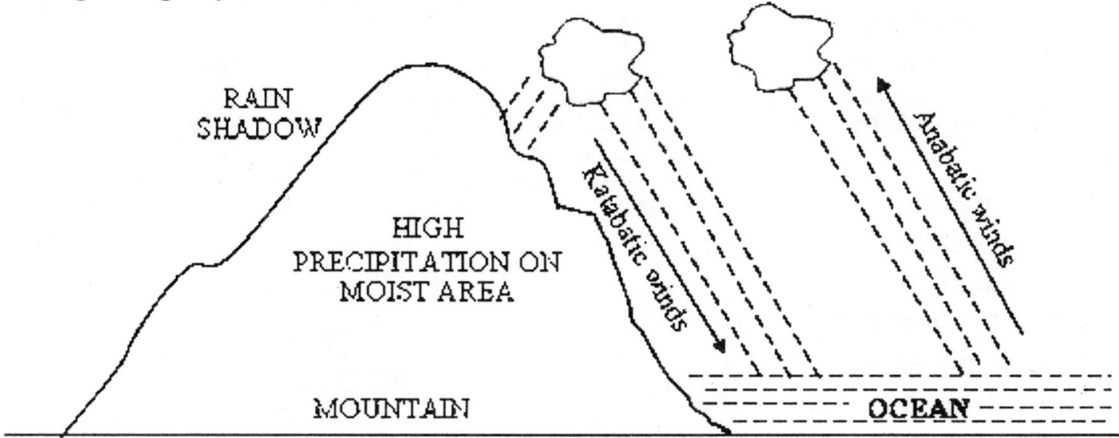

Fig. Mountain as climatic divider

BIOTIC FACTOR

Any activity of the living organisms which may cause marked effects upon vegetation in any way is referred to as "Biotic effect". The biotic effect may be both direct and indirect.

The plants live together in a community and influence one another. The communities interact with one another and shows following relationships-

1) **Neutralism**: There is no effect upon both the partners.

2) **Competition**: It is a relationship which exists between individuals of the some of different species, when the resources (e.g. space, light, CO_2, nutrient, water etc.) which they require are insufficient for the optimum growth of all.

   Competition between the same species is called intraspecific and between different species is called interspecific competition for e.g. Parthenium affects the growth of crop plants.

3) **Parasitism**: It is a relationship in which one organism called **"Parasite"**, obtains its nourishment either partially or completely from another living organism called **"host"**. The parasite may live inside (Endoparasite) or outside (Ectoparasite) the body of the host. It may or may not cause disease in the host.

   Most of the parasites belong to the group of fungi and bacteria. They usually parasitizes on higher plants but exceptions are found in cases where a fungus live on another fungus e.g. *Helicomyces* on *Pericormia*. Depending upon the organ of the host with which connection is established for obtaining nourishment, the angiospermic parasites are of two types-

(1) Stem parasite
(2) Root parasite

Type of interaction	Interaction is 'on'		Interaction is 'off'	
	Species A	Species B	Species A	Species B
1. Neutralism	0	0	0	0
2. Competition	-	-	0	0
3. Mutualism	+	+	-	-
4. Proto cooperation	+	+	0	0
5. Commensalism	+	0	-	0
6. Amensalism	0	-	0	0
7. Parasitism	+	-	-	0
8. Predation	+	-	-	0

where 0 : no effect, + : partners benefitted, - : harmful effect

Parasites, which are non green and obtain the whole nourishment form the host are called "**Total** or **holoparasites**". The ones, which are green and depend on the host for only water, minerals and some other chemicals, are called "**partial** or **semiparasites**". "**Hyperparasites**" are parasites that live on other parasites for eg. *Ciccinobolus cesatti* parasitise the powdery mildew fungi.

1) **Stem parasite**: The example of total parasite is *Cuscuta*. It live on number of angiospermic plants like- *Zizyphus, Citrus, Duranta* and so on. *Arceuthobium minutissimum* is a smallest dicot which is parasitic on *Pinus wellichiana*. The example of partial parasite is *Viscum* and *Loranthus*. Viscum attacking on several fruit plants like **Apple**, **Pears** etc. while *Loranthus* grows on **Mango** and *Acacia* etc.

2) **Root Parasite**: *Orobanche* is a total root parasite, which makes connection with the roots of the host by means of a number of suckers and obtain food in the form of sap.

   Sontalum album is a semiparsite, which obtain food by forming haustoria from *Dalbergia, Albizea* and *Eucalyptus* etc.

3) **Mutualism**: It is a mutually beneficial relationship between two organisms. Mutualism is also known as reciprocal symbiosis or simply simbiosis for eg. Lichen and Mycorrhizae etc.

4) **Protocooperations**: Partners, not completely depended upon each other. A type of interaction, in which both the partners are benefitted. But the

association is not obligatory. It can also be termed as non-obligatory mutualism for e.g. Rhizosphere and Phyllosphere. Microbes occur both on the surface as well as around the roots (Rhizoplane and Rhizosphere) and leaves (Phylloplane and Phyllosphere). Both roots and leaves excrete certain metabolic products like sugars and amino acids. They constitute an important source of nutrition for the microbes, the later protect the plant organs from the attack of pathogens. However, the association is not obligatory for both the partners.

5) **Commensalism**: It is the association between two organisms, in which one is benefited while the other is not harmed. Epiphytes and Lianas are examples of commensal plants.

6) **Amensalism**: It is a situation in which, one species is harmed or inhibited and the other is neither benefitted nor harmed by the association i.e.– In amensalism, one organism does not allow other organism to grow near it. The inhibition takes place by chemical substances, hence it is also called **Antibiosis** or **Allelopathy**, the inhibiting chemicals are also known as **Allelochemicals**. In some cases, **allelochemicals** are **autopathic** or **harmful** to the producing organisms as well as other organisms for eg. *Solidago*. In *Grevellia robusta*, the roots of older trees kill the seedling by allelochemics.

Roots of black Walnut (*Juglans nigra*) produce a chemical called **Juglone**, which is toxic to apple, tomato and alfa-alfa, while it stimulates the growth of some grasses like *Kentuchy* (Blue grass).

Salvia leucophylla and *Artemesia californica* give out volatile terpenes; similar volatile compounds have been recorded and identified by **Numata** (1978) in *Erigeron* and *Solidago*.

Certain species of algae produce antagonistic (toxic) substances, which adversely affect the other algal species. *Chlorella vulgaris* produces **Chlorellin** which is injurious to other algal species *Gymnodium venefium* by producing some antagonistic substances which cause the death of fishes. *Microcystis*, produces **hydroxylamine** toxin which causes mortality of fish and cattle.

A number of plants also produce toxic compounds harmful to microbes and animals. for e.g. *Tagetes erecta* (against nematodes), *Microcystis* (against aquatic animals), *Trichoderma* (against another fungus *Aspergillus niger*).

Predation: In predation, one free living organism called predator feeds on other organism. The predators are mostly animals. Amongst plants, they occur in some fungi.

Anthrobotrys entomophaga develops adhesive secretion at the tip of special hyphae for capturing insects.

Keystone species: Organisms whose presence is essential for the existence of a particular community is called as keytone species.

Fugitive species: Species of temporary habitat is called fugitive species.

Brood parasite: In this, one partner exploit the other partner for parental care and this is common in birds.

Cannibalism: It is a situation in which the predator and prey belong to same species.

Grazing and Browsing: Grazing means eating away of unharvested herbs as forage by animals for eg. eating away of grasses of goats, whereas browsing refers to a similar use of shrubs or trees by animals for eg. eating away of leaves and small twigs by camel.

MASTER STROKES

Q. 1. What is Bruckner cycle?

Ans. A cycle of climatic change, with an average (though irregular) periodicity of about 35 years. This was postulated by E.. Bruckner in 1890, who examined rainfall and temperature records going back to the 18th century.

Q. 2 What do you mean by arid climate?

Ans. Arid Climate A dry climate especially of deserts, where rainfall and available water cannot support much vegetation. The region is dry, because the total evaporation of the area is greater than the average rainfall. In other words aridity in determined by the amount of rainfall as well as the rate of 'evatranspiration' from vegetation and the rate of evaporation from land and water surfaces.

Q.3 What is Milankovitch cycles?

Ans. Milankovitch cycles named after Serbian scientist Milutin Milankovitch, who first described them in 1920s, are periodic shifts in the earth's orbit and tilt. The earth's elliptical orbit stretches and shortens in a 1,00,000 year cycle, while the axis of rotation tilts more or less at a 40,000 year interval. Furthermore, over a 26,000 year cycle, the axis wobbles like an out-of-balance spinning top. These variations change the distribution and intensity of sunlight reaching the earth's surface and consequently, global climate.

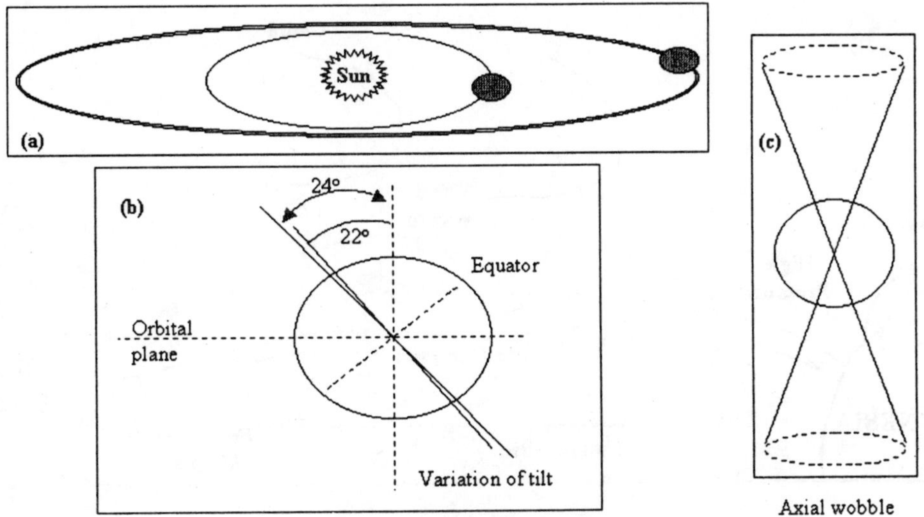

Fig. Milankovitch cycle

Milankowitch cycle, which affect long term climatic condition :

(a) Changes in the occupancy of the earth's orbit.

(b) Shifting tile of the axis.

(c) Wobble of the earth.

Q.4 Differenciate between La-Nina and El-Nino years?

Ans. During La-Nina years a large pool of warm surface water in the Western Pacific near Indonesia brings heavy rain to South-East Asia and drought to Western North and South America, while driving prevailing South-Easternly trade winds and ocean currents.

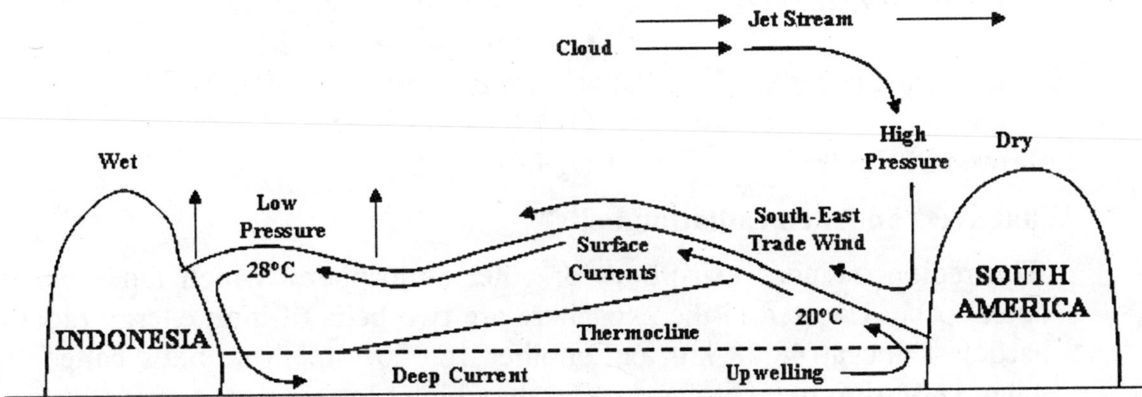

Fig. La-Nina years

In El-Nino years, this pool of warm water and its associated tropical low shift towards the Eastern pacific, bringing wet weather to the Americans and drought to Australia and Indonesia.

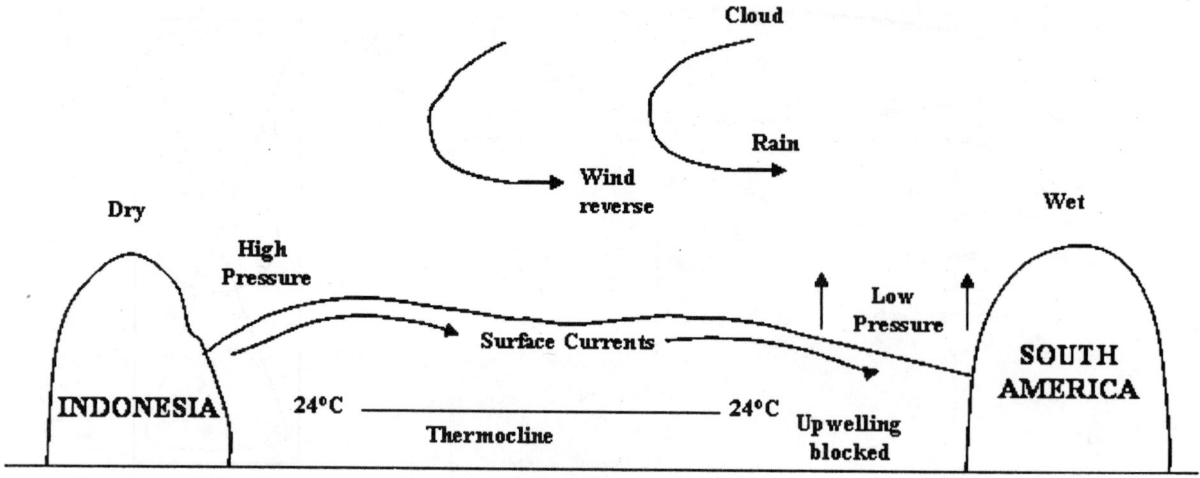

Fig. El-Nino years

Q. 1. What is normal environmental lapse rate?

Ans. One property that changes in the troposphere is the temperature. It drops to – 56^0C by the time an altitude of about 16 – 18km is reached. The rate of temperature drop is about 6.4^0C per 1000 meters. This is called normal environmental lapse rate.

Q. 2. What is Chapman mechanism?

Ans The oxygen is present in the atmosphere. It is seen that radiation of wavelength $\propto$ 242 nm promotes the reaction.

$$O_2 + hv \rightarrow O + O^*$$

The atomic oxygen produced react with O_2 to form ozone.

$$O^* + O_2 + M \rightarrow O_3 + M^*$$

And the ozone absorbs UV radiation of wavelength $\propto$ 340 nm;

$$O_3 + hv \rightarrow O_2 + O$$

Followed by the reaction: $O_3 + O \rightarrow 2O_2$

Q. 3. What are Van Allen radiation belts?

Ans The region includes **exosphere** or outer atmosphere, which fades out into outer space. Located in the exosphere are two belts of high energy radiation particles, one at 4,000 km and another 16,000 km. these belts called **Van Allen radiation belts** protects us by absorbing corpuscular energy of Sun.

Q. 4. What is atmospheric phenomenon?

Ans **Atmospheric Phenomenon :** Molecules of air scatter the shorter wavelength of light (which are at the blue end of spectrum of light) more than the longer wavelengths (at the red end) which is the reason the **sky appear blue in colour.** The cleaner the air, such as after a rain shower, the darker the shade of blue the sky appears. When the sun is near the horizon, its light passes though large dust particles which scatter longer wavelengths, giving a red sky at night and in the morning.

Wind broadly speaking is air in motion. In scientific sense, the airflow must be directed. Usually, only air movements in a horizontal direction are described as wind.

Wind arises as a result of differences in air pressure in the atmosphere, from which the air acquire velocity. The particles of air then follow the decline in air pressure from high pressure areas to low. The air particles also acquire acceleration from the divergent effect of **coriolis force** generated by earth's rotation.

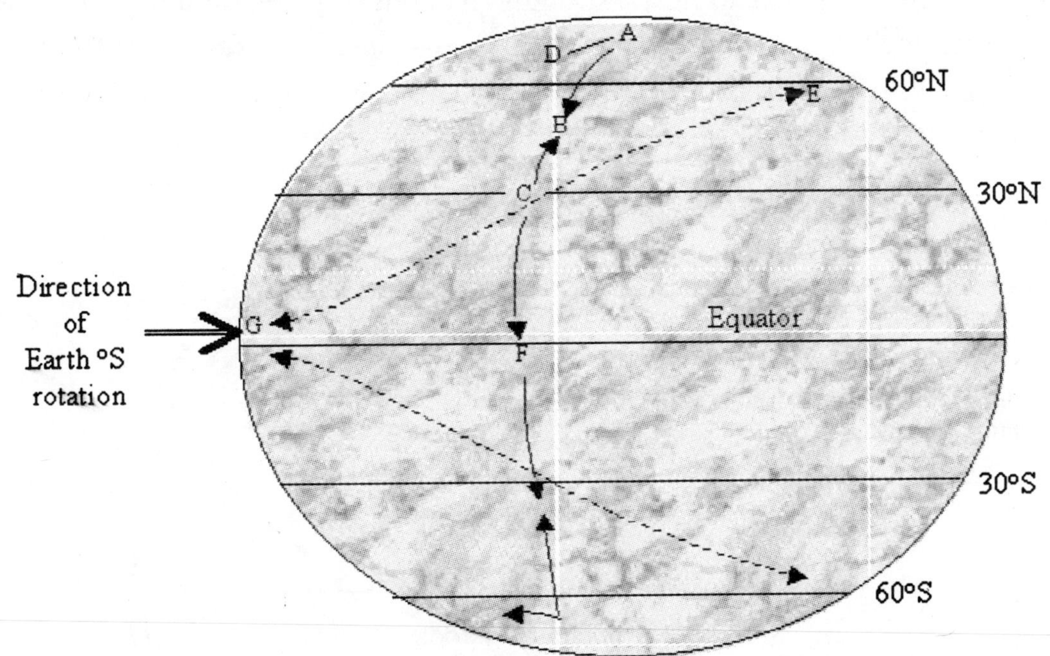

Fig. The coriolis force

Pressure Belts and Winds

Planetary or permanent winds blow round the year between different pressure belts. A low pressure belt covers the equatorial region, roughly between latitudes 10° N and S. This is called as **Doldrums**. It is characterized by weak, gentle winds and sudden storms. An **intertropical convergence zone** (ITCZ) is formed here as the air forms both the hemispheres converge here (as shown in figure). The warm moist air which accumulates here, rises up vertically resulting in cloud formation and thunderstorms at some places. The rising air moves toward the north and south poles. Part of the air sinks around latitude 30° North and South making **subtropical highs** (STH) in the two hemispheres. The wind is so much calm and weak that, in the historical past days, of sailships sailors had to throw their extra load and even horses to lighten their ships. The name **horse latitude** is thus came into use. Air diverges at the surface from the STH. The diverging air moves toward equator, undergoes deflection due to coriolis effect and becomes north-east trade wind and south-east trade wind and are called as **Easterlies**. Part of the diverging air from the STH moves toward poles and become **middle latitude westerlies** due to the same coriolis force.

The direction of westerlies is **south-west** in northern sphere and **north-west** in southern sphere. In southern sphere, maximum earth segment is covered by water, so the air becomes moistened and flows strongly without any restriction. They were named by the **sailors** as **roaring forties** (*गरजने वाला चालीसा*) , **furious fifties** (*क्रुद्ध पचासा*), and **shrieking sixties** (*चीखती साठा*) at 40°, 50° and 60° respectively. Due to these winds, heavy rainfall takes place in major cities of western region.

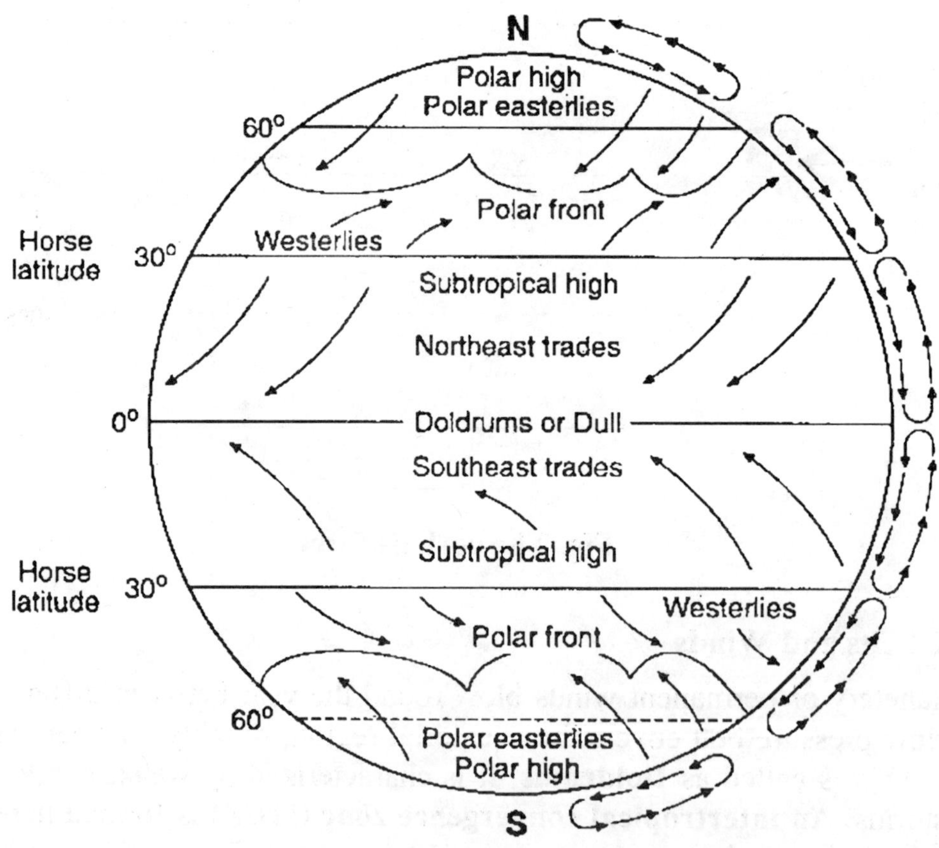

Fig. Pressure belts and winds

Beaufort Number

Wind speed is commonly represented on the **Beaufort** scale ranging from 0 to 12 as given below:

0. **Calm air :** Wind **speed** less than 1.5 km/hr
1. **Light air :** Wind speed 1.6-5 km/hr
2. **Light Breeze :** Wind speed 5-11 km/hr. Wind is felt on face, leaves and veins move gently.
3. **Gentle breeze :** Wind speed 12-19 km/hr. Leaves and twigs move constantly and a small flag is extended.
4. **Moderate breeze :** Wind speed 20-29 km/hr. Wind raises dust and small papers and moves small twigs and branches.
5. **Fresh breeze :** Wind speed 30-38 km/hr. Small trees begin to sway.
6. **Strong breeze :** Wind speed 39-49 km/hr. Telegraph wires starts whistling, umbrellas are difficult to control and large branches start moving.
7. **Strong wind :** Wind speed 50-61 km/hr. Walking become difficult and whole trees move.
8. **Fresh gale :** Wind speed 62-74 km/hr. Walking becomes very difficult. Tree twigs break off.
9. **Strong gale :** Wind speed 75-86 km/hr. Slight structural damage occurs, loose tiles and bricks come off.
10. **Whole gale :** Wind speed 87-100 km/hr., trees are uprooted.
11. **Storm :** Wind speed 101-120 km/hr. causes wide spread damage on land.
12. **Hurricace :** Wind speed higher than 120 km/hr. The sea surface is entirely white with air full of foam and spray.

ENTRANCE CORNER

1. The word ecology was first used by:
(a) Tailor (b) Ernst Haeckel
(c) A.G. Tansley (d) C.J. Krebs

2. Study of the Ecology of a single species or organism is called :
(a) synecoloyg (b) animal ecology
(c) autecology (d) plant ecology

3. Ecology takes into account :
(a) Environmental factor only
(b) effects of plants on environment
(c) plant adaptation
(d) relationship between organism and their environment

4. Synecology is the study of :
(a) animal community (b) individuals
(c) environment (d) none of these

5. Ecology of population is called :
(a) deme ecology (b) phytology
(c) phytosociology (d) phytogeography

6. Aristotle divided the earth climate into how many regions?
(a) One (b) Three
(c) Four (d) Five

7. La-Nina and EI-Nino are found between the countries:
(a) Indonesia and North America (b) Indonesia and South America
(c) Indonesia and China (d) None of the above

8. Habitat can be defined as :
(a) natural place of an organism (b) none-natural place of an organism
(c) both (a) and (b) (d) none of the above

9. The term niche was coined by :
(a) H.E. Hutchinson (b) J. Grinnell
(c) Putman (d) Charles Elton

10. Gause's principle was named as competitive exclusion principle by :
(a) G. F. Gause (b) E. P. Odum
(c) G. Hardin (d) H. E. Hutchinson

11. Ecological niche is an n-dimentional hypervolume. This concept was given by :
(a) O. Neill (b) C. J. Krebs
(c) R. H. Mac Arthur (d) G. E. Hutchinson

12. According to Odum, address of an organism is :
(a) habitat (b) niche
(c) species (d) none of these

13. According to Odum, profession of an organism is :
(a) species (b) habitat
(c) niche (d) none of these

14. The term autecology and synecology is coined by :
(a) Schroeter and Kirchner (b) Grinnell
(c) Putman (d) E. P. Odum

15. Deme ecology is :
(a) study of individual organism (b) study of ecology of population
(c) study of community (d) study of society

16. United Nation Conference on environment and development was first held in
(a) Rio de Janeiro (b) India
(c) USA (d) Johannesberg

17. The summit Riotlo (2002) held at the country :
(a) S. Africa (b) India
(c) USA (d) France

18. The concept of ecological niche was given in year :
(a) 1971 (b) 1917
(c) 1817 (d) 1931

19. Different organisms occupying similar niches in different geographical area :
(a) ecology (b) niche
(c) ecological equivalent (d) ethology

20. Basic unit of ecological hierarchy is :
(a) ecosystem (b) ecology
(c) population (d) individual

21. A unit of land delimited by natural boundary and having patches of different biotic communities is :
(a) ecosystem (b) ecology
(c) biome (d) landscape

22. A regional ecological unit having a specific climate is :
(a) biome (b) landscape
(c) ecosystem (d) both (b) and (c)

23. Equivalent to ecology is :
(a) bioecology (b) hexicology
(c) ethology (d) all of these

24. Autecology is the study of :
(a) population (b) community
(c) species (d) both (a) and (c)

25. Agriculture, animal husbandry and wild life management are part of :
(a) autecology (b) synecology
(c) applied ecology (d) system ecology

26. The highest level of ecological hierarchy is :
(a) ecosystem (b) biosphere
(c) biome (d) landscape

27. Population is :
(a) all animals of an area (b) all plant of an area
(c) both (a) and (b) (d) all individual of a species

28. Components of environments are :
(a) biotic (b) abiotic
(c) resource and regulatory factors (d) all of the above

29. Tropical zone extends from :
(a) 0^0–30^0 (b) 0^0–20^0
(c) 10^0–23^0 (d) 5^0–30^0

30. Mean annual temperature of tropical zone is :
(a) 10^0C (b) 16^0C
(c) 24^0C (d) 30^0C

31. Climate zone between 20^0–40^0 is :
(a) sub-tropical (b) temperate
(c) both (a) and (b) (d) temperate and sub-arctic

32. Winter are absent in climatic zone :
(a) sub-tropical (b) tropical
(c) alpine (d) arctic

33. A local variation of climate is called:
(a) niche (b) microclimate
(c) habitat (d) microhabitat

34. The role of an organism in ecological system is known as :
(a) habitat (b) herbivory
(c) niche (d) interaction

35. Niche of an organism denotes :
(a) habitat (b) microhabitat
(c) habitat as well as inter-relations (d) habitat as well as climate

36. Weather represents :
(a) short term properties of atmosphere
(b) seasonal changes in atmosphere
(c) average variations of atmospheric conditions
(d) all of the above

37. Study of inter-relationship between organisms and their environment is :
(a) ecology (b) ecosystem
(c) phytogeography (d) ethology

38. Study of inter-relationship between a species/individual and its environment in all states of its life cycle ....
(a) synecolgy (b) forestecology
(c) autecology (d) ecology

39. Ecology is connected with the study of :
(a) environmental factors (b) effect of plants on environment
(c) plant adaptations (d) all of the above

40. Study of inter-relationship between an entire community and its environment is :
(a) autecology (b) resource ecology
(c) species ecology (d) synecology

41. Synecology is study of inter-relationship between environment and :
(a) individual plant (b) population
(c) community (d) individual animal

42. Individuals of a species which occur in a particular area constitute :
(a) fauna (b) flora
(c) both (d) population

43. The term of ecology was actually coined by :
(a) Aristotle (b) Reiter
(c) Linnaeus (d) Odum

44. The sum total of the populations of the same kind of organisms constitute :
(a) colony (b) genus
(c) community (d) species

45. Ecology is study of relationship of :
(a) members of a family (b) man and environment
(c) organisms and environment (d) soil and water

46. Synecology is study of :
(a) human environment of a place
(b) biotic environment of a place
(c) abiotic component of a place
(d) biotic community in relation to environment of place

47. Which one is the famous plant ecologist of India?
(a) Charles Darwin (b) Ramdeva Misra
(c) Birbal Sahni (d) Jagdish Chandra Bose

48. Ecology was defined as "the reciprocal relationship between organism and their environment" by :
(a) Misra (b) Haeckel
(c) Odum (d) Lamarck

49. Plant and animals living in an area constitute :
(a) plantation (b) community
(c) population (d) ecosystem

50. Which of the following concerns synecology ?
(a) Same species (b) Different species
(c) Both (a) and (b) (d) None of these

51. Ecology deals with the study of :
(a) micro-organism
(b) relationship between living and non-living
(c) insects
(d) economic difference of countries

52. Ecology is the study of relationship of :
(a) man and environment (b) husband and wife
(c) soil and water (d) organism and environment

53. The study of interaction between living organism and their environment is called:
(a) phytogeography (b) ecology
(c) ecosystem (d) phytosociology

54. The term ecology was proposed by :
(a) Reiter (b) Aristotle
(c) Brown (d) Khorana

55. Ecology takes into account :
(a) environmental factor (b) plant adaptations only
(c) effect of plants on environment (d) all of the above

56. Biocenosis is the study of :

(a) living
(b) environment
(c) inter-relation between living and non-ling
(d) none of the above

57. The term biocenosis was proposed by :
(a) Tansley (b) Warming
(c) Karl Mobius (d) Non of these

58. Science which deals with the management of plants, animals, soil, water and minerals is :
(a) autecology (b) synecology
(c) resource ecology (d) phytosociology

59. The study of plant community structure is called :
(a) ecology (b) phytosociology
(c) autecology (d) ecosystem

60. Cytological study of species growing in different environment is termed :
(a) gene ecology (b) gynaecology
(c) ecotypes (d) cytoecology

61. Pedology is the study of :
(a) soil (b) crop disease
(c) rocks (d) locomotion of animal

62. The leaves of plants in warm, wet shaded micro-climates are :
(a) large and thin (b) small and thin
(c) large and thick (d) small and thick

63. The leaves of rainforest trees get rid of heat mainly by :
(a) radiation (b) conduction
(c) convection (d) evaporation

64. A tropical win or liana, is a growth from adapted for gaining access to :
(a) light (b) pollinators
(c) water (d) nutrients

65. An epiphyte is a growth form adapted for gaining access to :
(a) light (b) pollinators
(c) water (d) nutrients

66. A class of animals that is well adapted to warm, wet climates but not well adapted to other types of climates is :
(a) amphibians (b) reptiles
(c) birds (d) insects

67. A leaf adapted to a warm, dry climate is :

(a) large and thin (b) small and thin
(c) large and thick (d) small and thick

68. The leaves of a desert plant get rid of excess heat mainly by :
(a) radiation (b) conduction
(c) convection (d) evaporation

69. The deep lobes or toothed margins of leaves function to increase the :
(a) turbulence of air flow (b) smoothness of air flow
(c) conduction of heat (d) area for stomata

70. Desert annuals survive outside the rainy season as :
(a) leafless stems (b) dormant roots
(c) seeds (d) drought-resistant gametes

71. The silvery hairs on the surfaces of the leaves of desert plants and the bodies of desert insects function to :
(a) reflect solar radiation
(b) promote convection
(c) promote evaporation
(d) increase the ratio of surface area to volume for radiation

72. By growing close to the ground, a tundra plant :
(a) avoids the wind
(b) avoids herbivores
(c) attracts pollinators
(d) forms mutalistic relations with soil animals

73. The leaves of a tundra plant resemble the leaves of a plant in a :
(a) tropical rainforest (b) temperate deciduous forest
(c) northern coniferous forest (d) desert

74. The function of red pigments in tundra plants is to :
(a) expand the range of light used in photosynthesis
(b) warm the plants
(c) track the sun
(d) measure times

75. Transparent hairs on catkins and caterpillars function to :
(a) trap heat (b) trap moisture
(c) reflect light (d) collect dew

76. An insect survives the cold of winter in a dormant stage by converting its glycogen to :
(a) ATP (b) alcohol
(c) lactic acid (d) brown fat

77. What is best pH of soil for cultivation ?

(a) 3.4 – 5.4
(b) 4.5 – 5.5
(c) 5.5 – 6.5
(d) 6.5 – 7.5

78. Prolonged liberal irrigation of agricultural fields is likely to create the problem of : **(CBSE 2005)**

(a) acidity
(b) aridity
(c) metal toxicity
(d) salinity

79. More than 70% of world's freshwater is contained in : **(CBSE 2005)**

(a) antarctica
(b) mountains
(c) greenland
(d) polar ice

80. Which is not correctly matched ? **(CBSE 2005)**

(a) Laterite - Contains aluminium
(b) Terra rosa - Most suitable for roses
(c) Chrnozen - Richest soil
(d) Black cotton soil - Rich is calcium carbonate

81. In which one of the following pairs is the specific characteristic of a soil not correctly matched : **(CBSE 2004)**

(a) Chernozems - Richest soil in the world
(b) Black soil - Rich in calcium carbonate
(c) Laterite - Contains aluminium compound
(d) Terra rossa - Most suitable for roses

82. The pH of a fertile soil is usually around : **(CBSE 2001)**

(a) 2–3
(b) 6–7
(c) 8–10
(d) 11–12

83. If there was no CO_2 in the earth's atmosphere, the temperature of earth's surface would be: **(CBSE 1998)**

(a) as such
(b) less than the present level
(c) increase from present level
(d) dependent upon oxygen amount of the environment

84. The main role of bacteria in the carbon cycle involves : **(CBSE 1998)**

(a) assimilation of nitrogenous compounds
(b) photosynthesis
(c) chemosynthesis
(d) digestion or breakdown of organic compounds

85. Plants die from prolonged water logging because : **(CBSE 1997)**
(a) of dilution of soil minerals
(b) choking of roots due to oxygen lack
(c) dilution of cell sap in the plants
(d) leakage of nutrients due to excess water

86. The crucial atmospheric factor that facilitated the origin and evolution of animal species on earth is : **(CBSE 1995)**
(a) moisture content (b) availability of oxygen
(c) carbon dioxide (d) gaseous nitrogen

87. One of the following is linked with green-house effect : **(CBSE 1994)**
(a) CFC (b) Freon and CH_4
(c) CO_2 (d) all of these

88. Source of maximum sulphur is : **(CBSE 2003)**
(a) ocean (b) land
(c) rocks (d) lakes

89. Surface water of lake enrich in having : **(CBSE 2003)**
(a) organic substance (b) minerals
(c) inorganic substance (d) pollutants

90. Concentration of nitrogen remains constant by : **(CBSE 2001)**
(a) nitrogen cycle (b) thundering and light
(c) enzymes (d) both (a) and (b)

91. One of the following elements which is not much of importance to plants : **(CBSE 1999)**
(a) Ca (b) Zn
(c) I (d) Na

92. Ozone depletion is occurring due to : **(CBSE 1999)**
(a) PCB (b) CO
(c) PAN (d) None of these

93. Assertion : In tropical rain forest, O-horizon and A-horizon of soil profile are shallow and nutrient poor.
Reason : Excessive growth of micro-organisms in the soil depletes its organic content. **(CBSE 2006)**
(a) If both A and R are true and R is the correct explanation of A
(b) If both A and R are true but R is not the correct explanation of A
(c) If A is true but R is false
(d) If A is false but R is true
(e) If both A and R are false

94. Deforestation has an alarming effect on : **(AIIMS 2001)**
(a) increase in grazing area (b) sunlight
(c) weed control (d) soil erosion or desertification of habitat

95. The limiting factor in nitrification of soil is : **(AIIMS 2000)**
(a) pH (b) temperature
(c) light (d) air

96. An area of soil thoroughly wetted and allowed to drain till percolation stops will have a water content called : **(AIIMS 1995)**
(a) capillary water (b) storage water
(c) field capacity (d) gravitational water

97. Ozone hole is : **(BHU 1998)**
(a) absence of O_3 in troposphere (b) absence of O_3 in stratosphere
(c) deficiency of O_3 in stratosphere (d) deficiency of O_3 in troposphere

98. Large size of pinae in animals of warm region in comparison to animals of cold region is due to law of : **(CPMT 2006)**
(a) Dollo law (b) Gloger's law
(c) Cope law (d) Allen's law

99. Humus is present in : **(CPMT 2005)**
(a) horizon-A (b) horizon-O
(c) horizon-B (d) horizon-C

100. Maximum quantity of humus occurs in : **(CPMT 2003)**
(a) lowermost layer of soil (b) upper layer of soil
(c) middle layer of soil (d) same everywhere

101. Mechanical tissue is best developed in : **(CPMT 2003)**
(a) hydrophytes (b) halophytes
(c) xerophytes (d) mesophytes

102. Plants growing in saline and marshy conditions are called : **(CPMT 2002)**
(a) lithophytes (b) mesophytes
(c) halophytes (d) psammophytes

103. Soil transported by air is : **(CPMT 2002)**
(a) alluvial (b) colluvial
(c) glacial (d) eolian

104. The one to catch ultraviolet rays in the atmosphere is : **(CPMT 2004)**
(a) O_2 (b) O_3
(c) N_2 (d) CO_2

105. Humidity in atmosphere decreases rate of : **(DPMT 2005)**
(a) transpiration (b) photosynthesis
(c) glycolysis (d) growth

106. The conversion of NO_3 to N_2 is called : **(DPMT 2005)**
(a) nitrification (b) denitrification
(c) ammonification (d) nitrogen fixation

107. Mycorrhizae helps in : **(DPMT 2005)**
(a) nutrition uptaking (b) food manufacturing
(c) disease resistance (d) disease prevention

108. Field capacity consists of : **(DPMT 2002)**
(a) capillary water (b) gravitational water
(c) hygroscopic water (d) capillary and hygroscopic water

109. Soil rich in Fe and Al due to excessive leaching is : **(DPMT 2002)**
(a) alluvial (b) laterite
(c) loam (d) both (a) and (b)

110. Biogeochemical cycle f which element has atmospheric phase ? **(DPMT 2001)**
(a) Carbon (b) Sodium
(c) Phosphorus (d) Magnesium

111. Edaphology is connected with : **(DPMT 1999)**
(a) Plant and biosphere (b) Soil and living micro-organisms
(c) Animals and ecosystem (d) Soil and biosphere

112. Top soil is darker and : **(DPMT 1999)**
(a) contains more Na and Mg (b) is drier than subsoil
(c) contains more organic matter (d) is wetter than subsoil

113. The lowermost layer of atmosphere in which man and other living organisms exist, is called : **(Karnatka 2003)**
(a) mesosphere (b) stratosphere
(c) troposphere (d) thermosphere

114. Ozone layer is present in : **(Pb. PMT 2006)**
(a) stratosphere (b) thermosphere
(c) mesosphere (d) troposphere

115. Soil is a mixture of : **(Pb. PMT 2000)**
(a) sand and clay (b) sand and humus
(c) clay and humus (d) sand, clay and humus

116. Clay soil is obtained : **(Pb. PMT 2000)**
(a) in desert (b) around ponds
(c) on seashore (d) on rocks

117. Concentration of CO_2 on earth is : **(Pb. PMT 2000)**
(a) 0.3% (b) 0.03%
(c) 0.01% (d) 0.34%

118. One way cycle is : **(Haryana PMT 2003)**
(a) CO_2 cycle (b) H_2O cycle
(c) free energy cycle (d) O_2 cycle

119. Mangrove vegetation is found in : **(Haryana PMT 2005)**
(a) Dehradoon walley (b) Kullu valley
(c) Western ghats (d) Sundervans

120. Submerged hydrophytes have commonly dissected leaves for : **(Haryana PMT 2003)**
(a) decreasing surface area
(b) increasing surface area
(c) reducing effect of water currents
(d) increasing number of stomata

121. Humus is : **(Haryana PMT 2003)**
(a) completely decomposed organic matter
(b) partially decomposed organic matter
(c) partially decomposed inorganic matter
(d) completely decomposed inorganic matter

122. There is an increase in the green-house effect on earth due to : **(Haryana PMT 2000)**
(a) increasing CO_2 concentration (b) increasing SO_2 concentration
(c) hole in ozone layer (d) chlorofluorocarbon

123. Soil is composed of : **(Haryana PMT 1994)**
(a) mineral + water + air
(b) mineral + organic matter + water
(c) mineral + organic matter + air + water
(d) organic matter + water

124. Halophytes grow in saline soil having richnes of : **(AMU 2006)**
(a) $MgCl_2$, $CaSO_4$ (b) NaCl, $MgSO_4$
(c) $MgCl_2$, $MgSO_4$ (d) both (b) and (c)

125. Soil water available to roots is : **(AMU 2001)**
(a) surface water (b) hygroscopic water
(c) gravitational water (d) capillary water

126. Hibernation occurring in certain animals is : **(AMU 2000)**
(a) occasional (b) intermittent
(c) rhythmic (d) periodic

127. The life existing above and below the earth constitutes : **(AMU 2000)**
(a) population explosion (b) biome
(c) biosphere (d) noosphere

128. One of the very important agricultural practice 'the crop rotation' is usually adopted by our farmers : **(AMU 1996)**
(a) to harvest two different crops from the same field at the same time
(b) to increase to mineral elements in the same field
(c) to increase the nitrogen contents
(d) both (b) and (c)

129. One of these is a sedimentary cycle : **(Kerala PMT 2001)**
(a) phosphorus (b) hydrogen
(c) oxygen (d) nitrogen

130. Which of the following pair of bacteria is involved in two step conversion of NH_3 into nitrate ? **(MP PMT 2005)**
(a) Azotobacter and Nitrobacter (b) Nitrosomonas and Nitrobacter
(c) Azotobacter and Achromobacter (d) Pseudomoan and Nitrobacter

131. Mangroves are : **(MP PMT 2005)**
(a) xerophytes (b) hydrophytes
(c) halophytes (d) glycophytes

132. Which is not a part of atmosphere ? **(MP PMT 2005)**
(a) Light (b) Temperature
(c) Edaphic factor (d) Precipitation

133. The instrument which measures wind velocity is : **(MP PMT 2004)**
(a) lactometer (b) anemometer
(c) hydrometer (d) barometer

134. Hydrophytes are characterized by : **(MP PMT 2003)**
(a) thick and large leaf (b) delicate and mucilaginous stem
(c) short spinous stem (d) all of the above

One of the useful activities of most of the bacteria is : **(MP PMT 1998)**
(a) nitrogen fixation
(b) nitrification
(c) operation of biogeochemical cycles
(d) all of the above

135. The possibilities of presence of coal or fossil fuels in a particular area can be guessed by the study of : **(MP PMT 1998)**
(a) protozoans (b) pollen grains analysis (Paleobotany)
(c) both (a) and (b) (d) by studying sulphur cycle

136. H_2S adds foul smell to the water of tanks, ponds etc., is due to : **(MP PMT 1994)**
(a) anaerobiosis (b) aerobiosis
(c) biological magnification (d) nitrification

137. Control of re-radiating heart by atmospheric dust, O_3, CO_2 and water vapours is called : **(MP PMT 1994)**
(a) ozone layer effect (b) solar effect
(c) radioactive effect (d) green-house effect

138. The ozone layer which filters off U.V. radiation reaching the earth is present in : **(DY PATIL 2003)**
(a) lithosphere (b) ionosphere
(c) stratosphere (d) atmosphere

139. Which of the following soil is transported by air ? **(RPMT 2005)**
(a) Alluvial (b) Aerial
(c) Eolian (d) Glacial

140. Water holding capacity is highest in : **(RPMT 1995)**
(a) Sandy soil (b) silt soil
(c) Clay soil (d) loam soil

141. Soil water available to roots is : **(RPMT 1995)**
(a) surface water (b) hygroscopic water
(c) gravitational water (d) capillary water

142. The waxy surface of the floating leaves of the hydrophytes prevents : **(Orissa JEE 2005)**
(a) respiration (b) photosynthesis
(c) transpiration (d) clogging of stomata

143. Halophytes are : **(Orissa JEE 2004)**
(a) salt resistant (b) fire resistant
(c) cold resistant (d) sand loving

144. Rhizophora is a characteristic component of : **(Orissa JEE 2004)**
(a) marsh plants (b) swamp forests
(c) mangrove vegetation (d) salt swamp

145. Ozone occurs in the atmosphere in the region : **(JIPMRE 2003)**
(a) 5-15 km (b) 26-50 km
(c) 50-100 km (d) 100-300 km

146. Plants growing in acidic soils are known as : **(JIPMER 2002)**
(a) psammophytes (b) oxalophytes
(c) lithophytes (d) halophytes

147. Physical and chemical conditions of soil are studied under .... Factors : **(JIPMER 2000)**
(a) biotic (b) elimatic
(c) edaphic (d) topographic

148. Water-logging commonly occurs in : **(JIPMER 1996)**
(a) sandy soil (b) gravelly soil
(c) clay soil (d) loam soil

149. The percentage of CO_2 in atmosphere is : **(BVP Pune 2003)**
(a) 7.8% (b) 0.03%
(c) 0.04% (d) 0.01%

150. Root cap is absent in : **(Manipal 2005)**
(a) halophytes (b) hydrophytes
(c) xerophytes (d) homophytes

151. Characteristic feature of mangrove plants is : **(HP PMT 2005)**
(a) xivipary (b) heterospory
(c) parthenocarpy (d) apospory

ANSWER SHEET

1. (b)	**2.** (c)	**3.** (d)	**4.** (a)	**5.** (a)	**6.** (b)	**7.** (b)	**8.** (a)	**9.** (b)	**10.** (c)
11. (d)	**12.** (a)	**13.** (c)	**14.** (a)	**15.** (b)	**16.** (a)	**17.** (a)	**18.** (b)	**19.** (c)	**20.** (d)
21. (d)	**22.** (a)	**23.** (d)	**24.** (d)	**25.** (c)	**26.** (b)	**27.** (d)	**28.** (d)	**29.** (b)	**30.** (c)
31. (a)	**32.** (b)	**33.** (b)	**34.** (c)	**35.** (c)	**36.** (a)	**37.** (a)	**38.** (c)	**39.** (d)	**40.** (d)
41. (c)	**42.** (d)	**43.** (b)	**44.** (d)	**45.** (c)	**46.** (d)	**47.** (b)	**48.** (b)	**49.** (b)	**50.** (b)
51. (b)	**52.** (d)	**53.** (b)	**54.** (a)	**55.** (d)	**56.** (c)	**57.** (c)	**58.** (c)	**59.** (b)	**60.** (d)
61. (a)	**62.** (a)	**63.** (d)	**64.** (a)	**65.** (a)	**66.** (a)	**67.** (d)	**68.** (c)	**69.** (a)	**70.** (c)
71. (a)	**72.** (a)	**73.** (d)	**74.** (b)	**75.** (a)	**76.** (b)	77(c)	**78.** (d)	**79.** (d)	**80.** (b)
81. (d)	**82.** (b)	**83.** (b)	**84.** (d)	**85.** (b)	**86.** (b)	**87.** (d)	**88.** (c)	**89.** (a)	**90.** (d)
91. (c)	**92.** (d)	**93.** (a)	**94.** (d)	**95.** (a)	**96.** (c)	**97.** (c)	**98.** (d)	**99.** (a)	**100.** (b)
101. (c)	**102.** (c)	**103.** (d)	**104.** (b)	**105.** (a)	**106.** (b)	**107.** (a)	**108.** (d)	**109.** (b)	**110.** (a)
111. (b)	**112.** (c)	**113.** (c)	**114.** (a)	**115.** (d)	**116.** (b)	**117.** (b)	**118.** (c)	**119.** (d)	**120.** (c)
121. (a)	**122.** (a)	**123.** (c)	**124.** (d)	**125.** (d)	**126.** (d)	**127.** (c)	**128.** (d)	**129.** (a)	**130.** (b)
131. (c)	**132.** (c)	**133.** (b)	**134.** (b)	**135.** (d)	**136.** (b)	**137.** (a)	**138.** (d)	**139.** (c)	**140.** (c)
141. (c)	**142.** (d)	**143.** (c)	**144.** (a)	**145.** (c)	**146.** (b)	**147.** (b)	**148.** (c)	**149.** (c)	**150.** (b)
151. (b)	**152.** (a)								

Chapter 2
Ecosystem

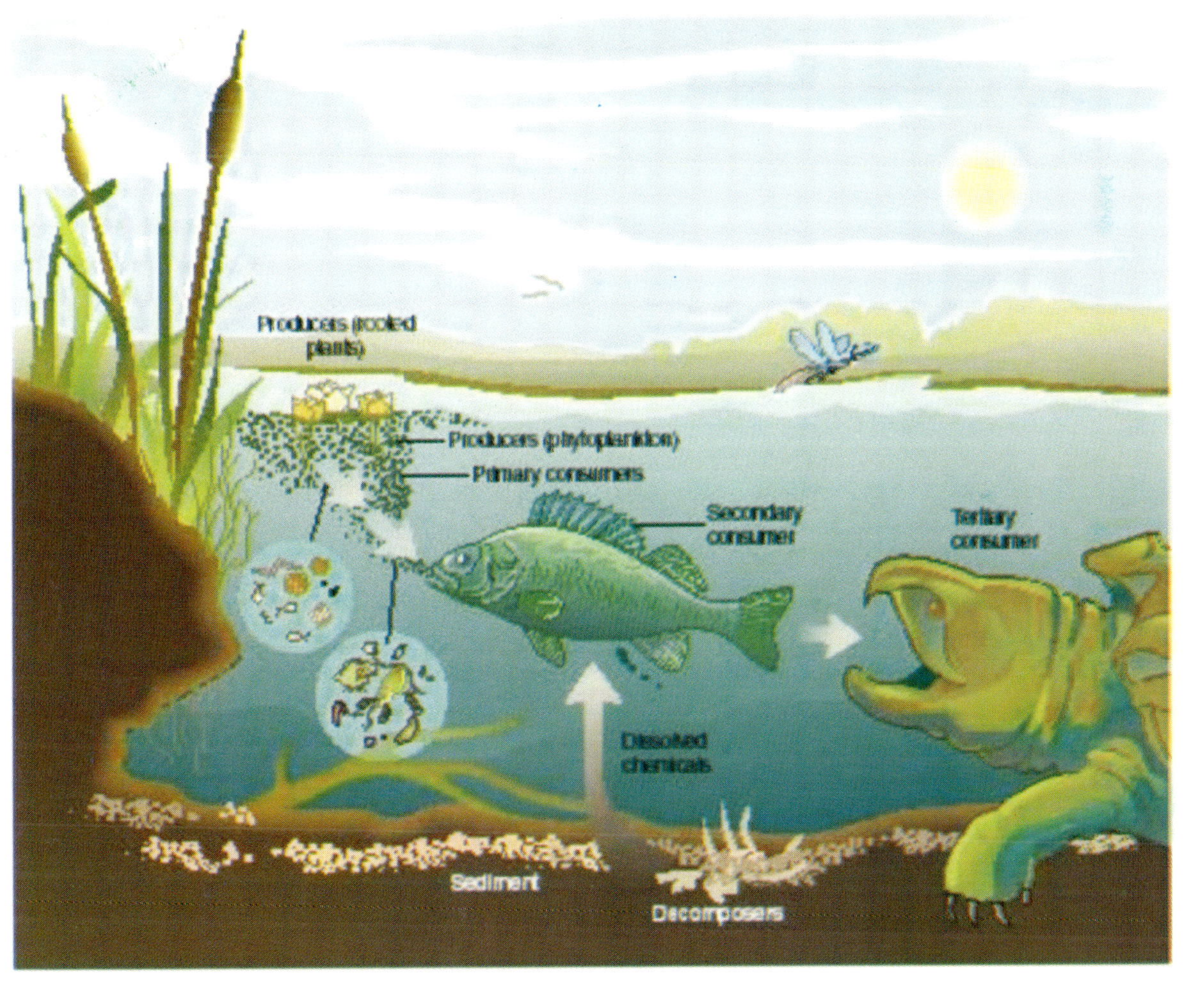

ECOSYSTEM

- The study of interrelation between biotic and abiotic component is called **ecosystem**.
- The term ecosystem was introduced by **A.G. Tansley** (1935).
- **Mobius** (1877) had used the term **biocoenosis**, while **Forbes** (1887) coined the term **microcosm** for the community of organisms.
- **Sukachev** (1944) employed the term **biogeocoenosis** (= geobiocoenosis) as synonym of ecosystem.
- Ecosystem may be as small as a drop of pond water called **microsystem**. It may be as large as ocean called **macrosystem**.
- Some workers consider the whole earth as ecosystem and call it **biosphere** (Vernadsky, 1929) or **ecosphere** (Cole, 1958).
- An ecosystem can be **temporary** (e.g., rain water pond) or **permanent** (e.g. river and ocean water).

Characteristics

1. Flow of Energy: All autotrophs obtain energy from the sun and convert them into chemical energy by the process of photosynthesis. Part of this energy is consumed by the autotrophs during their respiration, growth and other metabolic processes. The remaining chemical energy, called **net primary production**, is used in the body building of the autotrophic plants. It is the energy which is potentially available to the next trophic level.

2. Cycling of Nutrients: Number of inorganic nutrients are obtained by autotrophs from the environment, viz., C, N, P, K, Ca, Mg, Fe, etc. which become component of organic matter. From autotrophic plants, they are passed on to other living constituents of the ecosystem. With the help of decomposers, the nutrients are ultimately released to the environment again. In this way, the nutrients circulate in the ecosystem between the living and nonliving components.

3. Homeostasis: The process by which various components of a ecosystem maintains its functional balance or relatively stable state of equilibrium is called **homeostasis**. It is achieved through a number of controls or limitations called **cybernetics**. The term cybernetics is derived from the Greek work **Kybernetes** ("steers man" or "governer"). Cybernetics developed as the investigation of the techniques by which information is transformed into desired analogous in cybernetics. It found that any species does not increase in its number beyond a certain limit because environment has its own limitations to recycle wastes. Some populations regulate their densities themselves and the phenomenon is called self-regulation, e.g. *Salamander*. One living component of the ecosystem also helps to impose limitations on the other, through a **feedback system**. For example, phytoplanktons are autotrophs, they support zooplanktons which feed on the autotrophs. Phytoplanktons exert a **positive feedback** on zooplanktons while żooplanktons exert a **negative feedback** over them. Phytoplanktons have a high reproductive capacity and their number is maintained by zooplanktons. Any large scale increase in the number of zooplanktons will decrease the number of phytoplanktons to such an extent as to reduce the number of zooplanktons as well. Phytoplanktons will again multiply rapidly and ultimately increase the number of zooplanktons to a suitable level. Thus a system of regulating fluctuations occur in an ecosystem.

Rarely, there is large scale disturbance in one component of the ecosystem which upsets homeostasis and the system of controls, this phenomenon is called ecocatastrophe. In such situations, a new structural and functional balance is developed often by replacement of one type of component organism with another.

4. Inter-relations: The various components of an ecosystem are interwoven with each other and form a 3-dimensional complex, for e.g. plants provide shelter, food and oxygen to animals. The animals supply them with CO_2 and often help them in their pollination and dispersal. Ultimately all animals and plants die. Their bodies are broken down by a number of scavengers and decomposers which release the raw materials to the soil and atmosphere for re-use by autotrophic plants.

Components of Ecosystem

An ecosystem is an ecological unit consisting of both the biotic and abiotic factors forming a delicate balance. Although, the concept of ecosystem is the subject of controversy but according to Odum the ecosystem forms a purely functional stand point. The biotic components are living beings, they are often distinguished into autotrophs and heterotrophs. Autotrophic component is responsible for the fixation of light energy and the utilisation of inorganic constituents. Heterotroph are of four types of consumers (herbivore and carnivore), parasites, scavengers (detrivores) and decomposers. They are responsible for the utilisation, rearrangement and degradation of complex food materials synthesised by autotrophs. Those ecosystems which do not contain all the four basic components i.ė. abiotic substances, producers, consumers and decomposers are called **incomplete ecosystems** (Southwick, 1976), e.g. Abyssal depth of the sea and caves lack producers but contain only the consumers and decomposers, while those having four basic components are called **complete ecosystems**. The amount of living material present in an ecosystem at any time is called **standing crop**. The amount of inorganic materials is similarly known as **standing state. Biomass** is the term used in case standing crop is measured as weight.

Biotic Components

1. Producers: Producers or autotrophs (green plants) are those living components of the ecosystem which synthesise their own food from inorganic substances. Only producer have capability to convert one form of energy into other i.e. during photosynthesis they convert radiant energy into chemical energy and they should be called as **converters** or **transducers** (Kormondy, 1969). About 99% of the living mantle of the earth is made up of producers, the rest being consumers, parasites, scavengers, etc.

2. Consumers (Macroconsumers): They include heterotrophic organisms (animals) which depend on other organisms or particulate organic matter (food) produced by producers. They are also called **phagotrophs** *(phago-to* ingest). The consumers are of two types- **herbivores** and **carnivores**. Those which directly feed on plant matter are called **primary consumers** (Herbivores) while those feed on the flesh of other animals called **secondary consumers** or **primary carnivores**. The secondary consumers, in turn, are eaten by tertiary consumers and so on. The carnivores, which are not further preyed upon, are called **top carnivores**, *e.g.,* Vulture, Lion. Since organic matter is resynthesised by consumers, they are also called **secondary producers** in contrast to autotrophs which are called **primary producers**.

3. Parasites: Parasites are heterotrophs, they obtain food directly from other organisms of all trophic levels. They may also cause disease. Sometimes, a disease may spread in an epidemic form and destroy the host from a large area. Parasites, therefore, keep the population of other species under control.

4. Detrivores (Scavengers): Those small animals which feed upon dead bodies of other organisms are called scavengers. The common scavengers or detrivores are insects, beetles, termites and worms. Detrivores seem to be essential for quick breakdown of dead bodies of organisms.

5. Decomposers (Microconsumers): They include microbes viz., bacteria, actinomycetes and fungi. The decomposers secrete digestive enzymes which convert the complex organic substances into simple and soluble compounds. A number of minerals and raw materials are released during the process. The phenomenon is called **mineralisation**.

Both scavengers and decomposers are important organisms for the removal of detritus or dead bodies of organisms. They are, therefore, also called **reducers**. As compared to the term macroconsumers for phagotrophs, the reducers are also termed as **microconsumers**. The degree of removal depends upon the climate, type of vegetation, pH, number and kind of detritus feeders. The complete decomposition of organic matter takes two to a few months in tropical regions and 3-5 years in coniferous forests of temperate region (Ovington, 1962).

Detritus feeders may become food for a new group of consumers. Thus bacteria of decay become food for mosquito larvae. In this way the complete mineralisation can never occur though complete decomposition.

Abiotic Components

1. Inorganic Substances: They include water, minerals and gases. The inorganic substances required for the synthesis of organic substances are called inorganic nutrients or **biogenetic substances**. The nutrients keep on circulating in the ecosystem. They are picked up by the producers, passed on to the consumers and are transferred back to physical environment through the activity of detritus feeders.

2. Organic Substances: They are obtained by the excretion or by the degradation of dead bodies. The various organic substances are collectively called **organic detritus**. The important ones are amino acids, proteins, lipids, carbohydrates, etc. The organic substances link the biotic and abiotic components of an ecosystem.

3. Climate: The study of climate is called **climatology**. If the study of climate takes place for some period in geological past is called **palaeclimatology** while the study of climate in relation to organic life if called **bioclimatology**. Climate determines the flora and fauna of a place. It is a complex of hydrosphere, lithosphere and atmosphere which produce a number of variables like heat, light, rain, wind, dust storm, fire, fog, mist, snow, etc. and hence circulate nutrients.

Trophic Levels

The various steps through which food energy passes into an ecosystem are called **trophic levels**. All autotrophic plants or producers belong to the first trophic level or T_1. Primary consumers (herbivore) belong to the second trophic level or T_2, while the consumers of the second order or primary carnivores form third trophic level or T_3. There are 2-3 levels of carnivores and all of them are called predators. The ultimate carnivores which are not eaten by others are called **top carnivores**. They may belong to T_4 or T_5 trophic level. Decomposers form the ultimate or detritus trophic level (T_6). Root (1967, 1973) and Feinsinger (1976) have subdivided each trophic level into **guilds**. *A guild is a group of species which exploit a common resource base in a similar fashion* (Root 1967). For example, nectar feeding birds form a guild. They exploit a set or flowering plants. In each guild, species can replace one another to carry on the same functional role (functional equivalents). Competitive interactions are very strong amongst members of the same guild.

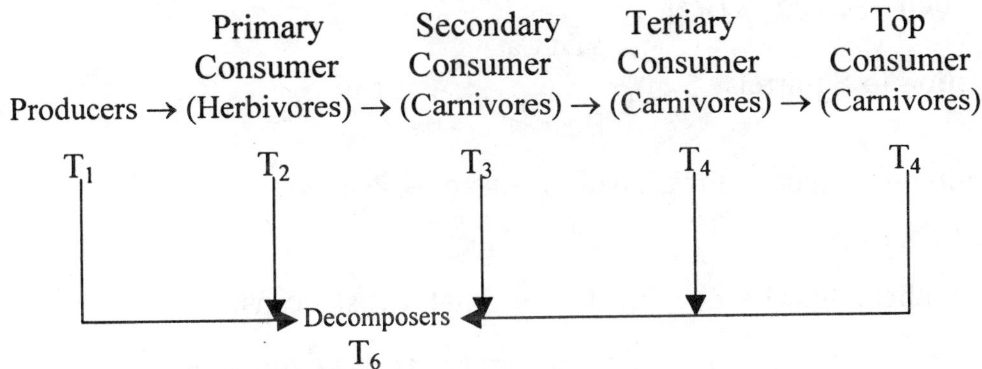

Fig. Various Trophic Levels in an Ecosystem

There arc certain organisms which do not have a fixed trophic level. For example, a top carnivore like lion may eat other carnivores or herbivores. Similarly man is omnivore, both herbivorous as well as carnivorous.

Food Chains

The transfer of food energy from the producer through a series of organisms (herbivores to carnibores to decomposers) with repeated eating and being eaten is known as food chain. There are three main types of food chains.

(i) Predator food chain (Grazing food chain)

(ii) Parasitic food chain (Auxillary food chain)

(iii) Saprophytic food chain (Detritus food chain)

(i) Predator food chains

In the predator food chains, organic food is manufactured by the autotrophic plants i.e. starting point of food chain is green plant. They are called **producers** and which obtain their food from producers are called **consumers**. The consumers which directly take their food from plants by eating their parts are known as **herbivores**. Herbivores are the first order of consumers. Herbivores are eaten by other animals known as **carnivores**. The carnivores which are not eaten by other animals called **top carnivores**.

Some of the predator food chains are as follows:

Land Food Chains

(i) Grass → Cattle → Man

(ii) Grass → Rabbit → Fox → Wolf → Tiger

(iii) Vegetation → Bees → Bear

(iv) Vegetation → Squirrels → Wild Cat / Bear → Tiger

(v) Vegetation → Insect → Frog/Toad → Snake → Peacock

Pond Food Chain

Diatom and other Algae → Zooplankton → Small Crustaceans

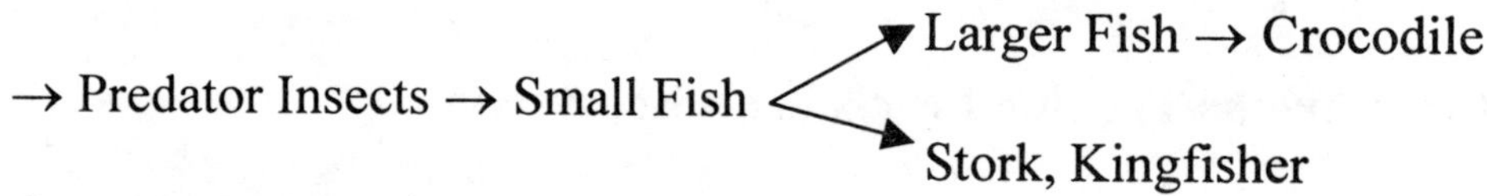

Oceanic Food Chain

Phytoplankton → Zooplankton → Small Fish → Larger Fish → Sharks, etc.

(ii) Parasitic Food Chain

In this case food energy passes through a series of parasite and hyperparasites. It goes from large organisms to smaller ones without outright killing as in the case of predators.

Tree (Producer)	→	Herbivores (Fruit eating birds)	→	Parasites on herbivores (Lice & Bugs)	→	Hyperparasites on Lice & bugs (Bacteria, Fungi & Actinomyceta)

(iii) Detritus Food Chain

In this type of food chain, starting point is dead and decaying organic matter. The dead organic matter (e.g., leaf litter, fallen twigs, decaying roots, corpses, dung, etc.) are acted upon by two types of organisms, detrivores and decomposers. Detrivores are organisms which directly feed on organic matter, e.g., termites, worms, beetles, insects (in terrestrial ecosystems), insect larvae, copepods, nematodes, amphipods etc. (in aquatic ecosystems). The decomposers include bacteria, actinomycetes and fungi. The latter are in turn preyed upon by several small animals including protozoans. These animals are consumed by smaller carnivores which in turn become the food of larger carnivores and so on.

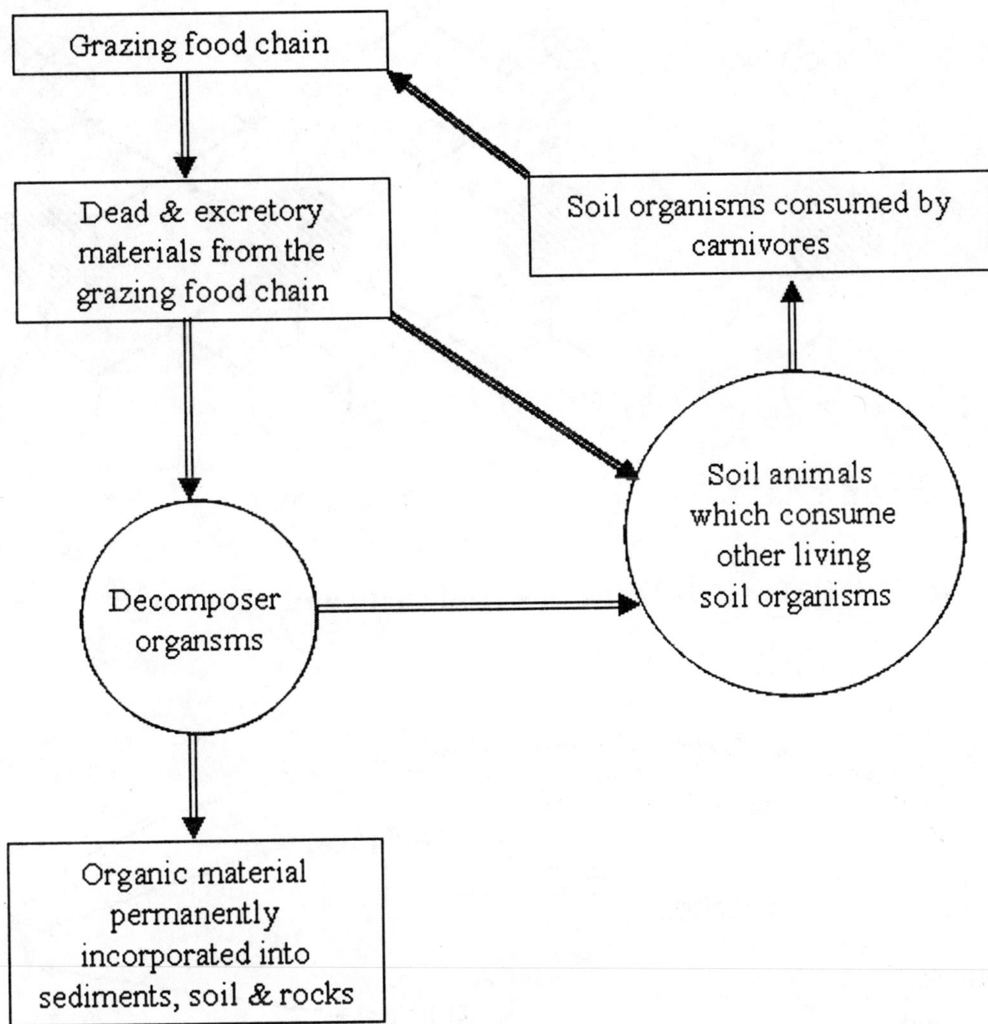

Fig. Detritus food chain

Food Webs

In natural ecosystem, the food chains do not operate independently. They are interconnected with each other forming a sort interlocking pattern which is called as **food web**.

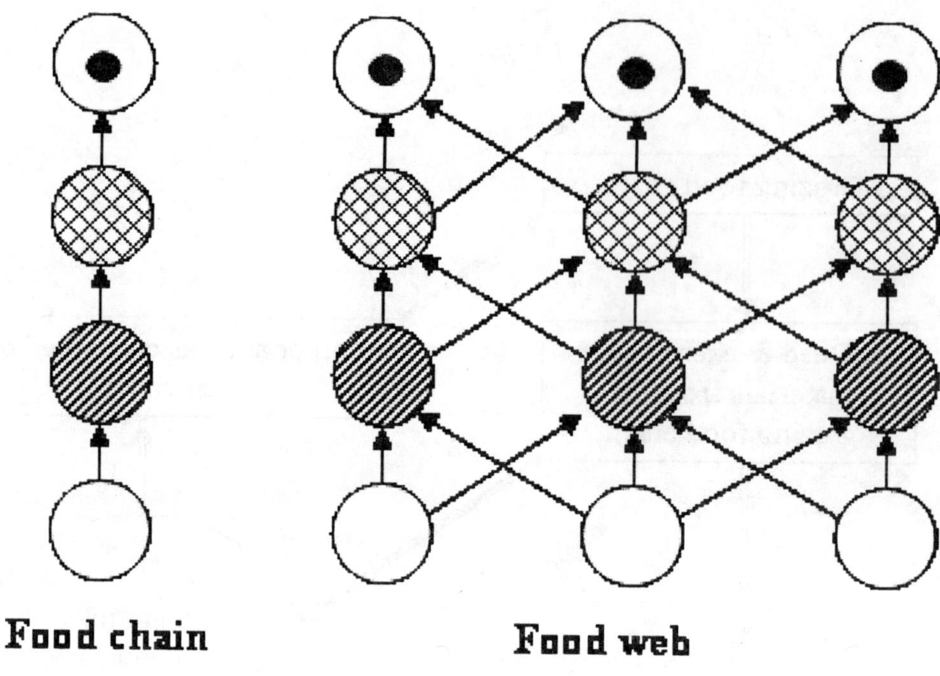

Fig. Differences between a food chain and a food web

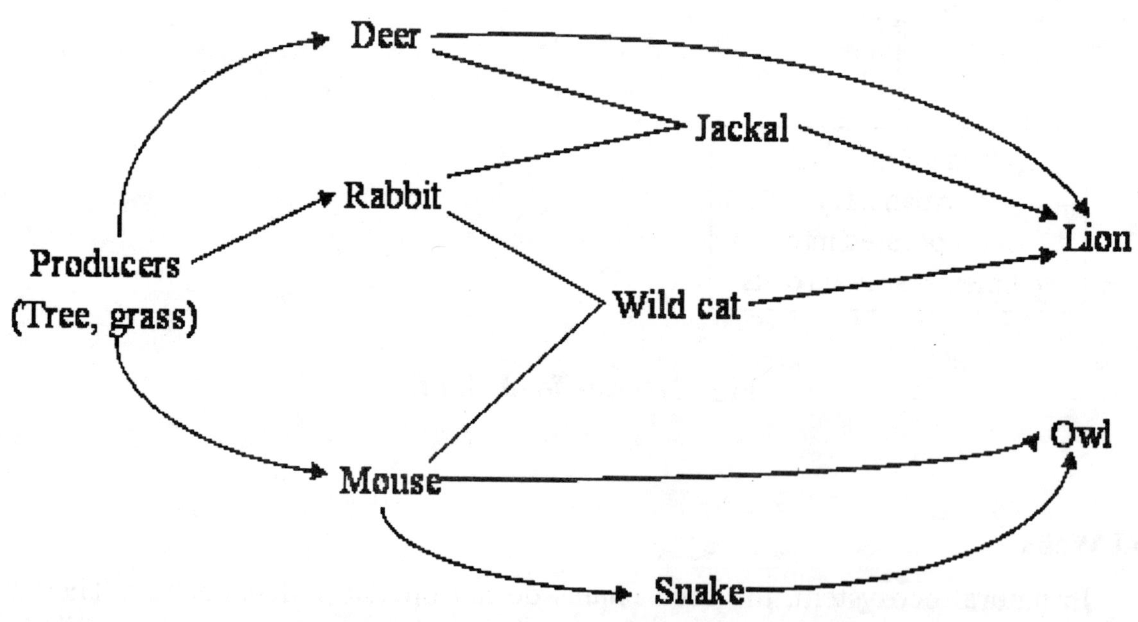

Fig. Food web in a forest

A food web opens several alternate pathways for the flow of food energy. It also allows an organism to obtain its food from two or more types of organisms of the lower trophic level. Thus a field mouse may be eaten by a wild cat, a snake or by owl. Similarly wild cats eat a number of herbivores like mouse, rabbit, etc. A wolf or jackal can eat both rabbit and deer. The primary consumers or herbivores compete with one another for the plant food. Their number is also governed by their natural predators. Excessive predation of a type of herbivore will reduce its number, resulting in the increase in the number of another herbivore. The latter, therefore, replaces the former as the prey for the predator. Meanwhile, the number of the former increases and a balance is re-established. In this way food web helps in maintaining the stability of the ecosystem. The larger the number of alternative pathways occurring in a food web, the more stable is the ecosystem. The alternatives also help in keeping the different species under check and maintain a state of equilibrium called **homeostasis**.

Ecological Pyramid

In any ecosystem, there occur a specific relationship between the various trophic levels of a food chain in terms of number, biomass and energy. If such relationship is represented in form of graphic diagrams, called ecological pyramids. Ecological pyramids were first devised by British ecologist Charles Elton (1927). They are also called **Eltonian pyramids**. It is of three types-

(1) Pyramid of number: Those pyramids which give the idea about the number of individuals of different trophic level are called pyramid of number. The base of the pyramid is constituted by the number of producer while the apex is formed by the number of top carnivores. The intermediate consumers are placed in between the two, depending upon their relationship in the food chain. In the majority of cases the pyramid of numbers is **upright** with a broad base formed by the number of producers. The number of individuals goes on decreasing from base to the apex. for e.g. grassland ecosystem, crop land ecosystem, pond ecosystem.

In forest ecosystem, the pyramid of number is **spindle** shaped because number is less than birds and again number of eagles are less but when parasitic food chain is considered pyramid become **inverted**.

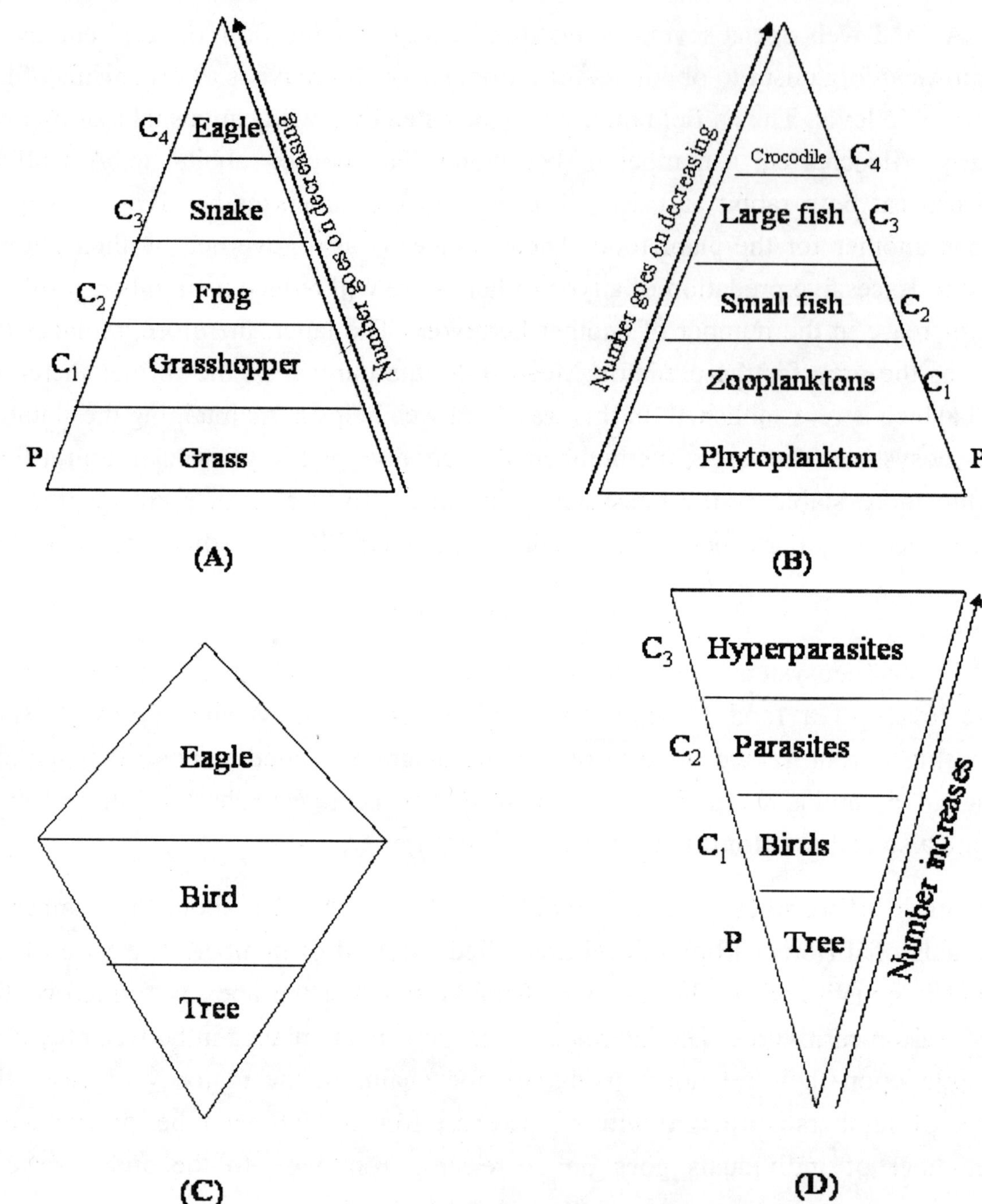

Fig. (A) Upright pyramid of a grass land ecosystem
(B) Upright pyramid of a pond ecosystem
(C) Spindle shaped pyramid of a forest ecosystem
(D) Inverted pyramid of a tree ecosystem

2. Pyramid of Biomass: Those ecological pyramids which gives the idea about biomass (total weight of organic matter) of different trophic levels is called pyramid of biomass.

The pyramid of biomass is also of two types. If a larger weight of producers support the smaller weight of consumers, results in **upright** pyramid and if a smaller weight of producers supports to the larger weight of consumers, an **inverted** pyramid of biomass results.

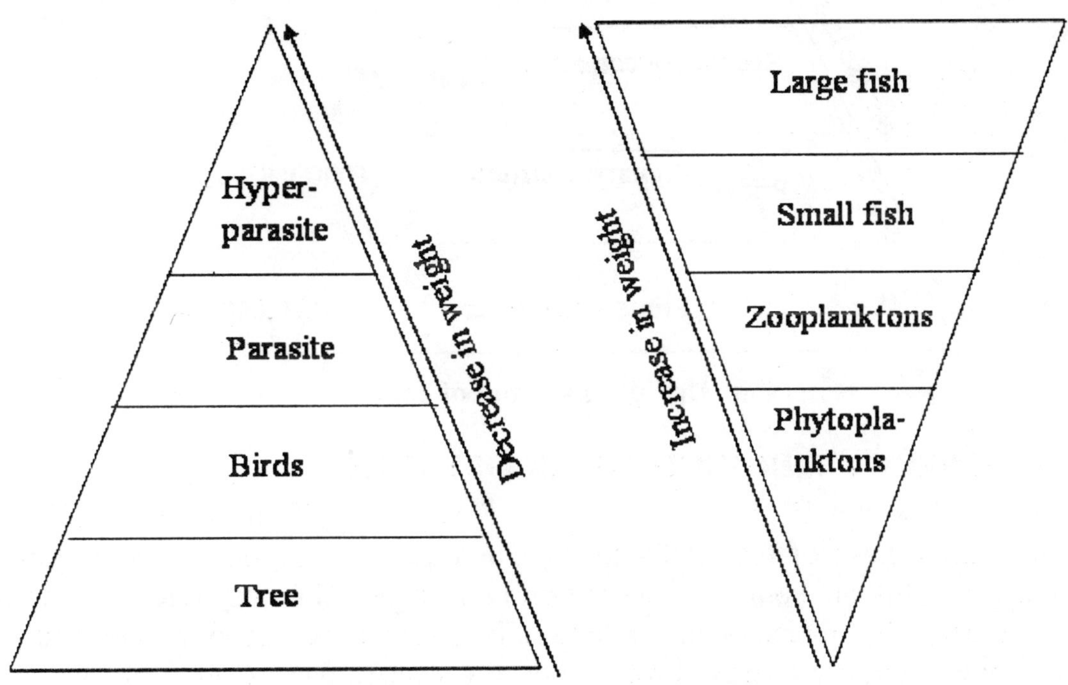

Fig. : An upright pyramid of biomass of a tree ecosystem

Fig. : An inverted pyramid of biomass of a pond ecosystem

(3) Pyramid of energy (Pyramid of productivity): Those pyramids which indicate the total quantity of energy used by different organisms at different trophic levels, called pyramid of energy. Pyramid of energy is **always upright**.

According to Lindman (1942) only 10% of energy is converted at each trophic level. For e.g., In case of pond ecosystem each 1000 calorie of energy produced in producers (algae etc.), 100 calorie may be converted in primary consumers, 10 calorie in the secondary consumers, 1 calorie in tertiary consumers. Thus the amount of energy gradually decreases at successive trophic levels hence it is also called as **law of 10%**.

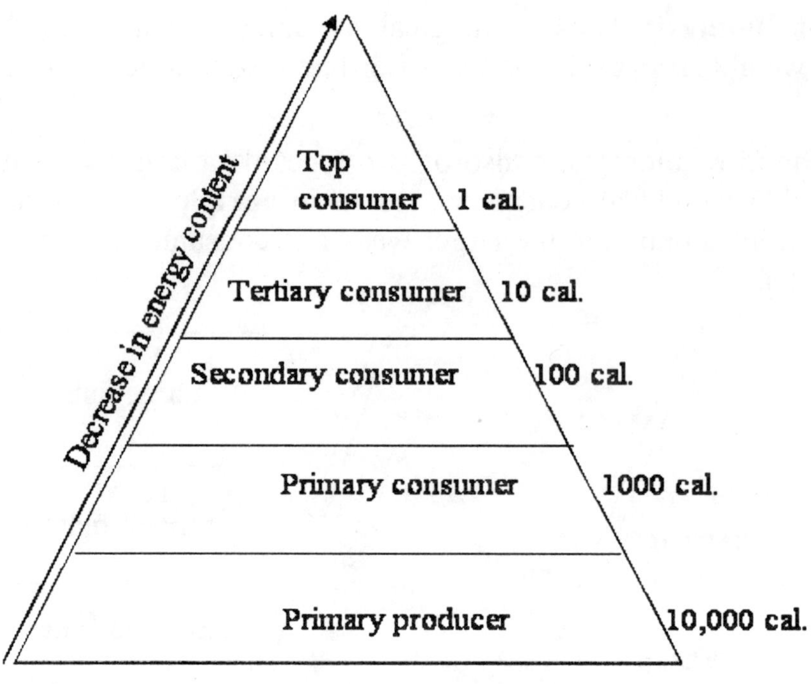

Fig. Upright pyramid of energy

FLOW OF ENERGY THROUGH THE ECOSYSTEM

Energy flow may be defined as the transfer of energy from one trophic level to other obeying the laws of thermodynamics. There is one-way flow of energy in the ecosystem. The ultimate source of energy on earth is sun. Photosynthesis is the only process on earth by which solar radiation is trapped by green plants and it is converted into chemical energy. From there, the energy passes on from one tropic level to the next. The flow of energy in the ecosystem is governed by two laws of thermodynamics. According to the first law of thermodynamics, energy can neither be created nor destroyed. According to the second law of thermodynamics every transformation of energy is accompanied by its dispersion. Therefore, one hundred percent transformation of energy from one form to the other or from one organism to the other is not possible. It is always accompanied by some dispersion or loss of energy in the form of heat, etc.

About 181,872 cal/cm^2/yr of solar radiations impinge upon the earth. Hardly 47-48% of them reach the surface of the earth. The quantity of radiations received at a place depends upon latitude and clarity of the atmosphere. It is 25000 cal/cm^2/yr in Britian (Phillipson, 1966), 47000 cal/cm^2/yr in Michigan (U.S.A), 60000 cal/cm^2/yr in Georgia (U.S.A) and 75000 cal/cm^2/yr in Varanasi, India (Singh, 1972). On an average, only 111 cal/cm^2/yr of energy is trapped by the autotrophs in producing organic food. The efficiency comes to only 0.1% (Lindeman, 1942). Only 0.2% of the incident solar radiations are trapped by aquatic ecosystems and 1% by terrestrial ecosystems. Herbaceous plants are more efficient than trees in trapping solar energy. The energy conserving efficiency has been recorded by Ambasht (1984) as 1.15% for grasslands, 0.9% for shrub savannah and 0.84% for a mixed forest. The modern crop plants can trap upto 5% of the radiations falling on them. The maximum efficiency is recorded for sugarcane. It is 10-12%.

The study of energy transfer between different trophic levels is called **bioenergetics** (Phillipson, 1966).

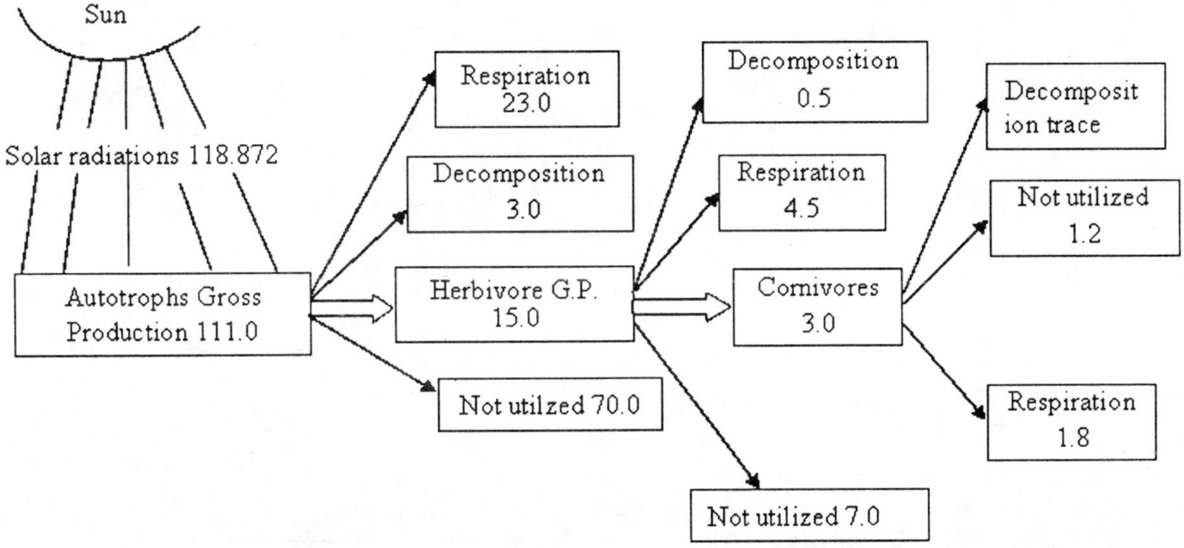

Fig. Energy flow sheet in a fresh water ecosystem (cal/cm^2/yr. After Lindeman 1942)

As shown in figure, out of total incoming radiation which is 118,872 gcal/cm^2/yr only 20.7% is consumed in biological reaction (Respiration etc.) of autotrophs. Herbivores consume 15 gcal/cm^2/yr as they feed or graze on autotrophy (13.5% of gross autotrophic production). The remainder 70 gcal/cm^2/yr (63.06%) is not used while 3 gcal/cm^2/yr (2.7%) is used in decomposition.

It is clear that much more energy is available for herbivorous than is consumed.

Lindemann has given 10% law regarding energy flow in various trophic level in an ecosystem. According to this law only 10% of the energy entering in a particular trophic level of organism is available for transfer to the next trophic level. All the energy transfers in a food chain follow the 10% law which means that the energy available at each successive trophic level is 10% of the previous level. Thus, there is gradual decrease in amount of energy available as we go from producer level to the higher trophic levels of organisms, for example- 1000 calorie of light energy is emitted by sun which falls on the green plants. Only 1% of light energy falling on them is converted into chemical energy of food. So, the energy which will be available in green plants as food will be only 1% of 1000 calorie which comes to 10 calorie. The remaining 990 calories of light energy which is not utilized by plants is reflected back into the environment. Now 10% of the 10 calorie of energy (which is 1 calorie) will be available for transfer at next trophic level, so that the herbivore will get 1 calorie energy at the second trophic level. The remaining 10-1 = 9 calorie will be lost to the environment. Again 10% of the 1 calorie (which is 0.1 calorie) will be transferred to the third trophic level of carnivores. The rest of it, 1-0.1 = 0.9 calorie will be lost to the environment.

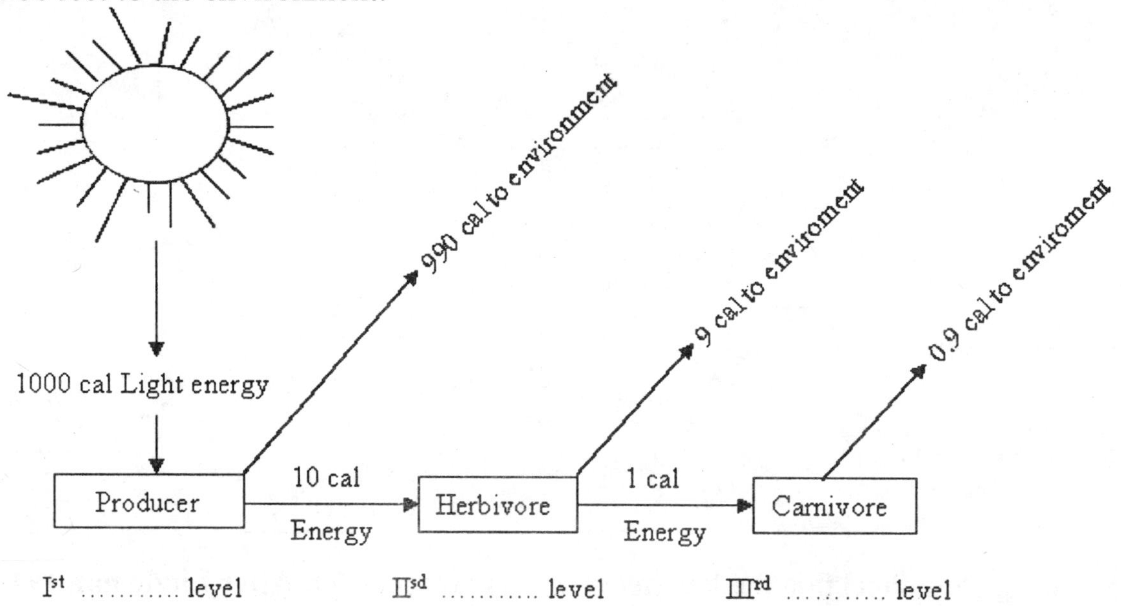

Fig. Energy flow in various trophic level (Lindemann's 10% law)

Ecological Efficiency

Ecological efficiency can be defined as the percentage of energy in the biomass produced by one trophic level i.e. incorporated into the biomass produced by the next higher trophic level. In other words, it is also called as **food chain efficiency** i.e. energy assimilated over the energy available between two trophic level is called **ecological efficiency**. Lindeman (1942) was the first to study the relationship between the energy consumed at the successive trophic levels. He called it as **progressive efficiency**. Kozlovsky (1968) has termed it as **productivity efficiency**. The ecological efficiency is 1-5% for producers and 10-30% for other trophic levels.

The ratio between the net production and energy assimilated at any trophic level is also called **ecological growth efficiency**. Its value lies between 10-50%. Ecological growth efficiency is higher in case of smaller or short-lived organisms as compared to larger or long lived organisms. In small organisms, a large part of the assimilated energy is used for growth while in case of large animals most of it is lost in maintenance of the body through respiration (97-99% in mammals).

The terms used to express various aspects of ecological efficiency are:

1. Photosynthetic efficiency: It is defined as the percentage of light energy assimilated by plants. It varies from 1 to 2%.

$$\text{Photosynthetic efficiency} = \frac{\text{Gross production}}{\text{Total incident light energy}} \times 100$$

2. Net production efficiency (NPE): The efficiency with which assimilated energy is incorporated into a production, growth and storage, is called the net production efficiency. It may be shown as :

$$\text{NPE} = \frac{\text{Production (reproduction and growth)}}{\text{Assimilation}} \times 100$$

In plants, NPE is defined as the ratio of net gross production. This varies between 30-85% depending on the habitat and growth pattern.

3. Gross production efficiency (GPE): This represents the overall efficiency of biomass production within a trophic level.

$$\text{GPE} = \text{Assimilation efficiency} \times \text{NPE} = \frac{\text{Production}}{\text{Ingestion}}$$

4. Assimilation efficiency: The entire food energy i.e. consumed is not assimilated. It may be shown as :

$$\text{Assimilation efficiency} = \frac{\text{Assimilation}}{\text{Ingestion}} \times 100$$

5. Exploitation efficiency: The efficiency by which the biological production of an entire trophic level is consumed.

$$\text{Exploitation efficiency} = \frac{\text{Ingestion of food}}{\text{Prey production}} \times 100$$

Productivity of Ecosystem

The rate of synthesis of energy containing organic matter by any trophic level per unit area in a unit time is described as its productivity. It is of the following types:

Primary Productivity: The rate at which energy is started up by the primary producers viz. green plants determines the basic or primary productivity capacity of an ecosystem. It is of two sub-types:

(a) *Gross Primary Productivity*: The total organic matter synthesised by the producers in the process of photosynthesis per unit time and area is known as gross productivity.

It includes the weight of organic matter added in the body of the producers plus the losses suffered by them due to respiration, grazing and other damages.

(b) *Net Primary Productivity*: The net productivity of energy is the gross productivity minus the energy used in metabolic processes.

Secondary Productivity: The rate at which the consumers synthesizes the energy yielding substances is termed as secondary productivity. It depends upon the loss, while transferring energy containing organic matter from the previous trophic level plus the consumption due to respiration. Respiration loss is about 20% for autotrophs, 30% for herbivores and upto 60% in case of carnivores. Therefore, net productivity decreases with each trophic level.

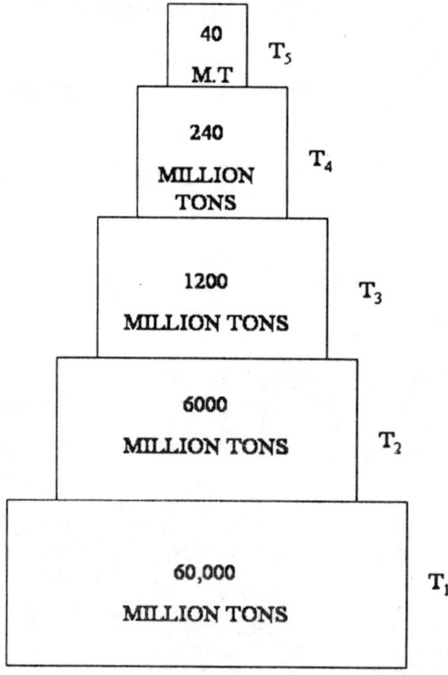

Fig. Annual biomass production/in million metric tons at various trophic levels in the oceans

Yield of an ecosystem depends upon the trophic level exploited by man. It is usually T_2 or second trophic level (herbivores) which is exploited on land for protein while in ocean, the trophic levels harvested by man are T_3, T_4 and T_5.

Productivity Levels

1. High Productivity Ecosystems: They mainly occur in tropic forests, flood plains, coral reefs, areas of upwelling, sugar cane fields etc. The net productivity is about 2-4 $kg/m^2/yr$. It depends upon the types of producers. The daily productivity ranges from 6-20 gm/m^2.

2. Average Productivity Ecosystems: The common agricultural crops and temperate forests belong to this category. The productivity is 1-2 kg/m^2/yr with a daily output of 3-5 gm/m^2.

3. Less Productivity Ecosystems: It contains land ecosystem and savannah. The annual productivity is 200-1000 gm/m^2 with a daily increment of 0.5-3.0 gm/m^2.

4. Low Productivity Ecosystems: Such kinds of ecosystem is found in deep sea and arid lands. The net productivity is less than 200 gm/m^2/yr. It is as low as 3 gm/m^2/yr in certain deserts.

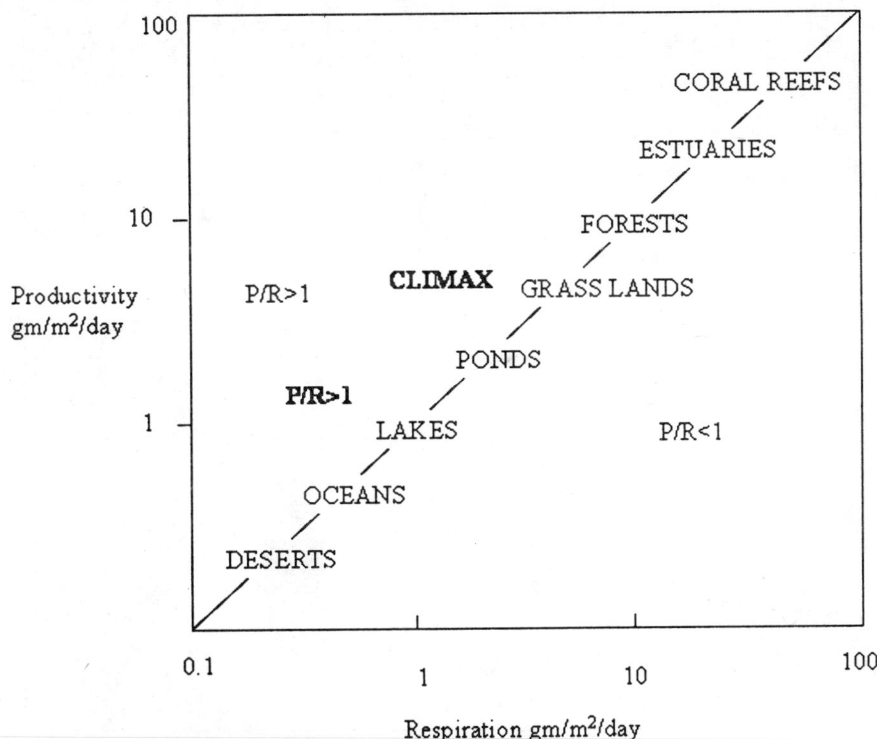

Fig. Magnitude of primary production in various ecosystems and climatic regions

Measurements of Primary Productivity

1. Harvest Method: It involves removal of vegetation periodically and weighing the material. In this vegetation of a unit area is harvested alongwith with its subterranean parts which is oven dried and weighed. The process is repeated at intervals. The change in biomass per unit area per unit time is depicted as ΔB.ΔB = B_2 (at time t_2) – B_1 (at time t_1). Net primary productivity is obtained by adding the loss due to death of plants or plant parts (L), herbivores and parasites (G).

P_n (Net Productivity) = ΔB + L + G.

2. Radio-Isotopic Method: In this, plant is illuminated in presence of radioactive $^{14}CO_2$ which form radioactive product. It is commonly used for the aquatic ecosystem. In this, two types of bottles i.e. light and dark containing phytoplanktons are suspended to the same depth of water. In the lighted chamber photosynthesis and respiration takes place simultaneously and the CO_2 coming out from the chamber is the unused gas of atmosphere + respiration of plant parts. In the dark CO_2 is due to respiration.

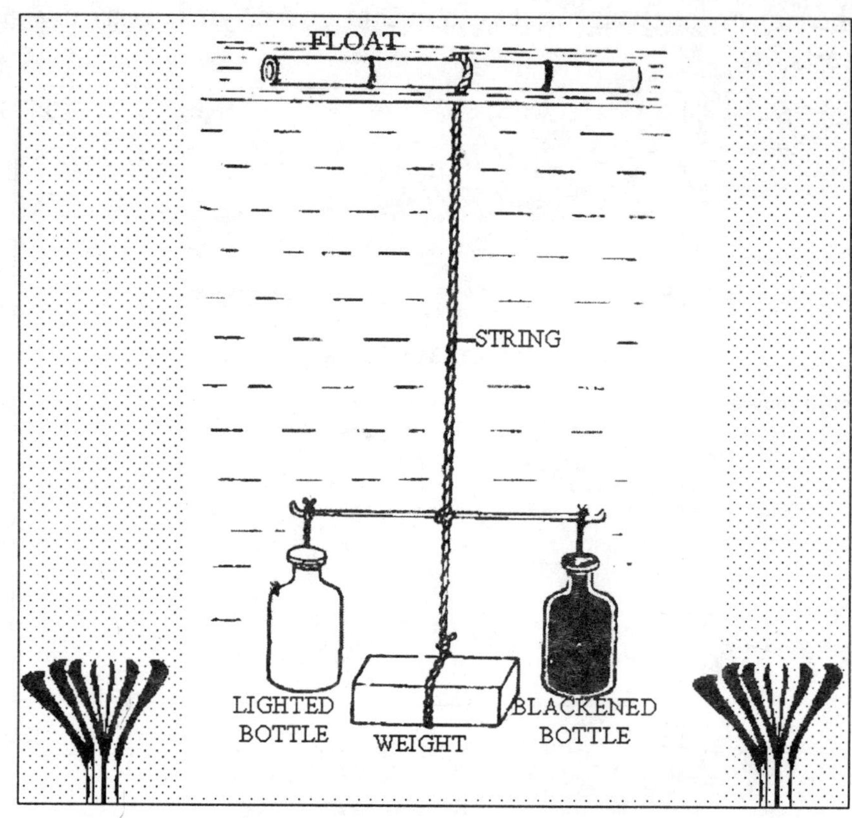

Fig. Method of light and dark bottles for estimating primary productivity through estimation of O_2 released during photosynthesis

3. Leaf Area Index (L.A.I.): It is calculated by the following formula:

$$LAI = \frac{TLA}{TGA}$$

where, TLA = Total leaf area

TGA = Total ground area found below the canopy of the plant

LAI = Indirect estimate of primary productivity upto a value of 4-5, beyond which the increase in productivity becomes stationary. It may rather show a decline (exception vertically oriented leaves).

Leaf area can be calculated by multiplying the number of leaves with the area of an average leaf. The latter is calculated by three methods: (i) Planimeter (ii) Graph paper (iii) Comparing weight of a square of a hard paper with that of a leaf replica.

4. Calorimetric Method: In this organic matter is oven-dried, weighed and taken in a crucible. The crucible is then placed in a bomb calorimeter. Dried organic matter of known weight is pressed into a tablet around the ignition wire. The bomb is filled with pure oxygen with a small amount of water. A known amount of water is poured into the kettle which is fitted with a stirrer and thermometer. Temperature of water is recorded. Electric current is now passed through the bomb. The organic matter ignites and generates heat, this raises the temperature of surrounding water. The temperature rise by ignition wire without organic matter is also noted. Caloric value is then calculated as below:

$$C.V. = \frac{W.\Delta t - \Sigma c}{G}$$

where W is the water equivalent value, Δt is the rise of water temperature, Σc is temperature rise due to ignition wire and G is the weight of dry organic matter.

BIOGEOCHEMICAL CYCLES

Term biogeochemical cycle was coined by **Vledimir Ivanovich Vernadasky** (1863-1945) A number of inorganic substances (C, H, O, N, P, K etc.) are required by the living organisms for their proper growth and development, called **biogenetic nutrients** or **essential elements**. The phenomenon of circulation of biogenetic nutrients between abiotic environment and biotic communities is called **biogeochemical cycling**. The amount of biogenetic nutrients presents in the abiotic environment per unit area at any time is known as **standing state**. Actually the biogenetic materials exist in two pools- **reservoir pool** and **exchange** or **cycling pool**. The reservoir pool is the large reservoir of the nutrients. However, it is largely unavailabie because of its very slow transfer to the cycling pool. The exchange or cycling pool is a small but more active pool of the nutrients which exchanges repeatedly between the abiotic and the biotic components of the ecosystem. The rate of this exchange is called **turn over rate**.

Biogeochemical cycles are of two types - **gaseous** and **sedimentary**. In the gaseous, the reservoir pool is largely atmosphere, though lithosphere may be a secondary pool e.g., carbon, oxygen, water, nitrogen. In the sedimentary cycle the elements are nongaseous and the reservoir pool occurs in the lithosphere, e.g., P, K, Ca, Mg, Cu, B, Zn, etc. In some cases an intermediate or temporary gaseous stage may occur in the biogeochemical cycle, e.g., SO_2 or H_2S in case of sulphur.

1. Carbon Cycle : Carbon occurs as carbon dioxide in atmosphere (6×10^{14} kg), bicarbonate, carbonic acid in hydrosphere ($1.3\text{-}5.0\times10^{15}$ kg) and in lithosphere

(2.8×10^{21} kg) as carbonate, graphite, shells, skeletons, fossil fuels (coal, petroleum and natural gas). Carbon dioxide is being added to the atmosphere through two types of processes i.e.

(i) **biological-** respiration of various organisms and decomposition of organic matter

(ii) **nonbiological-** combustion of carbon containing fuel. Burning of fossil fuel adds about 6 billion tons of carbon to the atmosphere. Use of fossil fuel is increasing day by day. This has increased the carbon dioxide content of the atmosphere by over 10% in the last 125 years. Other sources of increasing carbon dioxide concentration of the atmosphere are volcanic eruptions and hot springs. They mad add upto 100 million tons of carbon annually into the atmosphere. Weathering of rocks also releases carbon dioxide to the tune of another 100 million tons.

Carbon is being deposited into the lithosphere through several methods. The common ones are molluscan shells, protozoan tests, aquatic plants growing in alkaline waters, sedimentation and fossilisation of organic remains, seepage of carbon rich water into the interior of the earth, etc.

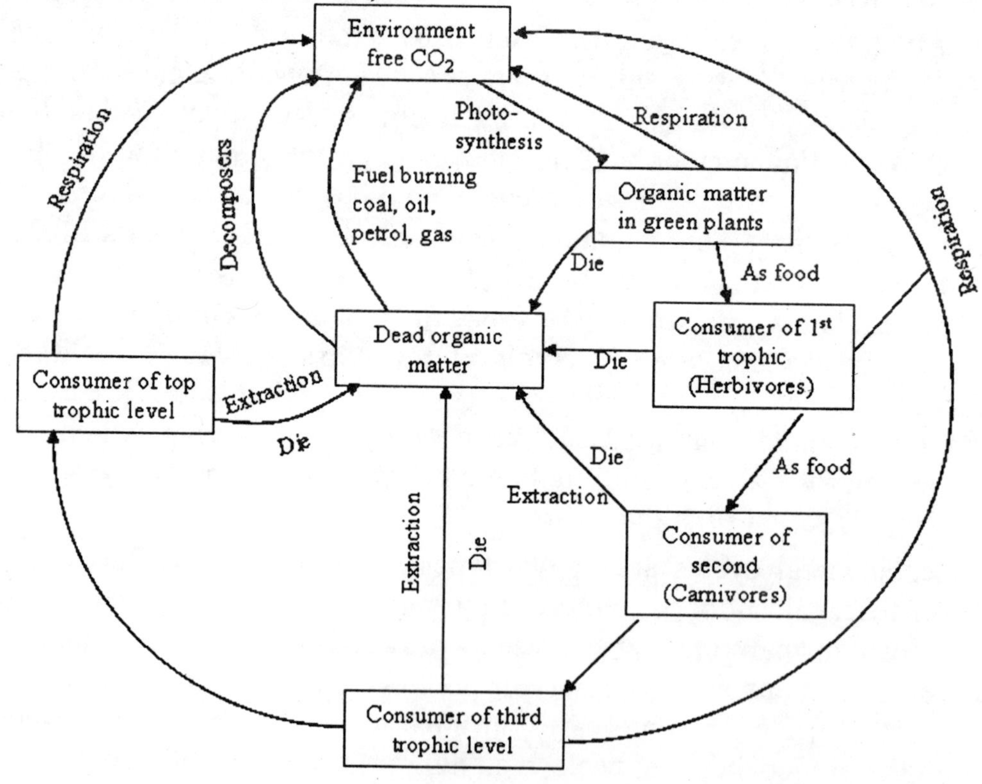

Fig. Carbon cycle

2. Oxygen Cycle : Oxygen occurs in the gaseous form in the atmosphere to the extent of 20-21% It also occurs as CO_2, H_2O (water) and a number of oxidised salts.

Oxygen is required by living beings for their respiration. The land organisms take it from the atmosphere while the aquatic forms obtain the same from their habitat, where atmospheric oxygen enters by diffusion. In the respiration, oxygen combines with hydrogen of the respiratory substrates of produce water.

$$C_6H_{12}O_6 + 6O_2 \rightarrow 6CO_2 + 6H_2O + 686 \text{ kcal}$$

In combustion, oxygen combines with organic matter to produce both water and carbon dioxide. Oxygen is also used in a number of nonbiological processes to produce oxides, sulphates, nitrates, etc. The same can also be got from microbial oxidation. Such substances are reduced both biologically and nonbiologically to release oxygen. However, the major part of oxygen replenishment is carried out by green plants in the process of photosynthesis.

$$6CO_2 + 12H_2O \xrightarrow[\text{Chlorophyll}]{\text{686 kcal}} C_6H_{12}O_6 + 6H_2O + 6O_2$$

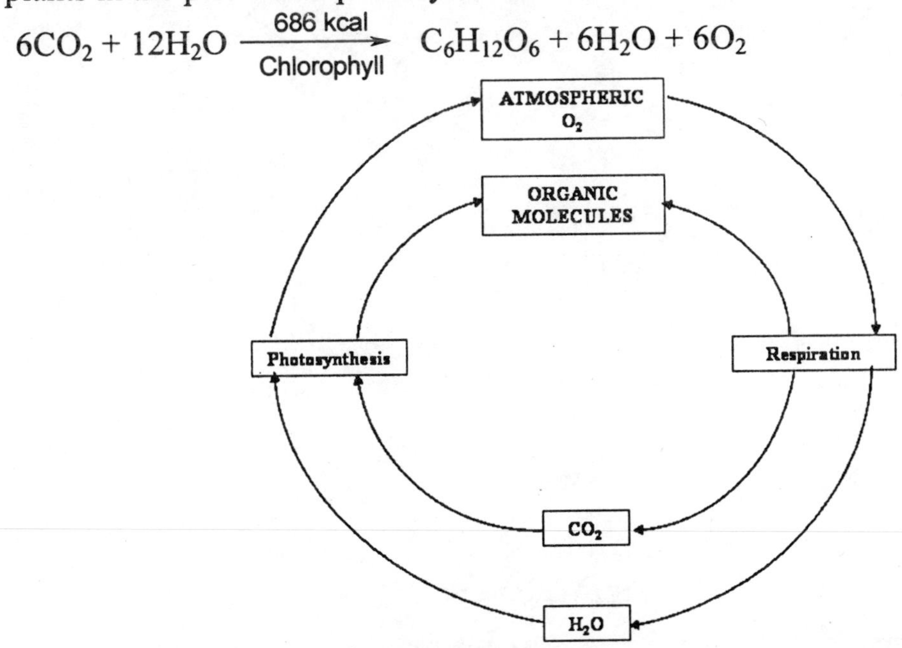

Fig. Oxygen cycle

Oxygen evolved during photosynthesis comes directly from the splitting of water. The process is called **photolysis** of water.

3. Water or Hydrological Cycle: Water is the essence of the living process. The content varies in different tissues of the organism and in different plants and animals. The hydrological cycle refers to the process whereby water gets converted from its liquid or solid state into its vapour state. As a vapour, the water has been capable of travelling considerable distances from its source prior to recondensing and returning to earth as precipitation. It constitutes 65-95% of the body weight of an organism. The living beings obtain regular supply of water through its circulation in the environment. The source of water in an ecosystem is precipitation of water vapours present in the atmosphere. It occurs in the form of rain, dew, hail, snow, etc. Water passes into the atmosphere by evaporation from the surface of moist earth, lakes, streams, oceans, etc. There is a circulation of water from sea to land and back. Plants obtain water from their surroundings. Part of this water becomes a constituent of food and protoplasm. The remaining is lost through transpiration. Animals get water through their food and drink. It is lost to the abiotic environment through sweat, excretion and decomposition.

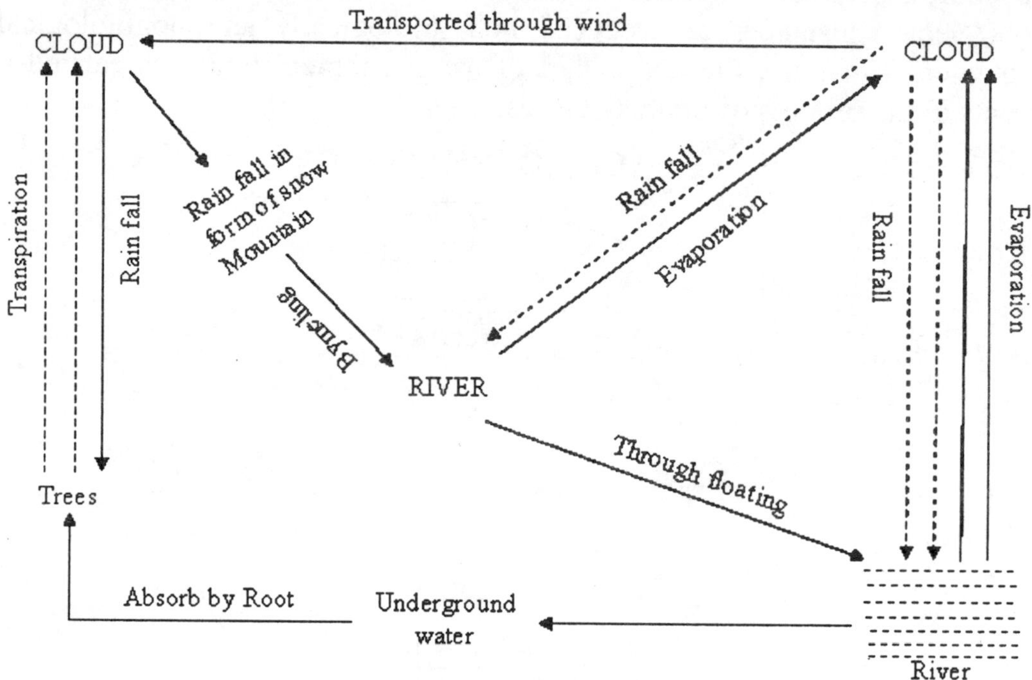

Fig. Water cycle

4. Nitrogen Cycle : Nitrogen is an essential element of all forms of life. In atmosphere, it is 79%. However, in the molecular form, it is never taken directly from atmosphere. The chief source of nitrogen for plants are nitrates in the soil. The atmospheric nitrogen is fixed symbiotically as well as asymbiotically by variety of microbes. Some of the atmospheric nitrogen is fixed by lightening also. In living organisms, nitrogen is found chiefly in proteins. Since protein is constantly being built up and broken down in metabolic activity, it is necessary to make available new source of nitrogen. Nitrogen is cycled from environment to organism back to organism back to environment by one of several paths.

Five types of bacteria are involved as key parts of the nitrogen cycle.

1. Nitrogen Fixing Bacteria – The chief symbiotic nitrogen fixers are *Rhizobium* bacteria found in the roots of leguminous plants. Asymbiotic nitrogen fixers are some blue green algae and species of *Azotobacter* and *Clostridium*. They have the ability to take free nitrogen gas from the atmosphere and convert it into soluble nitrates. Being soluble, nitrates can be taken into the roots of higher plants.

2. Putrefying Bacteria – These bacteria are found chiefly in the soil and in the mud. They break down animal and plant proteins converting them into ammonium compounds. Ammonium compounds are released into the soil where they can be used by other types of bacteria.

3. Nitrite Bacteria – These bacteria (e.g. *Nitrosomonas*) convert ammonium compounds into nitrites. They are soluble in water.

4. Nitrate Bacteria – These bacteria (e.g. *Nitrobacter*) convert nitrites into nitrates which are soluble in water.

5. Denitrifying Bacteria – These bacteria convert nitrates or ammonium compounds into free nitrogen.

The nitrogen cycle indicates that animals are not necessary for the successful operation of an ecosystem. Only the green plants and bacteria are essential.

1. Bacteria are needed because they can use the nitrogen in the atmosphere.
2. Plants are needed because they can use sunlight to synthesize organic compounds.

Animals only enter the cycle by feeding on plants. The nitrogen cycle, as well as the carbon cycle, would operate perfectly well without animals.

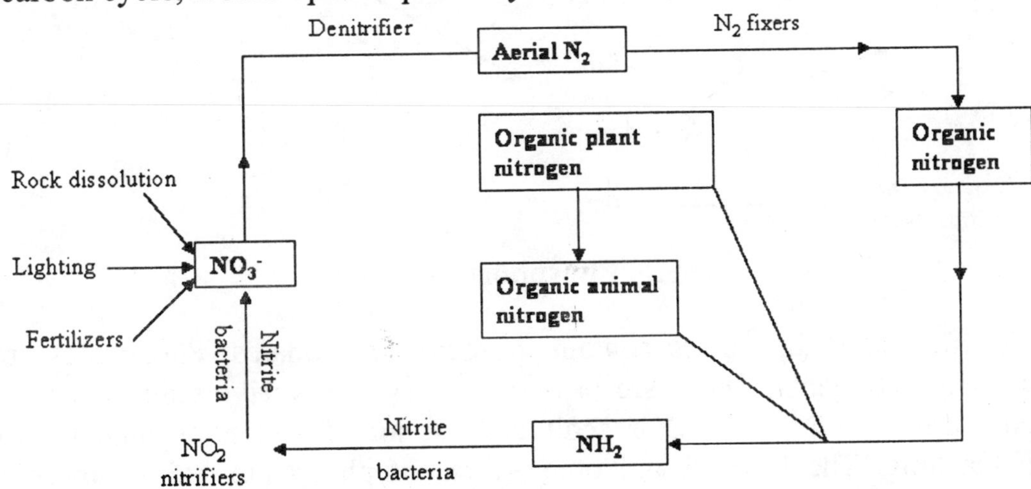

Fig. Nitrogen Cycle

5. Phosphorus Cycle : Phosphorus is a sedimentary cycle. Phosphorus is a component of nucleic acids, ATP, NADP, Phospholipids etc. It lacks atmospheric component. In soil, chief source of phosphates are rocks. Plant takes inorganic phosphates from the soil. The animals feed on plants taking in the organic phosphate, where it is mainly utilized in various biological processes, such as in the formation of teeth and bones. By the death and decay of plants and animals, or by the action of decomposers, the organic phosphates are subsequently returned to earth, where they are again available for recycling. The phosphorus cycle can be shown as follows:

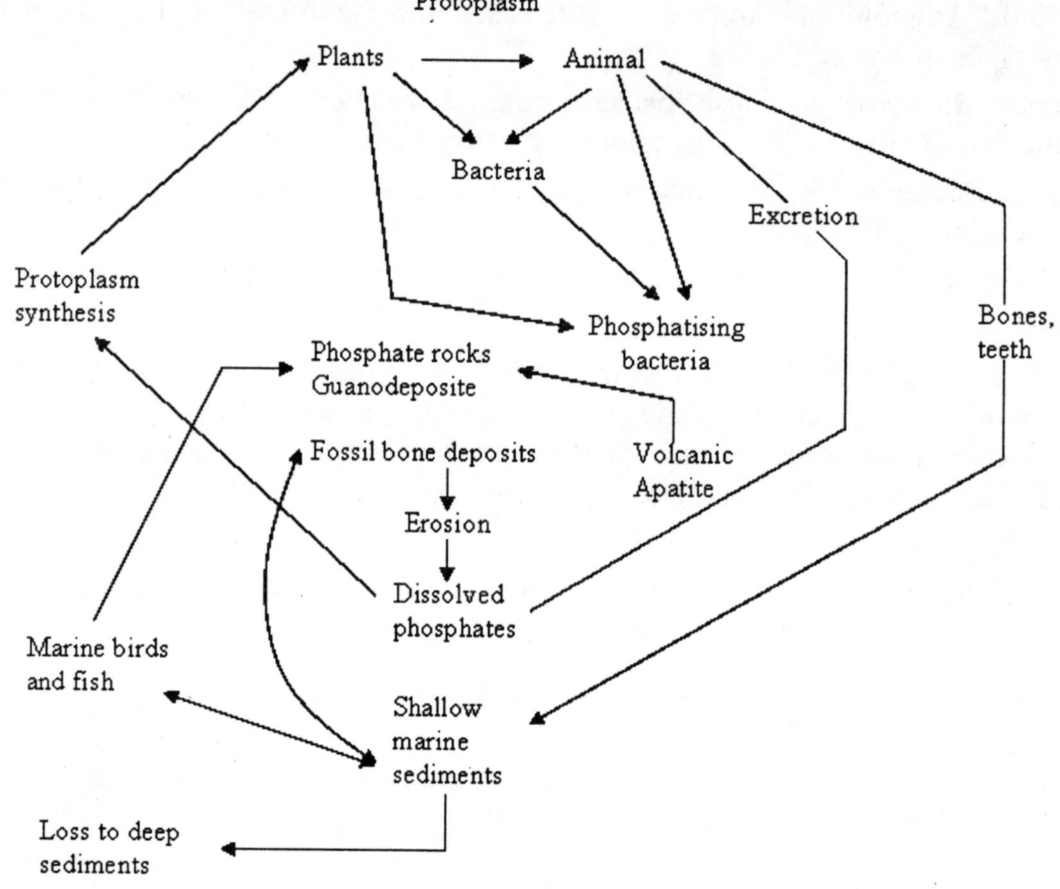

Fig. Phosphorus cycle

Several industrial wastes contain inorganic phosphates. Phosphates are also used in many detergents. These are passed to the streams, rivers and oceans where they settle down at the bottom as sediments. A lot of soil phosphate is also lost through leaching. The leached and other forms of phosphate sediments go out of circulation. They become available only after long geological periods when the sediments come to the earth surface as constituent of rocks.

6. Sulphur Cycle: This cycle provides an excellent example of the interaction and regulation that exits between different mineral cycles and demonstrates the complex biological and chemical regulation within such cycles. Sulphur cycle is a good example to illustrate linkage between the air, water and soil. Like the nitrogen cycle, it also illustrates the key role played by micro-organisms.

Plants and animals require a continuous supply of sulphur and its compounds in order to synthesize some amino acids and proteins. Only a few organisms gain their sulphur requirements in such organic forms as amino acid and cystein. Inorganic sulphate (SO_4) is the major source of biologically significant sulphur. Most of the biologically incorporated sulphur is mineralized by bacteria and fungi in ordinary decomposition by species of *Aspergillus* and *Neurospora*.

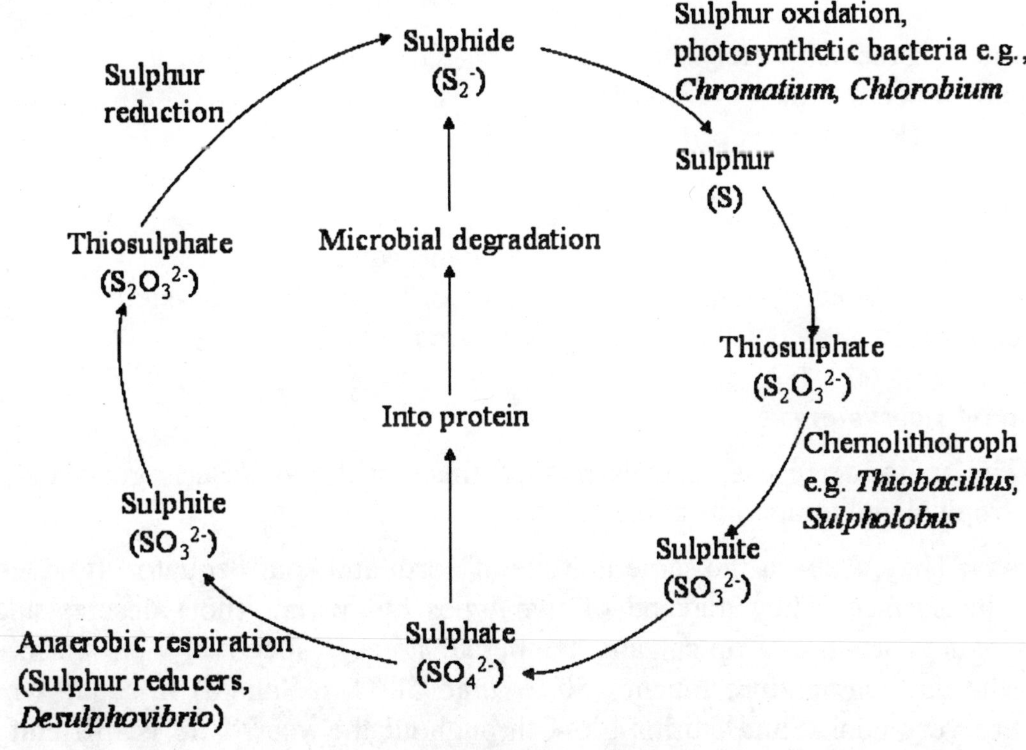

Fig. Sulphur Cycle

Plants absorb sulphur in the form of sulphate. Sulphate is produced in the soil from aerobic breakdown of proteins (e.g., *Penicillium, Neurospora*) and the microbial oxidation of H_2S.

$$2H_2S + O_2 \xrightarrow[\text{Thiobacillus}]{\text{Beggiotoa}} 2S + H_2O$$

$$2S + 2H_2O + 3O_2 \xrightarrow[\text{Thio-oxidants}]{} 2H_2SO_4$$

TYPES OF ECOSYSTEM

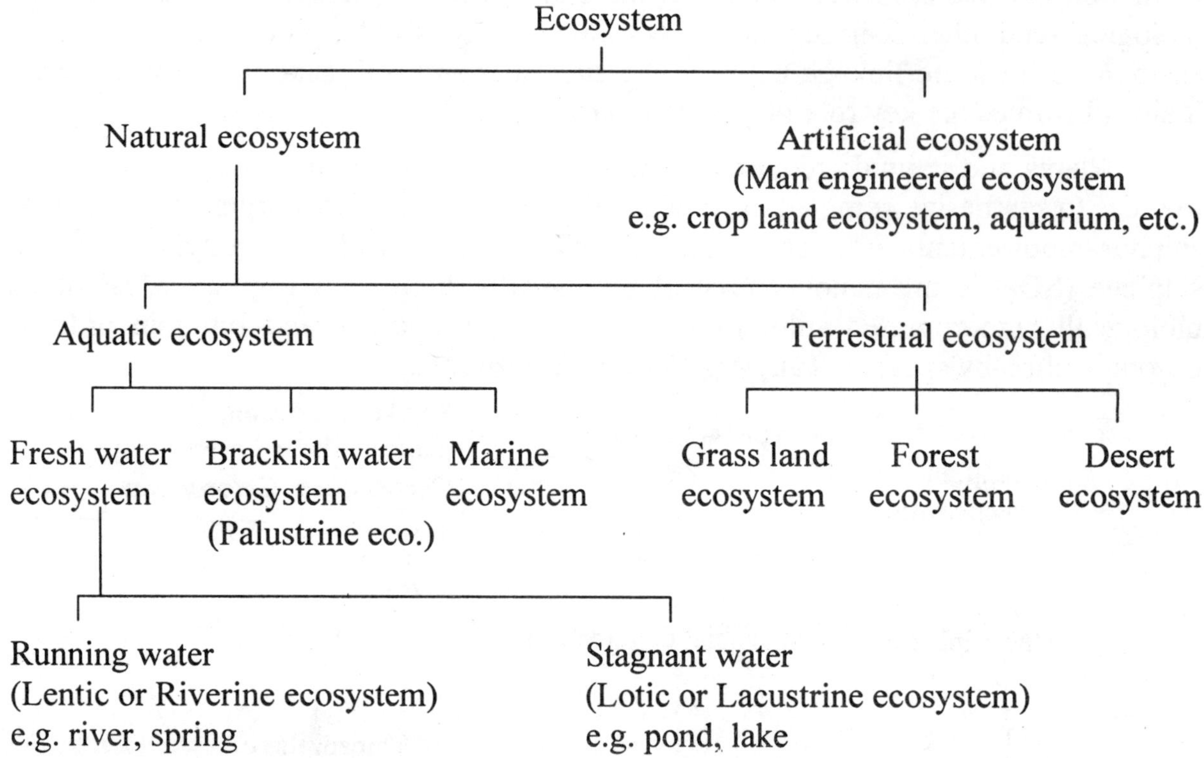

Terrestrial Ecosystems

The major terrestrial ecosystems of the world are desert, grassland, taiga, tundra, tropical and temperature forests, etc.

1. Deserts: They occur at the same latitude of north and south equator. It covers one-fifth of its surface. They may be of two types i.e. **warm** (hot) **deserts** and **cold deserts**. They occur in tropical and temperate areas respectively. In warm or hot deserts the day temperature touches 50°C (upto 58°C in Sahara) in shade while the nights are very cool. Sand storms blow throughout the year. Life is difficult in the deserts because of the very little soil moisture, dry atmosphere, strong winds and temperature extremes. A desert contains sand dunes, sand flats and dry rocks.

About 20% of the land area is occupied by deserts. The major deserts of the world are Death Valley (Great Western Desert, U.S.A.), Gobi (Central Asia), Arabian Desert,.Thar Desert (Indian subcontinent), Antacama (South America), Patagonia (South America), Kalahari (Africa), Sahara (Africa) and Australian Desert. The Thar desert of Indian subcontinent occurs in Rajasthan and adjoining areas of other states (including those of Pakistan).

Primary productivity of the desert is low (3 gm/m^2/yr). Desert fauna is consequently sparse. The desert animals are additionally faced with reduced availability of water and high temperature during the day time in warm deserts. Therefore, most of the desert animals live in burrows. They include lizards, snakes, rats and mice. Other animals are insects, insectivorous birds, hawks, foxes, etc. Some of the insects obtain their water supply from dew drops during the night. Amongst higher animals, camel is known to survive desert conditions. They are able to store water inside its body. The body cells of camel can survive a desiccation of upto 40% as against the limit of 15% in case of man. It has a mechanism to reduce sweat formation (Schmidtelson, 1964) by increasing body temperature during the day and lowering the same during night (upto 35°C) Desert or Kangaroo rat does not take water. It feeds on dry seeds. The only source of water for this animal is the one produced in respiration.

2. Grasslands

India does not have a true grassland. The grassland present in our country owe their origin to the destruction of one or the other type of natural forest vegetation by man, as well as due to other types of biotic pressure, such a grazing. Thus, grasslands are not climax formations in our country, but have developed secondarily. They occupy about 19% of the land surface. The grasslands are characterised by the occurrence of tall or short grasses and herbs. Natural grasslands occur in both temperate areas with an annual rainfall of 25-75 cm and tropical areas with a rainfall of less than 150 cm. Drought is a common controlling factor. At some places fire is known to be essential to maintain grasslands. The overgrazing has changed natural grasslands of Argentina into thorny scrub, because of the reduction of combustible matter and hence lower in frequency of fires. At some places grasslands have been converted into agricultural land. In some places forests have been degraded into grasslands by repeated human interference and overgrazing, rather 40% of the land is used for grazing animals.

Grasses are of two types- **Bunch grass** (forming clumps) and **Sod grass** (arising from rhizome). The common grasses include- *Cenchrus, Agropyron, Dichanthium, Sporobolus, Digitaria, Heteropogon* etc.

A number of herbs, shrubs, and trees occur scattered in the grasslands. Depending upon the size of the grasses and the abundance of non-graminaceous flora, a number of names have been given to grasslands. They are-

i) **Plains**: with short grass

ii) **Prairies**: with tall grass and shrubs

iii) **Pampas**: grassland of Argentina, charcterised by *Cortaderia sellona*. They are called **tussock** in New Zealand, **Veldt** in South Africa and **steppes** in **Eurasia**. Steppes are similar to prairies but with more scattered bushes. Veldt may contain few trees.

iv) **Savannah**: A grassland having scattered trees. One third of Africa is covered by grassland Savannah. The savannah are also common in other areas like- Australia, South Africa, India, Burma etc. Savannah are named after the dominant trees viz. Babul – *Accacia svannah, Eucalyptus savannah, Pine savannah.*

Grasslands are also classified according to their floristic composition for e.g. Whyte 1951 has described 8 types of grasslands-

1. **Schima-Dichanthium**: grow in black soil of Hyderabad, Bombay, M.P., Andhra Pradesh, Tamilnadu etc.
2. **Dichanthium-Cenchrus**: grow in sandy loam soil of Punjab, Delhi, Rajasthan, Gujrat, Cutch, Eastern U.P. and Bihar.
3. **Phragmites-Saccharum**: found in marshy areas of Terai of U.P., Bihar, Bengal, Assam, Sunderban etc.
4. **Bothriochloa**: found in high rainfall and low lying areas of Lonavala tract of Maharashtra.
5. **Cymbopogon**: Grow in low hills like Western Ghats, Vindhyas, Satpura, Aravali ranges etc.
6. **Arundinella**: Grow in high mountains like Western Ghats, Nilgiris, Himalayas, Eastern Punjab etc.
7. **Deyeuxia-Aurndinella**: Mixed temperate climate like Himalayas, Eastern, Punjab, Himachal etc.
8. **Deschampsia-Deyeuxia**: found in temperate alpine climate like Himalayas, Kashmir etc.

Primary consumers of a grassland ecosystem include- insects, termites, millipedes, grain eating birds and herbivorous mammals like rabbits, mice, buffalow, cow, goat, sheep, horse etc. Carnivorous animals are frogs, toad, snakes, hawks, owl, jackal, fox, wolf, tiger, lion etc.

3) Taiga

It occurs in the subarctic regions from North America to Eurasia. The summer is cool but the winters are snowy. The rainfall ranges between 40 cm to 100 cm per year. The ground is covered by litter, hence soil is acidic. The dominant vegetation of the ecosystem is coniferous therefore, taiga is also called **Evergreen** or **temperate coniferous forest**. The dominant trees are pine, fur, *Cedrus, Taxus*, *Quercus,* etc.

Insects, birds, hawks, squarrel, fox, musk, bear are the common animals.

4) Tundra

Tundra literally means treeless plains, and defines the community above the tree line on mountains, to the zone of perpetual snow and ice. The area lies between polar ice and the taiga in the arctic region. Tundra is found only in the northern hemisphere, because the corresponding region of the southern hemisphere is covered by Antarctic ocean. Tundra like ecosystem occurs in the alpine part of the mountains. It is often called **alpine tundra**. Tundra occupies about 8×10^6 km^2 area of the land mass.

Tundra area is covered by snow for most part of the year. The climate is extremely cold. Winter temperature is –30°C to –40°C. It is accompanied by strong winds and snow storms. The highest summer temperature is 10°C. Summer has rain and fog. It is unable to melt the snow except from top layers of the soil. The remaining part of the soil is covered with soil called **permafrost condition**. Vegetation is sparse i.e. **Elfine scrub**. Mosses and lichen grow easily. Anemony primula and Artemesia is the common animal.

5) Forest ecosystem

A type of terrestrial natural ecosystem dominated by trees. About one third of the earth's surface and about 14% of the land surface in India is covered by the forests. The forests are of several types which is as follows-

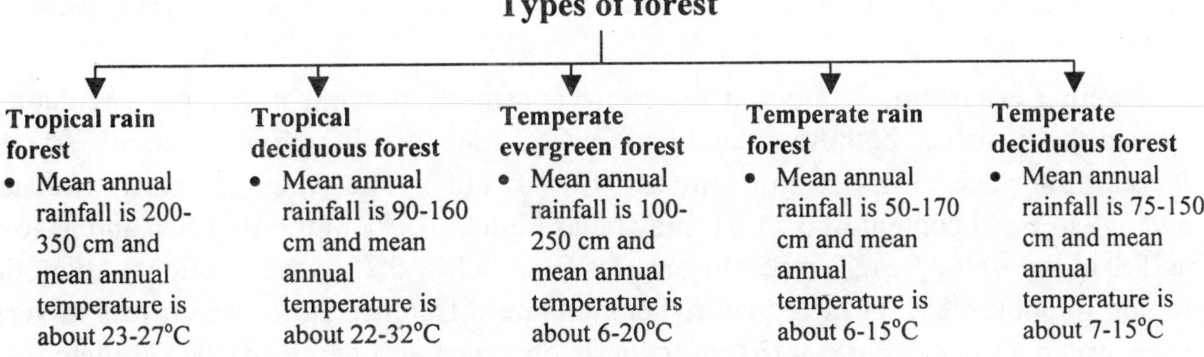

Aquatic Ecosystems

Water continues to be an essential requirement of life. We cannot imagine existence of life without water. Water has some unique features which are as follows:

- An excellent solvent and medium for chemical reactions.

- Liquid state between 4°C to 90°C which is suitable for existence of life.
- Good source of molecular oxygen and hydrogen in living cells.
- Protects organisms from thermal shock because it does not allow quick change in water.
- Water provides support to the aquatic organisms. This allows them to attain large size.

Light is deficient inside the aquatic habitat. Turbid and deep waters do not receive much light. Temperature changes are less as compared to land masses. Surface of water develops waves. Tides occur on the sea shore.

Depending upon their size and habit, the organisms found in the aquatic ecosystem are as follows:

(i) **Plankton**. Small, passively drifting surface organisms which comprise microscopic plants (**phytoplankton**) and animals (**zooplankton**).

(ii) **Neuston**. Large organisms found floating or swimming on the surface of water. They are of two types; **epineuston** (living on water surface) and **hyponeuston** (hanging from water surface, e.g., mosquito larvae).

(iii) **Nekton**. Organisms actively swimming inside water.

(iv) **Periphyton**. Organisms occurring attached to submerged vegetation above the bottom.

(v) **Benthos**. Organisms confined to the bottom.

Aquatic ecosystems are of several types- pond, lake, stream or river, estuary, ocean.

1. Ocean Ecosystem. The ocean water are constantly moving in a great circulatory system that involves both horizontal and vertical transfers. It comprises about 70% of the earth surface (361×10^6 km^2 out of 510×10^6 km^2). The water is saltish with an average mineral content of 3.5%. It may be as high as 4.5% in the Red Sea and as low as 1% in the Baltic Sea. Temperature varies from below 0°C in the Arctic or Antarctic region to about 28°C at the equator. Temperature difference is, however, less in very deep water. In hotter areas a **thermocline** is observed at a depth of 20-200 metre. At places cold or hot, water currents run in the ocean. Circulation water occurs due to wind stress, tides and temperature differences.

Ocean can be divided into four regions- continental shelf, continental slope, abyss and hadal regions.

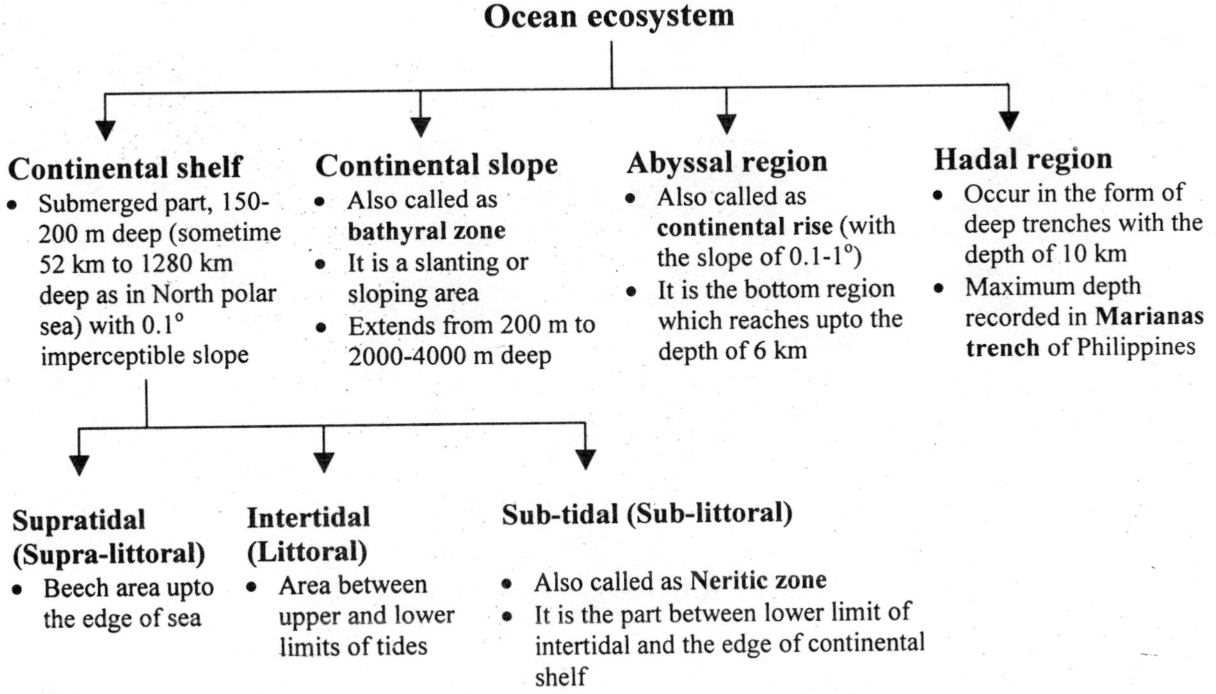

Hedgpeth (1957) has divided the environment of ocean in two parts i.e. **benthic** and **pelagic**

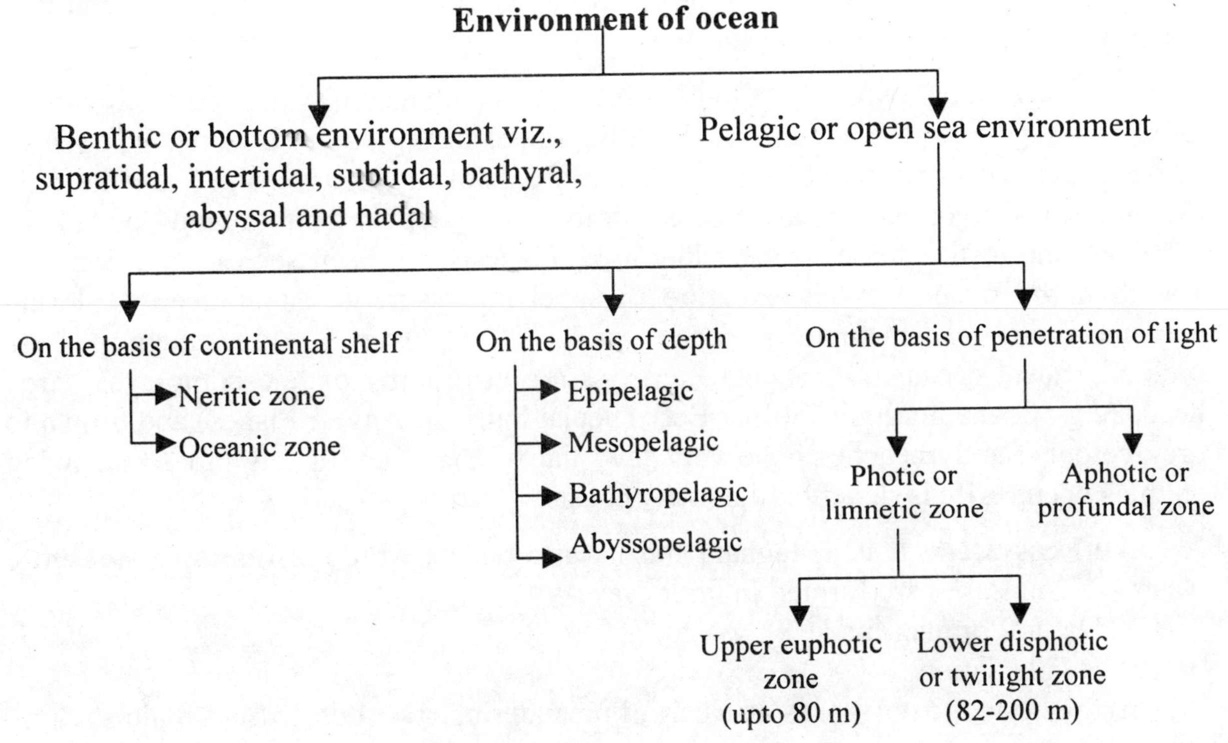

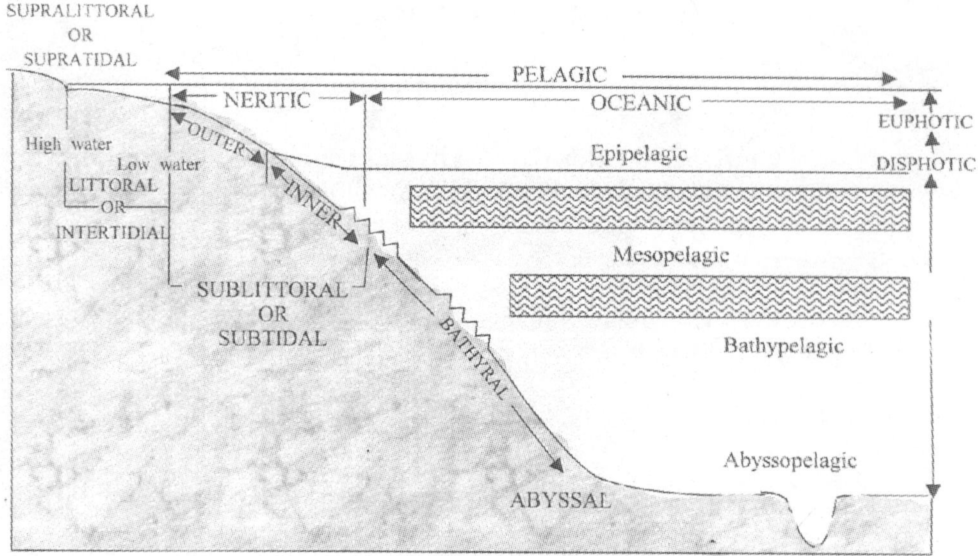

Fig. Marine Environment

Light penetration is sufficient in the euphotic region for the plants to carry out photosynthesis above compensation point. In the disphotic or twilight region light is sufficient for the plants to survive but not to perform photosynthesis above compensation point. Some light can also penetrate the ocean water upto 300 metres deep in the profundal or aphotic zone (total penetration 300 + 200 = 500 m) but it is insignificant for the plant life to survive.

The surface of water both in the open sea as well as along the coasts contains a large number of phytoplanktons (4-5 million/litre) in the euphotic zone. In open or oceanic area, phytoplanktons and the larger free floating marine algae are the only producers. Carnivorous animals pass down to greater depths from the mesopelagic or twilight zone to the depth of the hadal area. Fishes have been seen upto a depth of 10,300 m by means of **bathyscaphe** (a special pressure resistant deep exploring container) and caught upto a depth of 7,130 m. Over 90% of them possess bioluminescent organs. Other characteristics are protruding or telescopic eyes, large heads (e.g., Granadier Fish, Gulper Eel), special baits (in Angler Fishes) and brown to grey colour. Sandy beaches have very few plants. Crabs and a few other burrowing animals occur. Life is richer on the rocky coasts.

2. Lake Ecosystem. It is a stagnant fresh water bodies which also known as **Lentic** water system. Lakes are formed in three ways-

(i) Natural or man-made
(ii) By glaciation
(iii) Ox-bow or cutoff lakes as a part of meandering river cutoff from main stream.

Deep lakes show both thermal and photic zonation. Like ocean, the banks of the lakes constitute the **littoral area**. The depth to which light penetrates upto the compensation point is known as **photic** or **limnetic zone**. The deeper parts, where light is below compensation point, is termed as **profundal** or **aphotic zone**.

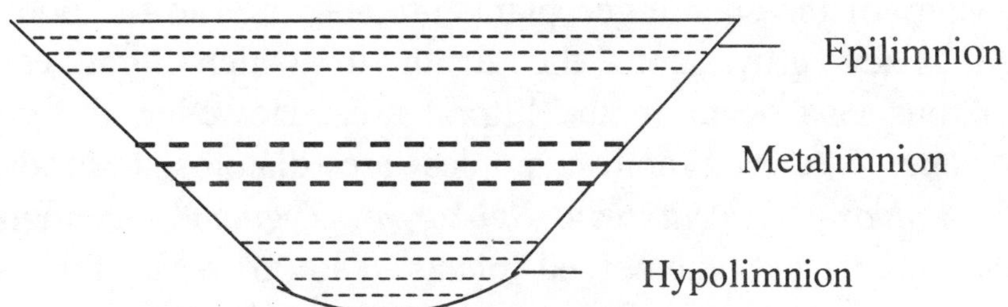

Fig. Thermal zonation in a deep lake

Three temperature strata are found in deep lakes-epilimnion, metalimnion and hypolimnion. **Epilimnion** is the upper stratum of water which is exposed to solar radiations. It is warm during the summer and comparatively cooler during the winter. Water keeps on circulating. **Hypolimnion** is the basal stratum where water is always cool. It does not circulate. **Metalimnion** is the transition zone between epilimnion and hypolimnion where temperature changes occur. The maximum temperature change occurs in the middle part of the metalimnion called **thermocline**. In temperate lakes, surface layers of water begin to cool off during autumn. As a result, temperature of the different strata becomes almost same. It causes mixing of water and helps to provide oxygen to deeper layers. In winter, temperature of epilimnion drops off and circulation stops. However, in spring the temperature rises again causing another circulation. Such lakes having twice circulation of temperature or water are called **dimictic**. Complete inversion or circulation in temperature stratification is absent in tropical lakes except for overturning of water once a year. Such lakes are termed as **monomictic**. Polymictic lakes show irregular circulation at irregular intervals.

On the basis of their productivity lakes are of two types- **oligotrophic** and **eutrophic**. Oligotrophic lakes have very little plant and animal life. Sambar lake of Rajasthan is one of them. It has a high salt content which makes it unfit for the growth of fresh water plants. Only a few blue-green algae grow happily in the lake, for e.g., *Anabaenopsis* and *Spintlina*. Amongst animals, brine shrimp and brine flies are common.

Eutrophic lakes are comparatively shallower, e.g., Dal lake of Srinagar (Kashmir). Some of the lakes possess floating soils, e.g., Khaziar lake near Chamba in

H.P., Dal lake of Kashmir. In the littoral area or bank of a lake are found amphibious plants (e.g., *Typha, Sagittaria*), free floating plants (e.g., *Pistia, Eichhornia, Lemna, Spirodela, Wolffia*), floating leaved hydrophytes (e g., *Nelumbo, Nymphaea, Cabomba, Potanlogeton* species), and submerged hydrophytes (e.g., *Potamogeton, Ceratophyllum, Zannichellia, Hydrilla, Vallisneria, Anacharis, Elodea, Chara, Nitella*). Many of the submerged plants are anchored to the bottom while others are suspended. They grow upto the depth to which light can penetrate. Some phytoplanktons also occur in the littoral area. However, they are abundant in the limnetic zone. The common examples are diatoms desmids, *Chlamydomonas, Chiarella, Volvox, Hydrodictyon, Spirogyra, Zygnema, Oedogonium, Cladophora*, etc. Some unattached submerged plants also grow in this area. At times the phytoplanktons and the free floating plants occur in blooms and over shadow other plants. No plant life occurs in profundal or aphotic zone.

The consumers of the littoral area are of several types. The zooplanktons include protozoa, water fleas, cyclopods, rotifers, ostracods, etc. The neuston organisms include water boatman and water spiders. The nekton or swimming forms are adults and larvae of diving beetles, hemiptera, larval stages of diptera, amphibians, tunks, water snakes, etc. The benthos are crabs, prawns, isopods, nymphs of several arthropods, mussels, clams, annelids, snails, etc. Fishes occur throughout the area.

The limnetic consumers are zooplanktons but their species are often, different from those of littoral area. Swimming animals or nektons are mostly fishes feeding on planktons as well as suspended plants. In the profundal area are found bacteria, fungi, bloodworms annelids and some clams. Many birds visit the lakes at various point to feed on fish and other animal life. They include ducks, crane herons and kingfishers.

3. Pond Ecosystem. It is also known as **lotic water ecosystem**. It is a small water reservoir which may or may not dry up during dry periods. In temporary ponds, living community is unstable but in permanent ponds a stable community can be found except for seasonal variations. Seasonal variations appear due to change in temperature, pH, rainfall and annual growth of some organisms. The various components of a pond ecosystem are as follows:

Abiotic Component. The quality and amount of water, frequency of rainfall, temperature, light penetration, pH, turbidity, O_2 content, CO_2 content, inorganic salts as well as organic matter are the constituents of the abiotic component of the pond ecosystem. They determine the productivity and richness of the biotic component.

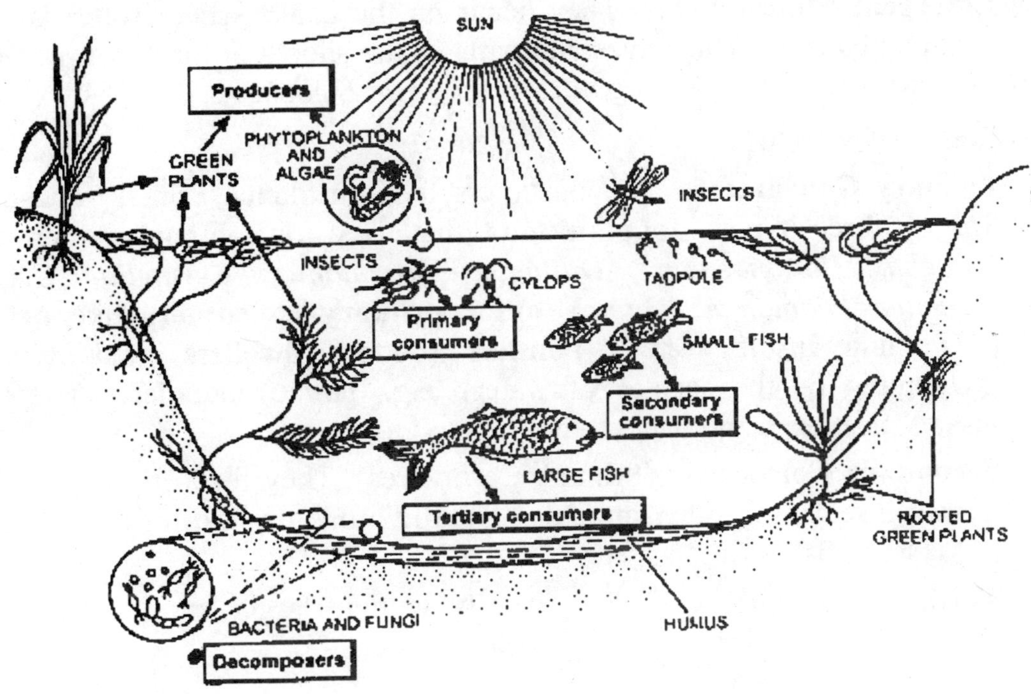

Fig. showing a Pond Ecosystem

Biotic Component. It consists of four types of organisms- producers, consumers, detrivores and decomposers.

(a) Producers. Producers are autotrophic plants that are able to trap radiant energy and manufacture organic food. The different producers of a pond ecosystem are-

(i) **Phytoplanktons**, e.g., Diatoms, *Euglena, Chlamydomonas*.

(ii) **Free Floating Macrophytes**, e.g., *Wolffia, Azolla, Salvinia, Spirodela, Lemna, Eichhornia, Pistia*. They may cover the entire surface of water by their growth and shade the remaining plant life to death. *Eichhornia* is regarded as an aquatic weed because it often grows in blooms and permanently change the entire ecosystem of a pond.

(iii) **Anchored Submerged Macrophytes**. They grow in deeper waters and include plants attached to the bottom, e.g., *Chara* (an alga), *Hydrilla, Vallisneria, Potamogeton* species, *Ceratophyllum, Zannichellia*. In later stages some of them detach from the bottom and become suspended.

(iv) **Floating Leaved Anchored Macrophytes**. They occur in less deep water. The leaves have long petioles which can adjust themselves in such a position as to keep the lamina floating on the surface of water. The upper surface of the leaf blade is unwettable, e.g., *Victoria, Nelumbo, Nymphaea*.

(v) **Emergent Macrophytes**. They occur on the banks where water is a few centimetres deep. The leaves and parts of the shoots of the plants come out of water, e.g., *Typha, Phragmites, Sagittaria, Eleocharis*.

(b) Consumers. They include-

(i) **Primary Consumers**. The consumers feed on plants. Primary consumers are of three types: **zooplanktons** that feed on phytoplanktons (e.g., *Amoeba, Paramaecium, Dilflugia, Brachionus, Keratella, Cyclops, Daphnia, Diaptomus, Moina*), **other primary consumers** like **nektons** (swimming inside water), **benthos** (bottom dwellers) and **neustons** (swimming on the surface of water), e.g., larvae, tadpoles, some bugs, insects, beetles, some fishes, etc.

(ii) **Secondary Consumers** (Primary Carnivores). They feed on zooplanktons, larvae, insects and other animals of herbivorous trophic level, e.g., frogs, bugs, some fishes, insects, etc.

(iii) **Tertiary Consumers**. They generally include larger carnivorous fishes, and carnivorous birds like King Fisher, Stork.

(c) Detrivores (Scavengers): They are animals which feed on the dead bodies of other organisms. The detritus feeders occur on the bottom of the pond ecosystem (benthos). The animals help in quick breakdown of the organic remains, e.g., *Chironomus* larva, some Water Skaters, worms, beetles, some small fishes.

(d) Decomposers: They are mainly bacteria, actinomycetes and fungi. The decomposers cause decomposition of organic matter by excreting digestive enzymes. The solubilised food is absorbed by the decomposers. In the process, a number of minerals are released. Like detrivores, the decomposers live on the bottom of the pond. As a result of their decomposing activity the bottom of the pond is rich in H_2S. Both decomposers and detrivores become food for the consumers.

4. Stream or River Ecosystem: It is a **lentic water ecosystem**. At high altitude, they flow rapidly hence very little plant growth exists while in plains the flow is less turbulent. However, in the rainy season most of the streams and rivers possess muddy water which does not allow the light to penetrate. The banks of the streams and rivers may possess marshy areas. Flowing water ecosystem has uniform oxygen tension with no thermal or chemical zonation. It is not completely dependent upon aquatic plants as producers. Detritus eaters and decomposers are comparatively more abundant. They obtain their food from the dead and living land organisms which happen to fall accidently or are brought down by rain water.

Producers are fewer due to seasonal increase in turbidity and constant flow of water. They are phytoplanktons, some attached algae (e.g., *Cladophora*), aquatic mosses and submerged seed plants. Reeds occur on the marshy banks. Larvae, snails and other animals typical of the pond ecosystem occur in the area. Besides the riverine ecosystem contains several types of fishes, crustaceans, crocodiles and some mammals. They do not show horizontal zonation. Instead longitudinal zonation may occur.

MASTER STROKES

Biomes or Life Zones

As the temperature and rainfall vary with geographic latitude and altitude, they divide the soil covered land surface into a number of distinct habitat zones or biomes. **A biome is, therefore, the largest terrestrial ecological unit, which is characterized by interaction of flora, fauna and abiotic substances.**

Of all the climatic factors, temperature and rainfall determine the nature of the biome to a large extent. For example, rain forests occur in tropical regions where temperature are warm and rainfall is plentiful throughout the year. Deserts, on the other hand, are found in tropical and temperate regions where rainfall is minimal. A biome gets its name from its climax or dominant plant form. For example, a grass biome is an area in which grasslands predominate; a coniferous biome is an area of conebearing evergreen trees. The various zoogeographic regions contain more than one biome. The oriental region, for instance, contain a tropical rainforest, a tropical deciduous forest etc., showing that the temperature and rainfall determine the nature of the biomes to a large extent.

Latitudinal Zones

There are three latitudinal zone on the earth:

(a) Tropical zone (Torrid zone): This zone is present between the tropic of cancer $\left(23\frac{1}{2}{}^{\circ}\mathrm{N}\right)$ and the tropic of Capricorn $\left(23\frac{1}{2}{}^{\circ}\mathrm{S}\right)$. In this zone, the sun's rays are vertically overhead during a part of the year. Therefore, this zone has maximum temperature on the earth.

(b) Temperate zones: These are the regions between the tropic of cancer and arctic circle $\left(23\frac{1}{2}{}^{\circ}\mathrm{N}\ \text{To}\ 66\frac{1}{2}{}^{\circ}\mathrm{N}\right)$ and the tropic of capricorn and the antarctic circle $\left(23\frac{1}{2}{}^{\circ}\mathrm{S}\ \text{To}\ 66\frac{1}{2}{}^{\circ}\mathrm{S}\right)$. In these zones, the sun's rays are never vertical during the year. The angle of incidence, the duration of sunshine are greater in summer than in winter. These regions are less hot than tropical zone.

(c) Polar zones (Frigid zones): These zones are the zones surrounding the poles. They extend upto the arctic circle in the Northern hemisphere and upto the Antarctic circle in the Southern hemisphere. In these zones, the sun's rays are not received during the long winter. The temperature is quite low even in summer.

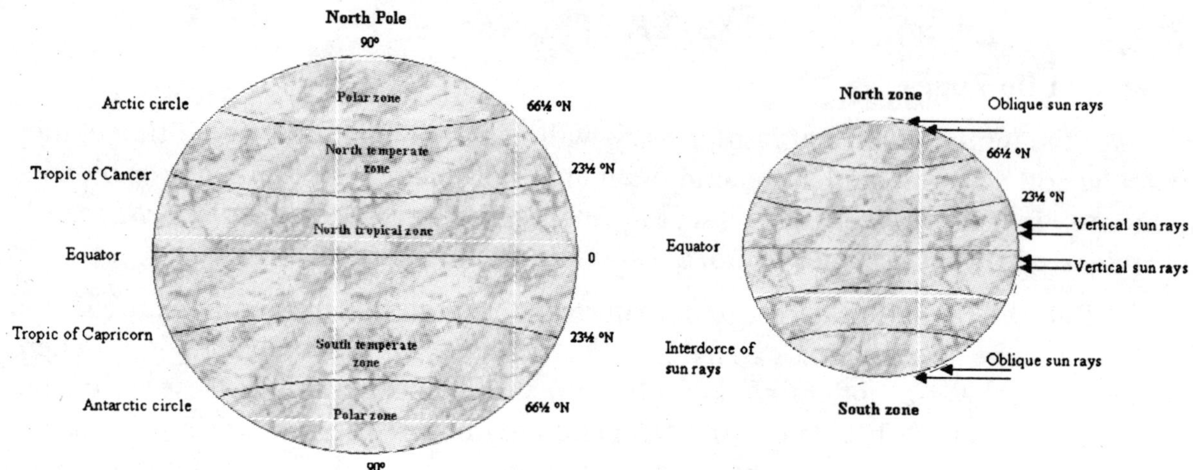

Fig. Temperature zone

Vegetation and annual rainfall:

(a) Evergreen tropical forest → 250-400 cm.
(b) Tropical deciduous forest → 100-200 cm.
(c) Taiga → 100-250 cm.
(d) Temperate deciduous forest → 75-150 cm.
(e) Chapparal → 50-75 cm.
(f) Grassland and Savanna → 25-750 cm.
(g) Desert → less than 25 cm.

ENTRANCE CORNER

1. Niche overlap indicates: **(CBSE 2006)**
(a) active co-operation between two species
(b) two different parasites on the same host
(c) sharing of one or more resources between the two species
(d) mutualism between two species

2. Which one of the following is not used for construction of ecological pyramids? **(CBSE 2006)**
(a) Dry weight (b) Number of individuals
(c) Rate of energy flow (d) Fresh weight

3. Which one of the following pairs is mismatched: **(CBSE 2005)**
(a) Tundra - permafrost
(b) Savanna - *Acacia* trees
(c) Prairie - epiphytes
(d) Coniferous forest - evergreen trees

4. Which of the following ecosystem has the highest gross primary productivity: **(CBSE 2004)**
(a) Grassland (b) Coral reef
(c) Mangroves (d) Tropical rain forest

5. Highest value in g/m^2/yr of a grassland ecosystem would be: **(CBSE 2004)**
(a) gross primary production (b) net primary production
(c) secondary production (d) tertiary production

6. A plant being eaten by a herbivorous which in turn is eaten by a carnivorous makes: **(CBSE 2002)**
(a) food chain (b) food web
(c) omnivorous (d) interdependent

7. Decomposers are: **(CBSE 2001)**
(a) Animalia and Monera (b) Protista and Animalia
(c) Fungi and Plantae (d) Bacteria and Fungi

8. The frog that feeds on an insect is a: **(CBSE 2000)**
(a) tertiary consumer (b) decomposers
(c) primary consumer (d) secondary consumer

9. Who proposed the concept of ecological pyramids? **(CBSE 2000)**
(a) Odum (b) Elements
(c) Tansley (d) Elton

10. Maximum biomass of autotrophs in oceans is made of: **(CBSE 2000)**
(a) benthic brown algae, coastal red algae and daphnids
(b) benthic diatoms and marine viruses
(c) sea grasses and slime moulds
(d) free floating micro-algae, cyanobacteria and nanoplankton

11. The organism feeding on herbivorous inset is : **(CBSE 2000)**
(a) primary consumer (b) secondary consumer
(c) tertiary consumer (d) top carnivore

12. Which part of the world has high density of organisms? **(CBSE 1999)**
(a) Grasslands (b) Savannas
(c) Deciduous forests (d) Tropical rain forests

13. The circulation or cycling of elements in an ecosystem is known as: **(CBSE 1999)**

(a) geological cycling (b) geo-chemical cycling
(c) bio-geochemical cycling (d) nutrient cycling

14. What is true of ecosystem ? **(CBSE 1998)**

(a) Primary consumers are least dependent upon producers
(b) Primary consumers out-number producers
(c) Producers are more than primary consumers
(d) Secondary consumers are the largest and most powerful

15. Role of bacteria in carbon cycle is: **(CBSE 1998)**

(a) photosynthesis
(b) chemosynthesis
(c) breakdown of organic compounds
(d) assimilation of nitrogen compounds

16. Plants such as *Prosopis, Acacia* and *Capparis* represents examples of tropical: **(CBSE 1998)**

(a) grassland (b) thorny deserts
(c) deciduous forests (d) evergreen forests

17. Relatively more abundant animals of desert ecosystem are: **(CBSE 1998)**

(a) arboreal (b) fossorial
(c) diurnal (d) aquatic

18. During adverse seasons 'therophytes' survive by : **(CBSE 1997)**

(a) bulbs (b) tubers
(c) rhizomes (d) seeds

19. Which of the following ecosystem has highest rate of gross primary production? **(CBSE 1997)**

(a) Grasslands (b) Mangroves
(c) Coral reefs (d) Equatorial forest

20. In any food chain the largest population is that of: **(CBSE 1996)**

(a) primary consumers (b) tertiary consumers
(c) producers (d) decomposers

21. The transfer of energy from a level to another is governed by law of thermodynamics. The average efficiency of energy transfer from herbivores to carnivores level is: **(CBSE 1996)**

(a) 10% (b) 25%
(c) 5% (d) 50%

22. In every ecosystem/food chain, the green plants are: **(CBSE 1996)**
(a) decomposers (b) producers
(c) consumers (d) none of these

23. The study of the interactions of the individual organism or a single species with the living and non-living components of its environment is: **(CBSE 1995)**
(a) autecology (b) autoecious
(c) autogamy (d) autonomic

24. Who are top carnivores ? **(CBSE 1995)**
(a) Insectivore plants
(b) Fox
(c) Hag fish
(d) Carnivores that are not eaten by animals

25. What are biotic components ? **(CBSE 1995)**
(a) The non-living components (b) The living components
(c) Both (a) and (b) (d) None of the above

26. The character of an ecosystem is determined by the environmental factor which is in shortest supply. This is the: **(CBSE 1994)**
(a) law of minimum (b) law of diminishing returns
(c) law of limiting factors (d) law of supply and demand

27. Bulk CO_2 fixation occurs in: **(CBSE 1994)**
(a) crop plants (b) oceans
(c) tropical rain forests (d) temperature forests

28. In grass-deer-tiger food chain, grass biomass is one tonne. The tiger biomass shall be: **(CBSE 1994)**
(a) 100 kg (b) 10 kg
(c) 200 kg (d) 1 kg

29. In a biotic community, the most important factor for survival of an animal is: **(CBSE 1994)**
(a) day length (b) soil moisture
(c) green food (d) predators

30. In a food chain, the largest population is that of: **(CBSE 1994)**
(a) producers (b) decomposers
(c) secondary consumers (d) primary consumers

31. Second most important trophic level in a lake is: **(CBSE 1994)**
(a) zooplankton (b) phytoplankton
(c) benthos (d) neuston

32. Tropical forests occur in India : **(CBSE 1994)**
(a) Jammu and Kashmir (b) Rajasthan
(c) Kerala and Assam (d) The forests do not occur in India

33. The order of organism in an aquatic food chain is: **(AFMC 2006)**
(a) Bacteria → Seal → Diatom → Fish → Crustacea
(b) Crustacea → Seal → Fish → Polar Bear → Diatom
(c) Polar Bear → Diatom → Seal → Crustacea
(d) Diatom → Crustacea → Fish → Seal → Bacteria

34. Deciduous forests have: **(AFMC 2004)**
(a) Variety of grasses (b) Broad-leaved trees
(c) Narrow-leaved trees (d) Variety of crocodiles

35. Broad-leaved oaks are found in: **(AFMC 2004)**
(a) tropical evergreen forests (b) temperature deciduous forest
(c) north coniferous forest (d) tropical deciduous forest

36. Nitrogen content of biosphere remains constant due to: **(AFMC 2001)**
(a) nitrogen fixation (b) nitrogen cycle
(c) industrial pollution (d) absorption of nitrogen

37. The organisms dwelling at the bottom of a lake are called: **(AFMC 2001)**
(a) Phytoplanktons (b) Zooplanktons
(c) Nektons (d) Benthos

38. An aquatic plant introduced from America to check pollution has become trouble some weed in India: **(AFMC 2000)**
(a) *Opuntia* (b) *Pistia*
(c) *Eichhornia* (d) *Aegilops*

39. One of the following exist in abundance in terms of genera and species: **(AFMC 2000)**
(a) herbivores mammals (b) carnivore mammals
(c) aquatic mammals (d) terrestrial mammals

40. Correct sequence in a food chain is: **(AFMC 1999)**
(a) grass→insects→birds→snake (b) grass→insects→insects→deer
(c) grass→wolf→deer→buffalo (d) bacteria→grass→rabbit→wolf

41. In a food chain, the total amount of living material, is depicted by: **(AFMC 1998)**
(a) pyramid of energy (b) pyramid of biomass
(c) pyramid of numbers (d) none of these

42. An ecosystem does not normally alter because it is in a state of: **(AFMC 1997)**
(a) homeostasis (b) imbalance
(c) deficient light (d) deficient components

43. 10% energy transfer law in food chain was first given by: **(AFMC 1997)**
(a) Lindeman (b) Tansley
(c) Elton (d) Raunkiaer

44. **Assertion**: Network of food chains existing together in an ecosystem is known as a food web.
Reason: An animal like kite cannot be a part of a food web. **(AIIMS 2006)**
(a) If both, the assertion and the reason are true and the reason is a correct explanation of the assertion
(b) If both, the assertion and reason are true but the reason is not correct explanation of the assertion
(c) If the assertion is true and the reason is false
(d) If both the assertion and the reason are false

45. Which one of the following is correct matching of a plant, its habit and the forest type where it normally occurs? **(AIIMS 2005)**
(a) *Prosopis*, tree, scrub
(b) *Saccharum*, grass, forest
(c) *Shorea robusta*, tree, coniferous forest
(d) *Acacia catechu*, tree, coniferous forest

46. Given below is one of the types of ecological pyramids. This type represents: **(AIIMS 2005)**

(a) pyramid of numbers in a grassland
(b) pyramid of biomass in a fallow land
(c) pyramid of biomass in a lake
(d) energy pyramid in a spring

47. Great barrier reef along east coast of Australia is a: **(AIIMS 2004)**
(a) population (b) community
(c) biome (d) ecosystem

48. Mr. X is eating curd/yoghurt. For this he is occupying trophic level: **(AIIMS 2003)**

(a) first (b) second
(c) third (d) fourth

49. **Assertion**: In a food chain members of successive higher levels are fewer in number.

Reason: Number of organisms at any trophic level depends upon the availability of organisms which serve as food at the lower level. **(AIIMS 2003)**

(a) If both, the assertion and the reason are true and the reason is correct explanation of the assertion
(b) If both, the assertion and reason are true but the reason is not correct explanation of the assertion
(c) If the assertion is true and the reason is false
(d) If both the assertion and the reason are false

50. **Assertion**: Tropical rain forests are disappearing fast from developing countries such as India.

Reason: No value is attached to these forests because these are poor in biodiversity. **(AIIMS 2003)**

(a) If both, the assertion and the reason are true and the reason is a correct explanation of the assertion
(b) If both, the assertion and reason are true but the reason is not correct explanation of the assertion
(c) If the assertion is true and the reason is false
(d) If both the assertion and the reason are false

51. Organisms found on the bottom of sea are: **(AIIMS 2002)**

(a) planktons (b) neuston
(c) benthos (d) phytons

52. The limiting factor in soil nitrification is: **(AIIMS 2000)**

(a) soil pH (b) light
(c) temperature (d) air

53. The flora and fauna in lakes or ponds are: **(AIIMS 2000)**

(a) Lentic biota (b) Lotic biota
(c) Abiotic biota (d) Field layer

54. In an ecosystem, trophic level of man is: **(AIIMS 1999)**

(a) producer (b) herbivore
(c) omnivore (d) carnivore

55. Moderate rain during summer produces: **(AIIMS 1998)**
(a) desert (b) grassland
(c) scrub forest (d) deciduous forest

56. Relationships in an ecosystem can be depicted through: **(AIIMS 1998)**
(a) pyramid of energy (b) pyramid of biomass
(c) pyramid of numbers (d) all of these

57. In India tropical wet evergreen rain forests are not found in: **(AIIMS 1998)**
(a) Tamil Nadu (b) Andaman
(c) West Bengal (d) Madhya Pradesh

58. Dense everygreen vegetation of broad sclerophyllous leaves and shrubs with fire resistant resinous plants is known as: **(AIIMS 1997)**
(a) chaparral vegetation (b) savanna vegetation
(c) steppe grassland (d) tundra vegetation

59. Deciduous forests have: **(AIIMS 1996)**
(a) variety of grasses (b) broad-leaved trees
(c) narrow-leaved trees (d) variety of crocodiles

60. Food chain is: **(AIIMS 1996)**
(a) number of human beings forming a chain for food
(b) animals near a source of food
(c) transfer of food energy from producers to consumers
(d) none of the above

61. Which one is the correct food chain? **(AIIMS 1996)**
(a) Eagle → Snake → Grasshopper → Grass → Frog
(b) Frog → Snake → Eagle → Grasshopper → Grass
(c) Grasshopper → Grass → Snake → Frog → Eagle
(d) Grass → Grasshopper → Frog → Snake → Eagle

62. The factors governing the nature of temperature, solutes, humus contents, nature of the soil etc., are considered as: **(AIIMS 1995)**
(a) biotic factors (b) edaphic factors
(c) climatic factors (d) niche

63. The storage of organic matter not used by heterotrophs is termed as: **(BHU 2006)**
(a) Gross primary production (b) Net secondary production
(c) Net production (d) Secondary production

64. Maximum absorption of rainfall water is done by: **(BHU 2004)**
(a) Tropical deciduous forest (b) Tropical evergreen forest
(c) Tropical savanna (d) Scrub forest

65. A man-made ecosystem is: **(BHU 2004)**
(a) less in diversity (b) more in diversity
(c) man does not make ecosystem (d) more stable than natural ecosystem

66. If the decomposers become extinct, the most severally affected would be: **(BHU 2003)**
(a) noncycling of minerals (b) damage of nitrogen fixation
(c) biomagnification (d) carnivores will be starved

67. The most stable ecosystem is: **(BHU 2002)**
(a) forest (b) mountain
(c) desert (d) ocean

68. In biotic community, primary consumers are: **(BHU 2002)**
(a) carnivores (b) omnivores
(c) herbivores (d) detrivores

69. Each successive trophic level has : **(BHU 2002)**
(a) less total energy (b) more total energy
(c) increased total energy (d) non-estimated energy contents

70. Pyramid of numbers in grass ecosystem is: **(BHU 2002)**
(a) linear (b) upright
(c) inverted (d) negative

71. Tansley for the first time coined the term: **(BHU 2002)**
(a) ecosystem (b) ecological pyramid
(c) ecology (d) biocoenosis

72. Generally the food chain has how many trophic levels ? **(BHU 2001)**
(a) One (b) Two
(c) Three or Four (d) Three

73. The soil salinity can be measured by: **(BHU 2000)**
(a) potometer (b) porometer
(c) calorimeter (d) conducting meter

74. Root cap is absent in : **(BHU 2000)**
(a) lithophytes (b) hydrophytes
(c) xerophytes (d) mesophytes

75. Transitional zone between two different vegetations of a region is: **(BHU 2000)**
(a) ecotone (b) ecodine
(c) eceeline (d) ecotype

76. Nitrogen fixation is: **(BHU 2000)**
(a) nitrogen $\rightarrow$ ammonia (b) nitrogen $\rightarrow$ nitrates
(c) nitrogen $\rightarrow$ amino acid (d) both (a) and (b)

77. Pyramid of energy in a pond ecosystem is always: **(BHU 1998)**
(a) upright (b) inverted
(c) both (a) and (b) (d) none of these

78. Tropical rain forests are found in: **(BHU 1998)**
(a) Andaman (b) Bihar
(c) Himachal Pradesh (d) Jammu and Kashmir

79. Deer in a forest is a: **(BHU 1998)**
(a) primary consumer (b) secondary consumer
(c) decomposer (d) none of these

80. Which is dominant in desert ? **(BHU 1997)**
(a) *Hyla* (b) Leopard
(c) Tiger (d) Lizard

81. Nitrogen is a critical element of the ecosystem because it is: **(BHU 1996)**
(a) essential element (b) abundant in atmosphere
(c) labile (d) fixed by microbes

82. Major communities of the world are: **(BHU 1996)**
(a) shrubs, jungles and gardens (b) ocean, freshwater and forest communities
(c) grassland and desert communities (d) both (b) and (c)

83. When a peacock eats snakes which eat insects thriving on plants, the peacock is: **(BHU 1995)**
(a) a final decomposer (b) the apex of the food pyramid
(c) a primary consumers (d) a primary decomposer

84. "Biosphere" is called as: **(Manipal 2006)**
(a) part of atmosphere where the livings cannot live
(b) part of atmosphere where the livings can live
(c) both (a) and (b)
(d) none of the above

85. Inverted pyramid is found in : **(Manipal 2005)**
(a) biomass pyramid of aquatic system
(b) energy pyramid of grassland
(c) biomass pyramid of grassland
(d) pyramid of number of aquatic system

86. Match the following and find correct combination **(Manipal 2002)**

A. Western Ghats — 1. Western and Eastern Himalays
B. Estuarine ecosystem — 2. Rajasthan, Punjab and parts of Gujarat
C. Indus Plain — 3. West Bengal and Andaman
D. Arctic Zone — 4. Cape Comorin

Codes:	A	B	C	D
(a)	4	2	3	1
(b)	4	3	2	1
(c)	1	2	3	4
(d)	3	4	2	1

87. Choose the correct sequence: **(Manipal 2002)**
(a) phytoplankton → zooplankton → crustacean → fish
(b) crustacean → fish → zooplankton → phytoplankton
(c) zooplankton → phytoplankton → crustacean → fish
(d) fish → crustacean → zooplankton → phytoplankton

88. Which of these is correct sequence for food chain ? **(Manipal 2002)**
(a) Insect-fish-zooplankton-phytoplankton
(b) Phytoplankton-zooplankton-insect-fish
(c) Fish-insect-zooplankton-phytoplankton
(d) Zooplankton-phytoplankton-insect-fish

89. Species diversity is lowest in ecosystem: **(Manipal 1997)**
(a) desert (b) tundra
(c) grassland (d) deciduous forest

90. Biomass of producers within specified area will be maximum in: **(Manipal 1997)**
(a) forest ecosystem (b) grassland ecosystem
(c) pond ecosystem (d) lake ecosystem

91. Most diverse organism of an ecosystem is: **(CPMT 2006)**
(a) producer (b) consumer
(c) decomposer (d) carnivores

92. In forest ecosystem, fungi is grouped as: **(CPMT 2003)**
(a) producer (b) consumer
(c) secondary consumer (d) decomposer

93. Energy enters a food chain through: **(CPMT 2003)**
(a) producers (b) decomposers
(c) herbivores (d) carnivores

94. Nitrates are transformed into nitrogen by: **(CPMT 2003)**
(a) ammonifying bacteria (b) nitrifying bacteria
(c) denitrigying bacteria (d) both (a) and (b)

95. Tropical forests are denser due to: **(CPMT 2002)**
(a) wild animals
(b) high temperature and less rainfall
(c) low temperature and excess rain
(d) high temperature and high rain

96. Source of energy in ecosystem is: **(CPMT 2002)**
(a) Sun (b) sugar produced in photosysnthesis
(c) green plants (d) ATP

97. Ultimate source of energy for living beings is: **(CPMT 2001)**
(a) carbohydrates (b) fats
(c) sunlight (d) ATP

98. The position of herbivores in the pyramid of biomass is: **(CPMT 2000)**
(a) first (b) second
(c) third (d) last

99. On the global basis the maximum productivity is shown by: **(CPMT 2000)**
(a) aquatic ecosystem (b) grasslands
(c) forests (d) deserts

100. Decomposers are: **(CPMT 1998)**
(a) autotrophs (b) heterotrophs
(c) autoheterotrophs (d) chemoautotrophs

101. Vegetation of Rajasthan is: **(CPMT 1996)**
(a) xerophytic (b) alpine
(c) arctic (d) deciduous

102. Which factor is responsible for poor vegetation in deserts? **(CPMT 1995)**
(a) Over grazing by animals (b) Low rainfall
(c) Poor fertility of soil (d) Native mankind

103. Overlapping region between two ecosystems: **(DPMT 2006)**
(a) biome (b) ecotone
(c) niche (d) photic zone

104. Study of periodical changes in plants is related to: **(DPMT 2006)**
(a) physiognomy (b) phycology
(c) phenology (d) photoperiodism

105. Which is an example of true pyramid in an ecosystem: **(DPMT 2006)**
(a) Pyramid of biomass (b) Pyramid of number
(c) Pyramid of energy (d) None of these

106. In ecotone some species become abundant and called: **(DPMT 2006)**
(a) edge species (b) endemic species
(c) rare species (d) threatened species

107. When spontaneous process occurs then free energy of system: **(DPMT 2003)**
(a) decrease (b) increase
(c) remains same (d) either can increase or decrease

108. Stratification is found in: **(DPMT 2002)**
(a) tropical rain forests (b) deciduous forest
(c) desert (d) both (a) and (b)

109. Which of the following is present in maximum amount in atmosphere? **(DPMT 2002)**
(a) Oxygen (b) Nitrogen
(c) Carbon dioxide (d) Hydrogen

110. Biochemical cycle with gaseous phase is: **(DPMT 2001)**
(a) carbon (b) sodium
(c) phosphorus (d) magnesium

111. Which one depicts nitrogen fixation? **(DPMT 1999)**
(a) $N_2 + 3H_2 \rightarrow 2NH_3$
(b) $2NH_3 \rightarrow N_2 + 3H_2$
(c) $2N_2$ + Glucose $\rightarrow$ 2 Amino acids
(d) $2NH_4^+ + 2O_2 \rightarrow 8e^- \rightarrow N_2 + 4H_2O$

112. Total amount of living material at various levels of a food chain is depicted by: **(DPMT 1999)**
(a) pyramid of energy (b) pyramid of numbers
(c) pyramid of biomass (d) all of these

113. An ecosystem contains: **(DPMT 1999)**

(a) producers (b) consumers
(c) decomposers (d) all of these

114. Biotic community alongwith its interacting physical environment comprises: **(DPMT 1997)**

(a) phytosociology (b) phytogeography
(c) ecosystem (d) ecology

115. Which one lies at the level of primary consumers? **(Karnataka 2004)**

(a) Fishes (b) Eagles
(c) Cattle and insects (d) Snakes and frogs

116. In a comparative study of grassland ecosystem and pond ecosystem it may be observed that: **(Karnataka 2003)**

(a) the abiotic components are almost similar
(b) the biotic components are almost similar
(c) both biotic and abiotic components are different
(d) primary and secondary consumers are similar

117. Which type of the following ecological pyramids are never inverted ? Pyramids of : **(Karnataka 2002)**

(a) biomass (b) number
(c) dry biomass (d) energy

118. Which one is inverted pyramid ? **(Karnataka 2001)**

(a) Pyramid of biomass in grassland
(b) Pyramid of biomass in pond ecosystem
(c) Pyramid of numbers in grassland ecosystem
(d) Pyramid of energy in a pond ecosystem

119. The term biocoenosis was coined by: **(Karnataka 2001)**

(a) Darwin (b) Haeckel
(c) Odum (d) Mobius

120. A rat feeding on potato tuber is: **(Karnataka 1999)**

(a) producer (b) primary consumer
(c) carnivore (d) decomposer

121. Lentic ecosystem occurs in : **(Karnataka 1999)**

(a) rain water (b) running water
(c) standing water (d) gravitational water

122. Match the items of column-I with those of column-II **(Karnataka 1999)**

Column-I	Column-II
A. Grass B. Grasshopper C. Frog D. Hawk	1. Decomposer 2. Secondary carnivore 3. Producer 4. Primary consumer 5. Primary carnivore

Codes:

	A	B	C	D
(a)	3	1	5	4
(b)	1	3	4	5
(c)	3	5	4	2
(d)	3	4	5	2

123. Phytoplanktons of a pond ecosystem act as: **(Karnataka 1997)**
(a) producers (b) primary consumers
(c) secondary consumers (d) decomposers

124. The organisms which attack dead animals are: **(Pb. PMT 2006)**
(a) first link of the food chain and are known as primary producers
(b) second link of the food chain and are herbivorous
(c) third link of the food chain and are tertiary consumers
(d) the end of food chain and are decomposers

125. Which creatures are direct or indirect food of all creatures on the ocean's surface? **(Pb. PMT 2005)**
(a) Protozoans (b) Phytoplankton
(c) Fish (d) Aquatic insects

126. If we completely remove the decomposers from an ecosystem, its functioning will be adversely affected because: **(Pb. PMT 2005)**
(a) herbivores will not receive solar energy
(b) mineral movement will be blocked very high
(c) the rate of decomposition will be very high
(d) energy flow will be blocked

127. Energy transfers or transformations are never 100% efficient. This is due to: **(Pb. PMT 2005)**
(a) entropy (b) homeostasis
(c) catabolism (d) anabolism

128. Decomposers are: **(Pb. PMT 2004)**
(a) autotrophs (b) heterotrophs
(c) organotrophs (d) autoheterotrophs

129. During food chain the maximum energy is stored in : **(Pb. PMT 2004)**
(a) producers (b) decomposers
(c) herbivores (d) carnivores

130. In the phosphorus cycle, phosphate becomes available by weathering of rock first to : **(Pb. PMT 2004)**
(a) consumers (b) producers
(c) decomposers (d) none of these

131. The major forest type found in India is: **(Pb. PMT 2003)**
(a) subtropical deciduous (b) tropical moist deciduous
(c) tropical deciduous (d) temperature deciduous

132. In tropic level an animal occupies which tropic level: **(Pb. PMT 2003)**
(a) 1st (b) 2nd
(c) intermediate (d) 3rd

133. Terai forest is : **(Pb. PMT 2000)**
(a) tropical (b) coniferous
(c) deciduous (d) temperate deciduous

134. Savanna is found commonly in : **(Pb. PMT 2000)**
(a) USA (b) USSR
(c) Australia (d) India

135. The graphic representation of the interrelationship between plants and animals is: **(Pb. PMT 1999)**
(a) ecological niche (b) ecological pyramid
(c) a trophic level (d) none of these

136. Ecosystem of a pond is: **(Pb. PMT 1999)**
(a) lotic (b) lentic
(c) benthic (d) xeric

137. In an ecosystem, which one shows one way passage? **(Pb. PMT 1999)**
(a) Free energy (b) Carbon
(c) Nitrogen (d) Potassium

138. Sal (*Shorea robusta*) forest is a: **(Pb. PMT 1998)**
(a) deciduous forest (b) evergreen forest
(c) rain forest (d) none of these

139. Severe winters with few months of summer are found in: **(Pb. PMT 1998)**
(a) Tundra (b) Arctic
(c) Taiga (d) Antarctic

140. 'Jhum' is : **(Pb. PMT 1998)**
(a) terrace farming (b) tribal farming
(c) shifting cultivation (d) farming on a slope

141. Which one is nature's cleaner ? **(Pb. PMT 1997)**
(a) consumers (b) producers
(c) decomposers and scavengers (d) symbionts

142. In plant ecosystem the rooted plants occur in : **(EAMCET 2001)**
(a) ionosphere (b) stratosphere
(c) limnetic zone (d) litterol zone

143. Which one has always a steeper vertical gradient ? **(Haryana PMT 2000)**
(a) Pyramid of mass (b) Pyramid of energy
(c) Pyramid of numbers (d) Pyramid of energy in aquatic system

144. Which one has evergreen vegetation and drought adapted animals? **(Haryana PMT 2000)**
(a) Chaparral (b) Savanna
(c) Tundra (d) Deciduous forest

145. Which of the following is the major adaptation for a plant to survive in the xerophytic conditions ? **(Haryana PMT 2000)**
(a) Long tap root (b) Lack of stomata
(c) Spines (d) Stipular leaves

146. Which of the following always show steeper slopes ? **(Haryana PMT 2000)**
(a) Pyramid of mass (b) Pyramid of number
(c) Pyramid of energy (d) Pyramid of energy in the aquatic biomes

147. The most productive zone in a pond ecosystem is : **(Haryana PMT 1998)**
(a) limnetic zone (b) profundal zone
(c) pedonic zone (d) littoral zone

148. The study of freshwater ecosystems is : **(Haryana PMT 1998)**
(a) Autecology (b) Hydrology
(c) Trichology (d) Limnology

149. Desert biome does not support much vegetation as it lacks: **(Haryana PMT 1994)**
(a) sufficient light (b) favourable temperature
(c) sufficient water (d) sufficient nutrients

150. Which is correct ? **(Haryana PMT 1994)**

(a) Least productive ecosystem is desert and deep lake
(b) Sugarcane is the most productive agroecosystem
(c) Most productive is coral reef system
(d) All of the above

151. Taiga refers to: **(AMU 2006)**

(a) Temperate deciduous forest
(b) Subtropical semi-deciduous forest
(c) Evergreen forest
(d) North temperate coniferous forest

152. In which part of the open sea producers are found? **(AMU 2005)**

(a) Aphotic zone (b) Abyssal zone
(c) Photic zone (d) None of these

153. Conversion of organic nitrogenous compounds into ammonium compounds is called : **(AMU 2002)**

(a) nitrification (b) denitrification
(c) ammonification (d) denaturation

154. Weathering of rocks makes phosphorus available to first : **(AMU 2002)**

(a) producers (b) decomposers
(c) consumers (d) none of these

155. An ammonifying bacterium is: **(AMU 2002)**

(a) *Clostridium* (b) *Azotobacter*
(c) *Nitrosococcus* (d) *Bacillus ramosus*

156. The biomass of each succeeding trophic level is: **(AMU 2002)**

(a) more than the one preceding (b) less than the one preceding
(c) constantly fixed (d) equal to next trophic level

157. Which ecosystem has the highest gross primary productivity ? **(AMU 2002)**

(a) Tropical rain forests (b) Coral reefs
(c) Mangroves (d) Grasslands

158. Path of energy flow in an ecosystem is : **(AMU 2002)**

(a) herbivores → producers → carnivores → decomposers
(b) herbivores → carnivores → producers → decomposers
(c) producers → carnivores → herbivores → decomposers
(d) producers → herbivores → carnivores → decomposers

159. Estuaries occur in : **(AMU 2001)**
(a) Orissa and Tamil Nadu
(b) Kerala and Tamil Nadu
(c) Kerala and Orissa
(d) Kerala, Tamil Nadu and Orissa

160. Which is correct food chain ? **(AMU 2001)**
(a) Phytoplankton → Fishes → Zooplankton
(b) Zooplankton → Phytoplankton → Fishes
(c) Phytoplankton → Zooplankton → Fishes
(d) Fishes → Zooplankton → Phytoplankton

161. Secondary producers are: **(AMU 1999)**
(a) herbivores
(b) heterotrophs
(c) carnivores
(d) green plants

162. Biome has organisms of various trophic levels **(AMU 1999)**
(a) reacting with abiotic environment
(b) constitue a sociological unit
(c) live a symbiotic life
(d) inhabit a desert

163. Which can be trophic level of a lion in forest ecosystem ? **(AMU 1998)**
(a) T_3 (b) T_2 (c) T_4 (d) T_1

164. Pyramid of number is stable when it is : **(AMU 1997)**
(a) upright
(b) linear
(c) downward
(d) variable

165. Plants that lack perennating buds and are annuals can be placed in one of the life form : **(AMU 1997)**
(a) phanerophyte
(b) chamacophyte
(c) cryptophyte
(d) therophytes

166. Most physiologically sensitive plants of saline soils are: **(AMU 1996)**
(a) psammophytes
(b) oxalophytes
(c) halophytes
(d) hygrophytes

167. The reason why the first trophic level in a food chain is always occupied by autotrophic plants is that : **(AMU 1995)**
(a) the number of herbivorous animals is larger than carnivorous animals
(b) plants produce organic food
(c) plants are very abundant
(d) plants occur everywhere

168. Transition zone between two ecosystems or vegetational regions is termed: **(Kerala PMT 2005)**

(a) Ecocline (b) Ecotone
(c) Ecad (d) Barrier

169. Which must be preserved in an ecosystem, if the system is to be maintained : **(Kerala PMT 2004)**

(a) producers and carnivores (b) producers and decomposers
(c) carnivores and decomposers (d) herbivores and carnivores

170. Rate of conversion of light energy into chemical energy of organic molecules in an ecosystem is called : **(Kerala PMT 2003)**

(a) net primary productivity (b) gross primary productivity
(c) net secondary productivity (d) gross secondary productivity

171. Tropical deciduous forest trees shed their leaves to : **(Kerala PMT 2003)**

(a) protect themselves from heat (b) save energy
(c) enhance rate of respiration (d) prevent loss of water

172. Grasslands of Asia are : **(Kerala PMT 2001)**

(a) savanna (b) pampas
(c) steppes (d) veldt

173. Which one is sedimentary cycle ? **(Kerala PMT 2001)**

(a) Oxygen cycle (b) Nitrogen cycle
(c) Hydrogen cycle (d) Phosphorous cycle

174. Under anaerobic conditions, denitrifying bacterium *Pseudomonas* changes: **(Kerala PMT 2001)**

(a) nitrate to molecular nitrogen (b) nitrate to ammonia
(c) nitrate to nitrite (d) nitrite to nitrate

175. Crystalline rocks are natural source of biogenetic element : **(Kerala PMT 2000)**

(a) calcium (b) phosphorus
(c) magnesium (d) calcium

176. Plains with snow, ice and frozen soil for most of the year are found in: **(Kerala PMT 2000)**

(a) chapparal (b) taiga
(c) tundra (d) savanna

177. A recently discovered ecosystem is : **(CET Chd. 2000)**

(a) crater (b) tundra
(c) floating iceberg (d) vent

178. Controlling factor in ecosystem is : **(CET Chd. 2000)**
(a) soil moisture (b) food
(c) predation (d) temperature

179. Inverted pyramid of biomass can be traced in one of the following ecosystems: **(CET Chd. 1998)**
(a) rain forest (b) desert
(c) ocean (d) tundra

180. Who proposed the concept of 'ecological niches' ? **(CET Chd. 1997)**
(a) Ransley (b) Richter
(c) Elton (d) Odum

181. The study of energy transfer in an ecosystem at different trophic levels is known as : **(CET Chd. 1997)**
(a) bioenergetics (b) geobiocoenosis
(c) holocoenosis (d) biosystem

182. The sphere of living matter together with water, air and soil on the surface of earth is: **(MP PMT 2004)**
(a) lithosphere (b) biosphere
(c) hydrosphere (d) atmosphere

183. A food chain starts with : **(MP PMT 2004)**
(a) nitrogen fixing organisms (b) photosynthesis
(c) respiration (d) decomposers

184. Plants and animals living in a particular area constitute : **(MP PMT 2002)**
(a) flora and fauna (b) community
(c) ecosystem (d) ecology

185. Ecosystem is : **(MP PMT 2002)**
(a) always open
(b) always closed
(c) both open and closed depending on biomass
(d) both open and closed depending on community

186. Pyramid of energy in a forest ecosystem is : **(MP PMT 2002)**
(a) always inverted
(b) always upright
(c) both upright and inverted depending on ecosystem
(d) first upright then inverted

187. In food chain initial organisms are : **(MP PMT 2002)**
(a) top consumers (b) secondary consumers
(c) primary consumers (d) photosynthates

188. A food chain can have trophic levels : **(MP PMT 2002)**
(a) three or four (b) three
(c) two (d) one

189. Energy stored at the consumer level is: **(MP PMT 2001)**
(a) gross primary productivity (b) net primary productivity
(c) net productivity (d) secondary productivity

190. Biotic components on an ecosystem include : **(MP PMT 2001)**
(a) producers, consumers and decomposers
(b) producers and consumers
(c) producers only
(d) consumers only

191. A man-made microecosystem is : **(MP PMT 2001)**
(a) plants grown in a pond
(b) crop field
(c) tank formed naturally in your court yard
(d) lake in a forest

192. Phytoplankton are destroyed. It will : **(MP PMT 2000)**
(a) cause spurt in population of primary consumers
(b) rapid growth of algae
(c) affect the food chain
(d) have no effect

193. The driving force for an ecosystem is : **(MP PMT 2000)**
(a) biomass (b) producers
(c) carbohydrates in producers (d) solar energy

194. A treeless biome is : **(MP PMT 2000)**
(a) tundra (b) grassland
(c) desert (d) all of these

195. Top most members of most of Eltonian pyramid are : **(MP PMT 1999)**
(a) producers (b) decomposers
(c) carnivores (d) herbivores

196. Zooplanktonic forms are : **(MP PMT 1999)**
(a) primary producers (b) carnivores
(c) primary consumers (d) secondary consumers

197. Which of the following occurs in abiotic component of ecosystem ? **(MP PMT 1999)**
(a) Flow of energy
(b) Cycling of materials
(c) Consumers
(d) Flow of energy and cycling of materials

198. First link in food chain is green plant because : **(MP PMT 1999)**
(a) it alone can synthesis food
(b) it is fixed at one place
(c) it can pick up everything
(d) it can eat everything

199. Ecosystem has two components : **(MP PMT 1999)**
(a) animals and humans
(b) plants and animals
(c) biotic and abiotic
(d) weeds and trees

200. A habitat not suitable for primary productivity is : **(MP PMT 1999)**
(a) cave
(b) pond
(c) meadow
(d) forested river bank

201. Energy transfer from organism to organism in a natural community develops: **(MP PMT 1999)**
(a) biological control
(b) food chain
(c) natural barriers
(d) food web

202. Biological equilibrium is equilibrium amongst : **(MP PMT 1998)**
(a) producers and decomposers
(b) producers and consumers
(c) producers, consumers and decomposers
(d) consumers and decomposers

203. Number of individuals of a species in a particular ecosystem at a given time remains constant due to : **(MP PMT 1998)**
(a) predators
(b) parasites
(c) man
(d) food

204. Importance of ecosystem lies in : **(MP PMT 1998)**
(a) cycling of materials
(b) flow of energy
(c) both (a) and (b)
(d) none of these

205. A lake ecosystem is : **(MP PMT 1997)**
(a) artificial
(b) abiotic
(c) natural
(d) hydrological

206. Food levels of an ecosystem are known as: **(MP PMT 1996)**
(a) producers levels
(b) herbivore levels
(c) consumer levels
(d) trophic levels

207. Tropical dry deciduous forests occur in India in : **(MP PMT 1996)**
(a) Andamans (b) Eastern Himalayas
(c) Madhya Pradesh (d) Kerala

208. Number of primary producers per unit area would be maximum in : **(MP PMT 1995)**
(a) forest ecosystem (b) pond ecosystem
(c) desert (d)

209. A food chain begins with : **(MP PMT 1995)**
(a) photosynthetic organisms (b) nitrogen fixing organisms
(c) decomposers (d) consumers

210. Temperate evergreen forests are found in: **(MP PMT 1994)**
(a) Hamalayan ranges (b) Western ghats
(c) Aravalli ranges (d) Assam

211. In the pyramid of food the primary producers lies at: **(CMC Ludhiana 2003)**
(a) base (b) apex
(c) near the position of base (d) near the position of apex

212. Biome is : **(CMC Ludhiana 2003)**
(a) all organisms on earth
(b) fauna on earth
(c) flora and fauna
(d) a major ecosystem by delimited climate and geographic conditions

213. Sea-weed belongs to trophic level : **(CMC Ludhiana 2002)**
(a) primary consumers (b) secondary consumers
(c) decomposer (d) primary producer

214. Ecosystem is : **(Orissa JEE 2005)**
(a) open (b) closed
(c) both open and close (d) neither open nor closed

215. The pyramid that cannot be inverted in a stable ecosystem, is pyramid of : **(Orissa JEE 2005)**
(a) number (b) energy
(c) biomass (d) all of these

216. Biosphere reserve programme started in India : **(Orissa JEE 2005)**
(a) 1986 (b) 1984
(c) 1982 (d) 1988

217. An ecosystem is not : **(Orissa JEE 2004)**
(a) open system (b) closed system
(c) variable system (d) none of these

218. The scientist who coined the term ecosystem was ? **(Orissa JEE 2003)**
(a) Ernst Haeckel (b) E.P. Odum
(c) A.G. Tansley (d) Charles Elton

219. An aquatic ecosystem with maximum primary productivity is : **(Orissa JEE 2003)**
(a) ocean (b) pond
(c) estuary (d) river

220. Tiger lies at the apex of food chain because : **(Orissa JEE 2002)**
(a) tiger has maximum biomass
(b) tiger feeds on a number of herbivores which feed on a number of plants
(c) tiger has many enemies
(d) tiger is omnivorous

221. Sea weed is : **(CMC Vellore 2002)**
(a) primary producer (b) primary consumer
(c) both 'a' and 'b' (d) decomposer

222. Bottom dwellers where do light occurs is found in: **(CMC Vellore 2002)**
(a) benthic zone (b) abyssal zone
(c) pelagic zone (d) lotic zone

223. Plants which grow on rocks are : **(CMC Vellore 2002)**
(a) lithophytes (b) halophytes
(c) oxylophytes (d) psammophytes

224. First link in any food chain is a green plant because : **(RPMT 1995)**
(a) green plants can alone synthesize food from sunlight
(b) they can eat everything
(c) they are fixed at one place
(d) none of the above

225. Flow of energy in the biosphere is : **(GGSIPU, N. Delhi 2002)**
(a) unidirectional (b) multidirectional
(c) diffuse (d) linear

226. The most favoured habitat of a plant is called : **(JIPMER 1999)**
(a) ecological habitat (b) microhabitat
(c) ecotype (d) ecocline

227. An ecosystem must have continuous external source of : **(JIPMER 1994)**
(a) food (b) minerals
(c) energy (d) all of these

228. Which must be preserved in an ecosystem if the system is to be maintained? **(JIPMER 1994)**
(a) Producers and carnivores (b) Producers and decomposers
(c) Carnivores and decomposer (d) Herbivores and carnivores

229. Which is the correct sequence in the food chain in a grassland ? **(HP PMT 2005)**
(a) Grass → wolf → deer → buffalo
(b) Bacteria → grass → rabbit → wolf
(c) Grass → insect → birds → snakes
(d) Grass → snake → insect → deer

230. In a food chain, lion is a : **(MH CET 2003)**
(a) secondary consumer (b) primary consumer
(c) tertiary consumer (d) secondary producer

231. Grasshopper is : **(BVP Pune 2003)**
(a) primary consumer (b) secondary consumer
(c) decomposer (d) primary decomposer

ANSWER SHEET

1. (b)	**2.** (a)	**3.** (c)	**4.** (d)	**5.** (a)	**6.** (a)	**7.** (d)	**8.** (d)	**9.** (d)	**10.** (d)
11. (b)	**12.** (d)	**13.** (c)	**14.** (c)	**15.** (c)	**16.** (b)	**17.** (b)	**18.** (d)	**19.** (c)	**20.** (c)
21. (a)	**22.** (b)	**23.** (a)	**24.** (d)	**25.** (b)	**26.** (a)	**27.** (b)	**28.** (b)	**29.** (c)	**30.** (a)
31. (a)	**32.** (c)	**33.** (d)	**34.** (b)	**35.** (b)	**36.** (b)	**37.** (d)	**38.** (c)	**39.** (d)	**40.** (a)
41. (b)	**42.** (a)	**43.** (a)	**44.** (c)	**45.** (b)	**46.** (c)	**47.** (d)	**48.** (c)	**49.** (d)	**50.** (c)
51. (c)	**52.** (a)	**53.** (a)	**54.** (c)	**55.** (d)	**56.** (d)	**57.** (d)	**58.** (a)	**59.** (b)	**60.** (c)
61. (d)	**62.** (b)	**63.** (c)	**64.** (b)	**65.** (a)	**66.** (a)	**67.** (d)	**68.** (c)	**69.** (a)	**70.** (b)
71. (a)	**72.** (c)	**73.** (d)	**74.** (b)	**75.** (a)	**76.** (d)	**77.** (a)	**78.** (a)	**79.** (a)	**80.** (d)
81. (c)	**82.** (d)	**83.** (b)	**84.** (b)	**85.** (a)	**86.** (b)	**87.** (a)	**88.** (b)	**89.** (b)	**90.** (a)
91. (c)	**92.** (d)	**93.** (a)	**94.** (c)	**95.** (d)	**96.** (a)	**97.** (c)	**98.** (b)	**99.** (a)	**100.**(b)
101.(a)	**102.**(b)	**103.**(b)	**104.**(c)	**105.**(c)	**106.**(a)	**107.**(a)	**108.**(b)	**109.**(b)	**110.**(a)
111.(a)	**112.**(c)	**113.**(d)	**114.**(c)	**115.**(c)	**116.**(c)	**117.**(d)	**118.**(d)	**119.**(d)	**120.**(b)
121.(c)	**122.**(d)	**123.**(a)	**124.**(d)	**125.**(b)	**126.**(b)	**127.**(a)	**128.**(b)	**129.**(a)	**130.**(b)
131.(c)	**132.**(b)	**133.**(a)	**134.**(c)	**135.**(b)	**136.**(b)	**137.**(a)	**138.**(a)	**139.**(a)	**140.**(c)
141.(c)	**142.**(d)	**143.**(b)	**144.**(a)	**145.**(a)	**146.**(c)	**147.**(a)	**148.**(b)	**149.**(c)	**150.**(d)
151.(d)	**152.**(c)	**153.**(c)	**154.**(a)	**155.**(a)	**156.**(b)	**157.**(a)	**158.**(d)	**159.**(d)	**160.**(c)
161.(b)	**162.**(a)	**163.**(c)	**164.**(a)	**165.**(d)	**166.**(c)	**167.**(b)	**168.**(b)	**169.**(b)	**170.**(b)
171.(d)	**172.**(c)	**173.**(d)	**174.**(a)	**175.**(b)	**176.**(c)	**177.**(a)	**178.**(c)	**179.**(c)	**180.**(d)
181.(a)	**182.**(b)	**183.**(b)	**184.**(a)	**185.**(a)	**186.**(b)	**187.**(d)	**188.**(a)	**189.**(d)	**190.**(a)
191.(b)	**192.**(c)	**193.**(d)	**194.**(d)	**195.**(c)	**196.**(c)	**197.**(b)	**198.**(a)	**199.**(c)	**200.**(a)
201.(b)	**202.**(c)	**203.**(a)	**204.**(c)	**205.**(c)	**206.**(d)	**207.**(c)	**208.**(b)	**209.**(a)	**210.**(a)
211.(a)	**212.**(d)	**213.**(d)	**214.**(a)	**215.**(b)	**216.**(a)	**217.**(b)	**218.**(c)	**219.**(c)	**220.**(b)
221.(a)	**222.**(a)	**223.**(a)	**224.**(a)	**225.**(a)	**226.**(b)	**227.**(c)	**228.**(b)	**229.**(c)	**230.**(c)
231.(a)									

Chapter 3
Plant Communities

PLANT COMMUNITIES

Group of different population growing in a particular area is called **Community**. According to Oosting (1976), "Community is an aggregate of living plants having mutual relation among themselves and with the environment". Grubbs (1987) defined community as "collection of plant population found in habitat type in one area and integrated of degree of dependents". The study of plant community is called **Phytosociology** or **Plant Sociology**".

Characteristics of Community

1) Species Diversity: Each community is made up of different species of plants, animals and microbes that differ taxonomically from each other called species diversity. There are two levels of species diversity i.e. **regional diversity** of a large continent within which different communities exists, and **local diversity** of a given particular area where different communities exist at different latitudes.

Ratio between the number of species and "Importance values" (numbers, biomass, productivity etc.) of individuals are called **species diversity indices**.

Each species of the community has its own definite range of tolerance limits against physical or biological environmental conditions called **ecological amplitude**.

2) Species Dominance: In any community, one or a few species dominate either in number or in biomass over the other species called species dominance for e.g. in **grassland community**, grasses are the dominant species.

3) Coexistance: Species occurring in their particular habitat do not live in complete isolation as pure cultures, but they coexist in mutual adjustment called **Coexistance**.

4) Interdependency: All the individuals of the community have ability to live under the conditions of the habitat and they are interdependent upon one another to some extent. It is called dependency.

5) Stratification: The division of community into definite vertical strata is called **Stratification**. In stratification, communities exhibit a specific structure. Structurally it can be divided into two divisions, i.e.-

i) **Horizontal division** or **Zonation** e.g.: Pond communities
ii) **Vertical** or **Stratification division** e.g. forest community.

The community of a pond or a lake ecosystem horizontally shows three important zones as shown in figure below-

i) Littoral zone
ii) Limnetic zone
iii) Profundal zone

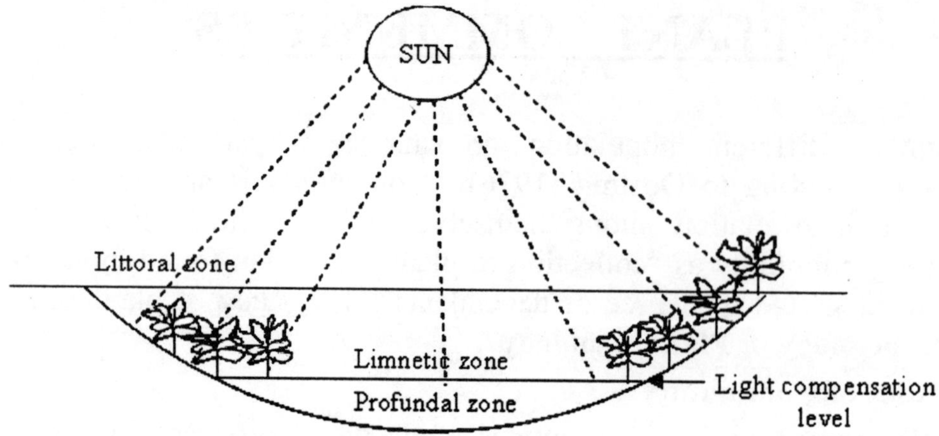

Fig. Diagrammatic sketch showing three major zones of a fresh water body as lake

In forest vegetation, the stratification reaches its greatest complexity. As shown in figure five vertical subdivisions may be present-

i) Subterranean zone
ii) Forest floor
iii) Ground vegetation or herbaceous vegetation
iv) Shrubs and
v) Trees

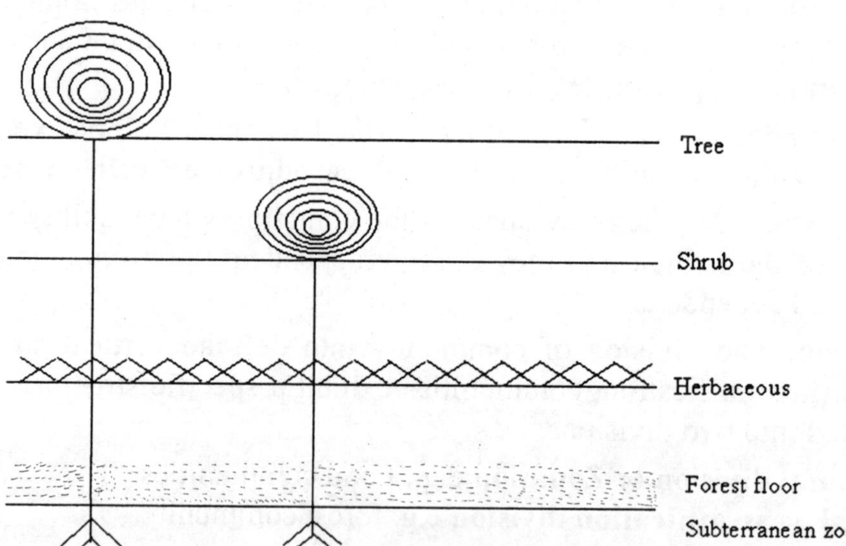

Fig. showing stratification of a forest vegetation

6) Ecotone: The transition zone between two biotic communities or ecosystems is called **ecotone**. It is characterised by the presence of species of both the communities. The ecotone contains more species and often a denser population than either of the neighbouring communities and this is known as the **principle of edge effect**. For example, the transition zone between forest and grassland or the intermediate zone between any two major land or aquatic communities are the examples of **ecotone**.

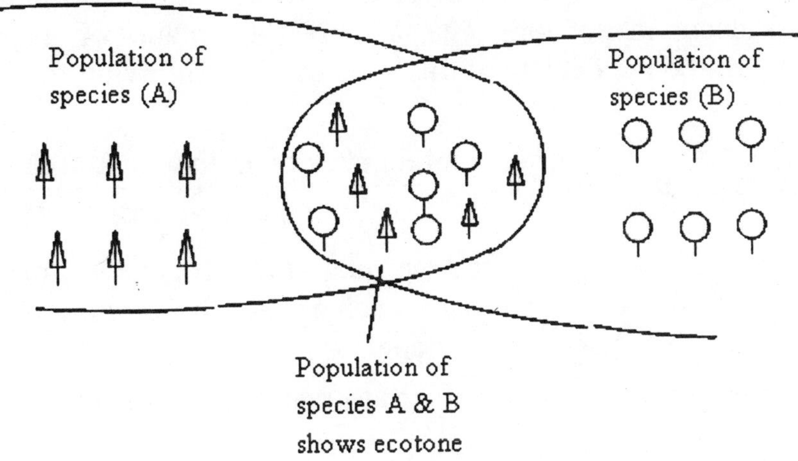

Fig. showing ecotone characterised by edge effect

Study of Community

The field of community ecology, which deals with the study of plant community that make up vegetation, their origin, formation, composition and structure is known as **plant sociology** or **phytosociology**. These various characters used for such purposes are broadly classified into following two categories-

1) Analytical characters
2) Synthetic characters

Analysis of community

Analytical method

Quantitative Characters
- Frequency
- Density
- Abundance
- Cover
- Basal cover

Qualitative characters
- Life form
- Physiognomy
- Stratification
- Phenology & Periodicity
- Sociability
- Vitality

Synthetical method
- Presence of constance
- Fidelity
- Dominance
- IVI
- Other methods

Quantitative character

1. Density: The density of an individual refers to the number of individuals occuring in per unit area while the *crude density* refers to the number of individuals of a particular species occur in per unit area, e.g., if 1000 plants of *Argemone* species occur in per acre, will be the density of this species. The following formula is used for calculating density:

$$\text{Density of a species per unit area} = \frac{\text{Total number of individuals of the species}}{\text{Total number of quadrats studied}}$$

In some cases, e.g. grass and vegetatively propagated plants, the term individual creates difficulty. In such cases, each aerial tiller or shoot arising out from the soil is generally regarded as one individual.

The proportion of density of a species to that of stand as a whole is referred to as relative density.

The following formula is used for calculating relative density (RD) of a species:

$$\text{Relative density (RD) of a species} = \frac{\text{Total number of individuals of a species}}{\text{Total number. of individuals of all species}} \times 100$$

Density of species in a field is determined by the method given in Table.

Name of species	Number of individuals in different quadrats each of 1 square metre size										Total No. of individuals	Density
	1	2	3	4	5	6	7	8	9	10		
1. *Cynodon dctylon*	×	7	5	4	2	2	7	4	7	6	44	$\frac{44}{10} = 4.4$ plants/m^2
2. *Evolvulus* sp.	6	×	7	8	3	×	3	2	×	5	34	$\frac{34}{10} = 3.4$ plants/m^2
3. *Argemone maxicana*	×	×	×	2	2	4	1	4	5	×	18	$\frac{18}{10} = 1.8$ plants/m^2
4.												
5.												
6.												

2. Frequency: Frequency refers to the degree of dispersion of individual species in an area and is usually expressed in terms of percentage. The frequency of a species is determined with the help of the following formula:

$$\text{Frequency} = \frac{\text{Total number of quadrats in which species has occured}}{\text{Total number of quadrats studied}} \times 100$$

Suppose, species 'X' occurred in 5 quadrats out of total ten quadrats studied, the frequency of species X will be $\frac{5}{10} \times 100 = 50\%$.

If the line or belt transect method is used for sampling then each line or belt is recognized as one quadrat for the purpose of frequency calculation.

If point frame method of sampling is used, the frequency is calculated by the following formula:

$$\text{Frequency of a species} = \frac{\text{Total number of species occur in all quadrats} \times 100}{\text{Total number of quadrats studied}}$$

For calculating the frequency, the following pattern is adopted:

Name of the species	Number of sample plots										Frequency percentage
	1	2	3	4	5	6	7	8	9	10	
1. *Parthenium hysterophorus*	×	×	-	-	×	×	×	×	×	×	$\frac{8}{10} \times 100 = 80\%$
2. *Argemone maxicana*	-	-	×	×	×	×	×	×	-	-	$\frac{6}{10} \times 100 = 60\%$
3. *Cynodon* sp.	×	×	×	×	×	×	×	-	-	-	$\frac{7}{10} \times 100 = 70\%$
4. *Euphorbia* sp.	×	-	-	-	-	×	×	×	-	×	$\frac{5}{10} \times 100 = 50\%$
5. *Boerhaavia* sp.	-	-	-	-	×	×	-	×	-	×	$\frac{4}{10} \times 100 = 40\%$
6.											

(In the above table, the occurrence of species is marked by ×)

If the individuals of a species are evenly distributed over the area, they may occur in all the sample plots and thus the frequency of species will be 100%. Poorly dispersed species will occur only in a few quadrats and their frequency will be low. This indicates that higher the frequency value of a species in the area, the greater will be the uniformity in the spread.

Raunkiaer recognised five frequency classes of plant species in the community on the basis of their frequency percentages. These are as follows:

Class A-1 to 20% frequency
Class B-21 to 40% frequency
Class C-41 to 60% frequency
Class D-61 to 80% frequency
Class E-81 to 100% frequency

Raunkiaer (1934) suggested that the number of species in frequency class A is greater than that of class B; B is greater than class C, class C is greater, or equal or lesser than class D; and D is lesser than class E.

$A > B > C \underset{<}{\overset{>}{=}} D < E$. This is also known as **Raunkiaer's 'frequency law'**.

From the above frequency law, it is apparent that the species with low frequency values are higher in number than the species with higher frequency value in most natural communities.

The dispersion of species in relation to that of all the species is termed as relative frequency of a species. Relative frequency (RF) is determined by the following formula:

$$\text{Relative frequency (RF) of a species} = \frac{\text{Frequency of the species in stand } \pi}{\text{Sum of the frequencies for all species in stand x}} \times 100$$

3. Abundance: It usually gives the numerical strength of a species. To determine abundance, the sampling is done by quadrat or other methods at random at many places and the number of individuals of a species is added for all the quadrats studied. The abundance is determined by the following formula:

$$\text{Abundance} = \frac{\text{Total number of individuals of the species}}{\text{Total number of quadrats in which the species has occurred}}$$

Abundance also have 5 classes- rare (1-4), occasional (5-14), frequent (15-29), abundant (30-99) and very abundant (100). For calculating the abundance, the following pattern is adopted:

Abundance Classes

Name of species	**Number of individuals in different quadrats**										**Total No. of individuals**	**Density**
	1	**2**	**3**	**4**	**5**	**6**	**7**	**8**	**9**	**10**		
1. *Cynodon dctylon*	×	7	5	4	2	2	7	4	7	6	44	$\frac{44}{9} = 4.8$ plants/m^2
2. *Evolvulus* sp.	6	×	7	8	3	×	3	2	×	5	34	$\frac{34}{7} = 4.25$ plants/m^2
3. *Argemone maxicana*	×	×	×	2	2	4	1	4	5	×	18	$\frac{18}{6} = 3.0$ plants/m^2
4.												
5.												
6.												

Abundance refers actually to **density** of population in those quadrats in which a given species occurs. In low form of vegetation like grasslands, abundance can be recorded by uprooting the plants.

4. Cover: The cover implies the area covered or occupied by the leaves, stem, flower and fruits, as viewed from the top. It is of two types-
(i) Crown (Canopy) cover
(ii) Basal (Stem base) cover

Area of cover is calculated by multiplying the average area of radius with πr^2. In forest vegetation, the stratification reaches to its greatest complexity, here strata are well marked and each layer of vegetation is considered separately for measuring the coverage. The coverage can be measured by quadrat method, transect method and point method of sampling. Basal coverage of tree is measured at 4.5 feet above the ground called diameter of **breast height**, while in case of grasslands, estimate of total spread of foliage has little meaning. Basal area in such cases refers to the coverage of ground 1 inch above the ground surface by stems and leaves. It is also called **herbage** cover. Basal area of tree is calculated by the following formula-

$$\text{Basal area per tree} = \frac{\text{Total basal area}}{\text{Number of trees}}$$

The area of coverage is used to express the dominance. The higher the coverage area the greater is the dominance. The average basal area is calculated out of average cross section area of stem penetrating the soil. The average area of one stem multiplied by the density (no. of individuals per unit area) gives the basal cover per unit area.

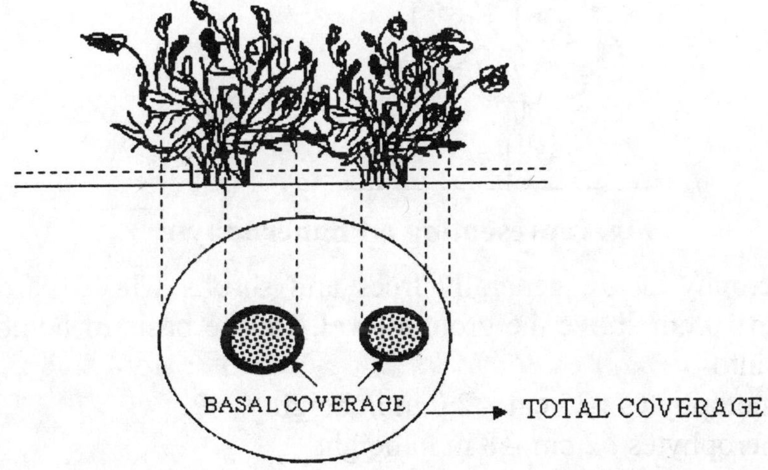

Fig. showing basal and total coverage of two clumps of grasses

Relative dominance (R.D.) is the proportion of the basal area of a species to the sum of the basal coverage of all the species in the area.

$$\text{Relative dominance of a species} = \frac{\text{Total basal area of the species in all the quadrats}}{\text{Total basal area of all the species in all the quadrats}} \times 100$$

The 3/2 thinning law : Most aspects of growth of population are density related. One important generalisation applied is 3/2 thinning law. If we plot the relationship between the dry weight and density of shoots in plant population, the line related weight of each individuals to density has a slope of -1.5 or -3/2. The slope would be 1, if increasing density has been exactly compensated by reduction in weight of individuals. Thinning is normally inversely density dependent, but does not always occur in the growth of the plants is extremely plastic. The 3/2 is universal and applied well in a wide variety of plants from mosses to trees. Exact reason of its occurrence is still not clear.

Group effect or phase phenomenon : In insects, when some individuals are separated from their society, they live for shorter time than their conspecific left in the group. This phenomenon has been described as group effect. It has been observed in many fishes, birds, rodents and even apes. **Uarow** (1921) has used the term phase phenomenon for group effect like situation in insects.

Life form

Life form may be defined as "The vegetative appearance of plant based on its size, shape, branching, crown and life span". Also known as physiognomy.

Various types of life forms have been recognized, out of which the most important and accepted rule is given by "**Raunkiar 1935**". According to him, life form is defined as "Life form is the sum of the total adaptation of plants".

He classified phanerogames on the basis of position of perrenating organ and degree of their protection during adverse condition.

1) Phanerophytes

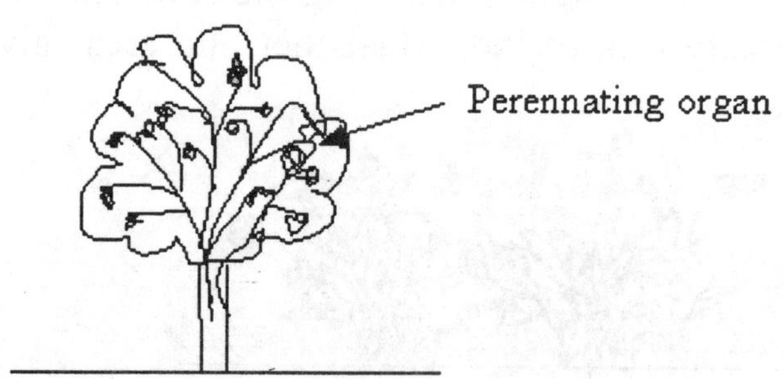

Fig. representing a Phanerophyte

The phanerophytes are generally trees and shrubs whose parrenating organ situated more than 10 cm above the ground level. On the basis of height of trees it is further classified into-

a) Nannophanerophytes : 25 cm – 2.0 m in height
b) Microphanerophytes : 2 cm – 8 m in height
c) Mesophanerophytes : 8 m – 30 m in height
d) Megaphanerophytes : More than 30 m in height.

To these categories, three more types have been added by later workers they are epiphytes, lianas and stem succulents:

1) Epiphytic phanerophytes
2) Stem succulent phanerophytes
3) Liana phanerophytes e.g. *Bauhania vahlii*

2) Chaemophytes

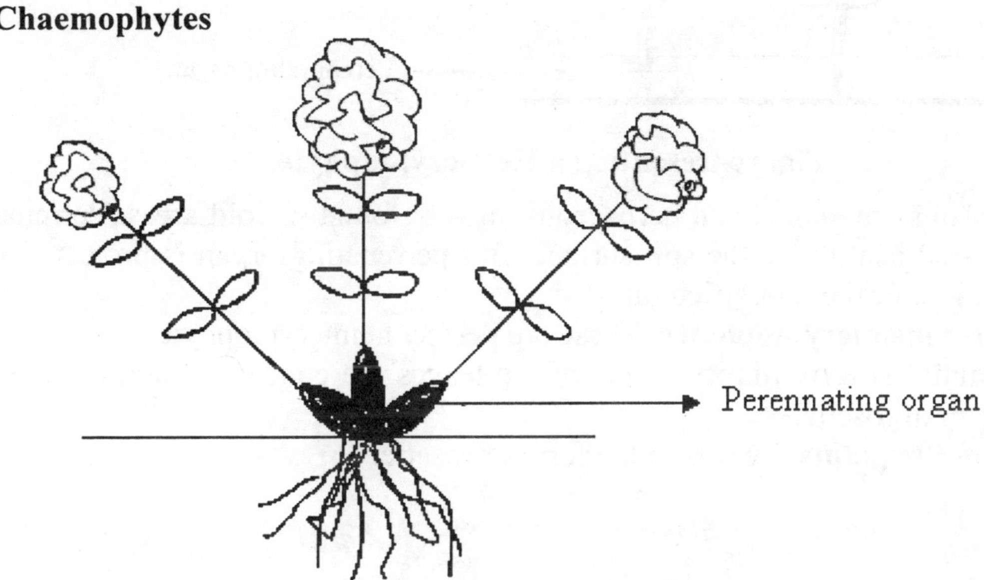

Fig. representing a Chaemophyte

It is commonly found in cold areas. Plants are usually less than 25-30 cm in height. The perrenating buds are shoot apices lying above the soil but close to ground. Rest of the plants lie in unfavourable season. The dead shoot covers and protects the perrenating buds or shoots. For example- *Trifolium repens.* Chaemophytes are of 4 types-

i) **Suffruticose chaemophytes**: Plant is short and perrenating organ at the base.

ii) **Passive chaemophytes**: Plants are short but perrenating organs found on horizontal branches.

iii) **Active chaemophytes**: Aerial erect branches are totally absent, only horizontal branches are found. They behave like perrenating organs.

iv) **Cushion chaemophytes**: They constitute a transitional form between chaemophytes and hemicryptophytes. Shoots are small and form cushions.

3) Hemicryptophytes

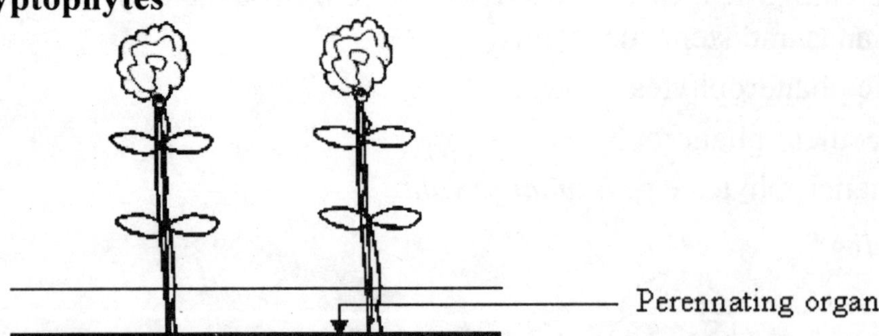

Fig. representing a Hemicryptophyte

The plants are annual and herbaceous, mostly found in cold areas. Perrenating organ is situated just above the soil surface. But perrenating organ is protected with soil or snow. It is further classified into-

i) **Protohemicryptophytes**: These are perfect hemicryptophytes.

ii) **Partial rosette plants**: Some of the leaves are condensed and arranged in form of rosette.

iii) **Rosette plants**: Leaves in the form of rosette.

4) Cryptophytes

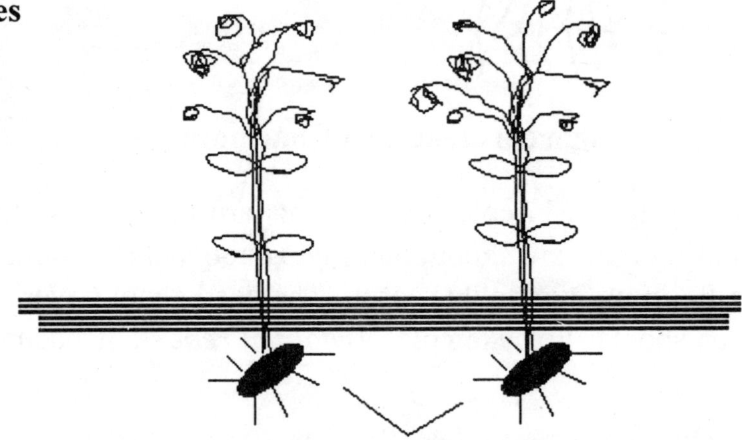

Fig. represents cryptophytes

In these plants, perrenating organ is found protected beneath the soil surface in the form of rhizome, tubers and corck etc. It is also classified into 3 parts-

i) **Geophytes**: Land plants

ii) **Heliophytes**: Grown in marshy places

iii) **Hydrophytes**: Submerged or aquatic floating plants.

5) Therophytes: These are annual plants, where entire plant die each year and perrenation takes place by seeds.

Raunkiaer also gives another system of life forms on the basis of size of leaves-

i) **Leptophylls**: Less than 25 mm^2
ii) **Nanophylls**: 25-225 mm^2
iii) **Microphylls**: 225-2025 mm^2
iv) **Mesophylls**: 2025-1825 mm^2
v) **Macrophylls**: 1825 mm^2 – 164025 mm^2
vi) **Megaphylls**: More than 164025 mm^2

Stratification and Aspectation

In order to make maximum utilization of the available physical and physiological requirements, different species of plants in a community are arranged in different vertical strata or layers is called **stratification**. Stratification of vegetation is another tool for the study of community. A forest community may have five strata viz. sub-terranean, forest floor, herbaceous, shrubs and trees. The grassland community may be distinguished into two or three vertical strata. When the flora of a community is periodically observed round the year, changes in the appearance of vegetation may be apparent with the change in season. This is known as **aspectation**. For this, phenology (study of periodicity phenomenon of plants, e.g. germination, vegetative growth, flowering, fruiting, seed settings etc. in relation to climate) of species in relation to different seasons of the year is recorded.

Periodicity and Phenology

The various physiological process of plants like germination of seeds, growth, flowering, fruiting an seed formation occur in a seasonal manner. This is known as **periodicity**; while the systematic study of periodicity phenomenon of plants in relation to their climate is called **phenology**. Different species have different periods of seed germination, vegetative growth and flowering.

Sociability

Sociability denotes the space relationship of plants to one another. Plants generally grow as isolated individuals, in patches, colonies or groups. Some plant species grow better when nearer to each other and produce thick population. Others become weak or die due to hard competition and as such they cannot form big populations. Nature of reproduction and fruit and seed dispersal are greatly affected by the way in which they are aggregated. Species with same density, may differ in sociability values. Thus sociability expresses the degree of association between species. Braun-Blanquet (1951) has recognized following five sociability classes which accommodate different types of species.

Class 1 = Shoots grow singly.
Class 2 = A group of 4-6 plants at one place.
Class 3 = Many smaller scattered groups at one place.
Class 4 = Several bigger groups of many plants at one place.
Class 5 = A large group of plants occupying larger area.

High degree of sociability is seen in those plants which shows high power of productivity.

Vitality or Vigour

It is concerned with the health of plants i.e. the capacity of plants by which they complete their life cycle. In plants, stem height, root length, area of the leaf, number of leaf, number of flowers, number and weight of fruits and seeds etc. determine the vitality.

SYNTHETIC CHARACTER

It based on the data obtained during the analysis of plants. Synthetic characters are actually computed from analytical characters. The synthetic characters are determined in following terms:

a) Presence: It indicates the occurrence of a species in a stand. **Braun Blanquet** has classified it into 5 classes-

Rare: Present in 1 to 20% of the sampling unit.
Seldom Present: Present in 21 to 40% of the sampling unit.
Often Present: Present in 41 to 60% of the sampling unit.
Mostly Present: Present in 61 to 80% of the sampling unit.
Constantly Present: Present in 81-100% of the sampling unit.

(b) Constance: It is the degree of presence in unit area. It is generally determined from frequency and the following 5 constance classes have been recognised.

Constance 1 = 1 – 20% Frequency
Constance 2 = 21 – 40% Frequency
Constance 3 = 41 – 60% Frequency
Constance 4 = 61 – 80% Frequency
Constance 5 = 81 – 100% Frequency

(c) Fidelity

The term **fidelity** refers to faithfulness of the species to its community. Fidelity is the degree with which a species is restricted in distribution to one kind of community. Such species are sometimes called as **indicator species**. Fidelity is the measure of ecological amplitude (Hanson, 1950). Fidelity of a species is expressed in relation to a particular community. A species may have high fidelity for one community and a low fidelity for another. **Braun Blanquet** recognised the following five fidelity classes of the species:

Fidelity Class	**Characteristics**
Class-1	**Strange species**: It is the rarely occurring species which either have arrived from other communities or are relics from other earlier stages of succession.
Class-2	**Indifferent species**: Species which occur in several kinds of communities without exhibiting any preference to any particular kind of community.
Class-3	**Preferential species**: Species which occur in several kinds of communities but are predominant in one.
Class-4	**Selective species**: Species occurring in one kind of community but rarely in other communities.
Class-5	**Exclusive species**: Species which are completely restricted to one kind of community.

(d) IVI (Importance Value Index)

In a highly heterogenous plant community data of frequency, density and cover of species do not yield total picture of ecological importance independently. The overall picture of ecological importance of a species in relation to the community structure can be obtained by adding the values of relative density, relative dominance and relative frequency. This total value out of 300 is called IVI of the species.

The IVI was first pointed out by **Curtis** and **Mc, Intosh (1451)**. It gives complete picture of sociological character but does not give the dimension of relative density, relative dominance and relative frequency. For the preparation of phytograph, a circle is made and then the circle is divided into four equal quarters by two radial lines lying at right angles to each other. Three radii from centre to circumference are divided into 100 equal divisions (segments) and the fourth radius is divided into 300 equal parts to mark the IVI value. On 0-100 scale in the three radii, the values of relative frequency, relative density and relative dominance are marked, and on 0-300 scale value of IVI is marked. All these points on different radii are joined by lines. Thus a phytography illustrating the sociological characters and IVI of individual species is obtained.

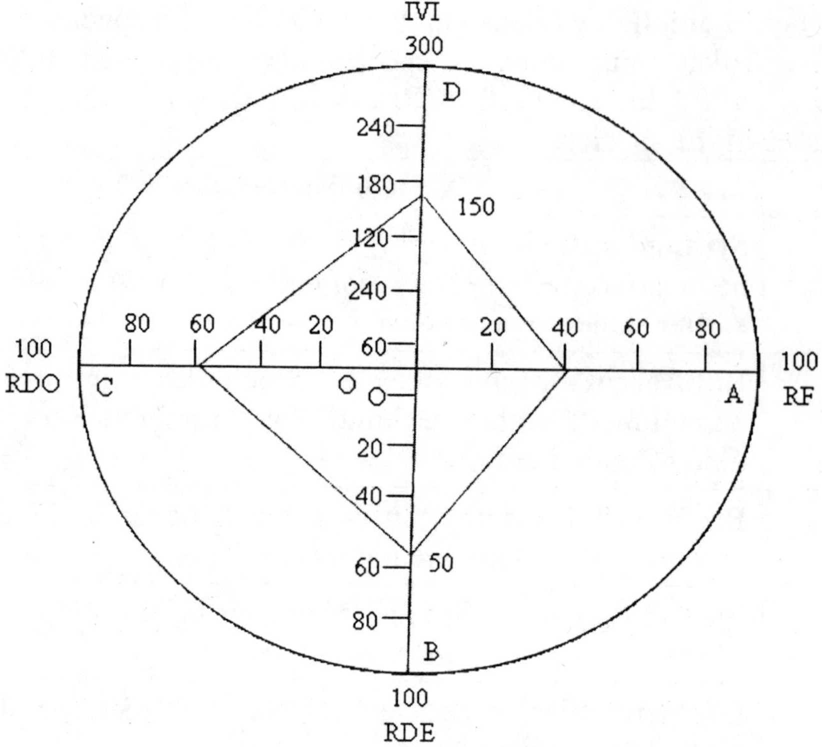

Fig. Phytograph of a species with relative frequency (RF) 40%, relative density (RDE) 50%, relative dominance *RDO) 60%, and IVI value 150.

MASTER STROKES

Q. What do you mean by age pyramid?

Age pyramid: Age distribution is an important characteristic of population which influences natality and mortality. The ratio of the various age groups in a population determines the current reproductive status of the population, thus anticipating its future. **Bodenheimer** (1958) divided the individuals of population into three major ecological ages (age groups). These are –

1. Pre-reproductive
2. Reproductive
3. Post-reproductive

The individuals of pre-reproductive group are young, those of reproductive group are mature and those in post-reproductive group are old.

The model representing geometrically, the proportion of different age groups in the population of any organism is called **age pyramid**. Age pyramids are of following three types-

a) Triangular age pyramid: This pyramid shows that the number of pre-reproductive individuals is high, number of reproductive individuals is moderate while post-reproductive individuals are fewer.

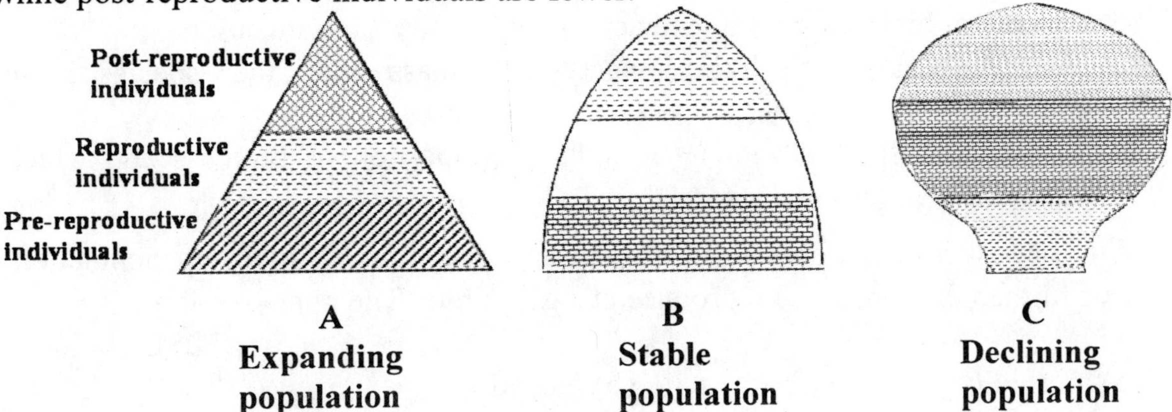

A
Expanding population

B
Stable population

C
Declining population

Fig. showing age structure in different types of population

b) Bell shaped age pyramid: This pyramid shows that the number of pre-reproductive and reproductive individuals is almost equal. The reproductive individuals are comparatively fewer. The population size is remain stable, neither growing nor diminishing.

c) Urn shaped age pyramid: It shows increased numbers of middle aged and old organisms as compared to young ones in the population. It is indicative of contracting or diminishing population.

ENTRANCE CORNER

1. The formula for exponential population growth is: **(CBSE 2006)**
(a) dt/dN = rN (b) dN/rN = dt
(c) rN/dN = dt (d) dN/dt = rN

2. Which of the following is not true for a species? **(CBSE 2005)**
(a) Members of a species can interbreed
(b) Variations occurs among members of a species
(c) Each species in reproductively isolated from every other species
(d) Gene flow does not occur between the population of a species

3. There exists a close association between the alga and the fungus within a liches. The fungus : **(CBSE 2005)**
(a) fixes the atmospheric nitrogen for the alga
(b) provides protection, anchorage and absorption for the alga
(c) provides food for the alga
(d) releases oxygen for the alga

4. A keystone species is the one which has : **(CBSE 2004)**
(a) a small proportion of total biomass but has a huge impact on community's organization and survival
(b) a plenty of biomass but low impact on community's organization
(c) a rare species with little impact on biomass and other species in te community
(d) a dominant species which has a large proportion of biomass and which affects many other species

5. Why opposite forces operate in growth and development of every population. One of them has ability of reproduce at a given rate. The opposing force is : **(CBSE 2003)**
(a) morbidity (b) fecundity
(c) biotic potential (d) environmental resistance

6. Mycorrhiza is an example of : **(CBSE 2003)**
(a) symbiotic relationship (b) ectoparasitism
(c) endoparasitism (d) decomposers

7. Which of the following utilizes inorganic materials ? **(CBSE 2002)**
(a) Autotrophs (b) Decomposers
(c) Saprophytes (d) Heterotrophs

8. The most important character of speciation is : **(CBSE 2002)**
(a) reproductive isolation (b) ethological isolation
(c) behavioural isolation (d) geographical isolation

9. The interdependent evolution of flowering plants and pollinating insects together is known as : **(CBSE 2002)**
(a) mutualism (b) co-evolution
(c) commensalism (d) co-operation

10. Association between entomophilous flowers and pollinating agent is :
(a) commensalisms (b) co-evolution
(c) mutualism (d) co-operation

11. In exponential phase, the sequence of events in bacterial multiplication is : **(CBSE 2002)**
(a) stationary, lag, log, decline (b) stationary, log, lag, decline
(c) lag, log, decline, stationary (d) decline, log, lag, stationary

12. An unrestricted reproductive capacity is called : **(CBSE 2002)**
(a) birth rate (b) biotic potential
(c) carrying capacity (d) fertility

13. In an exponential bacterial culture, the number of cells/ml is plotted on halflog graph. The graph is : **(CBSE 2002)**
(a) sigmoid (b) hyperbola
(c) ascending straight line (d) descending straight line

14. If in any given population progeny is less adapted than its ancestors, it is due to : **(CBSE 2001)**
(a) migration (b) mutation
(c) genetic drift (d) gene flow

15. A force that acts against maximum population growth is : **(CBSE 1998)**
(a) population pressure (b) carrying capacity
(c) saturation point (d) environmental resistance

16. Genetic drift operates in only..........populations : **(CBSE 1998)**
(a) island (b) smaller
(c) larger (d) Mendelian

17. After about every 5 to 8 years there occurs warm ocean in peru Currents and the East Pacific of south America, it is widely known as : **(CBSE 1998)**
(a) Magnox (b) El-Nino
(c) Gulf stream (d) Aye ayes

18. The epecies which are different from each other and does not interbreed are : **(CBSE 1998)**
(a) sympatric (b) congenic
(c) sibling (d) parapatric

19. Species occurring in different geographical areas are called as : **(CBSE 1998)**
(a) neopatric (b) sibling
(c) sympatric (d) allopatric

20. The types of animals which are relatively more abundant in desert grassland are : **(CBSE 1998)**
(a) auroral (b) diurnal
(c) fussorial (d) quatic

21. The most important factor for success of any animal population is : **(CBSE 1997)**
(a) natality (b) unlimited food
(c) adaptability (d) inter-species activity

22. The present population of the world is about : **(CBSE 1997)**
(a) 5000 million (b) 100 million
(c) 6 billion (d) 15 billion

23. Closely related, morphologically similar sympatric populations but reproductively isolated are designated as : **(CBSE 1996)**
(a) niche (b) ecotone
(c) sibling (d) clines

24. At the present rate of growth the human population growth in India : **(CBSE 1996)**
(a) is to follow a sigmoid curve as in case of many other animals species
(b) is to reach a zero population growth as in case of some animals species
(c) can be decreased by allowing the natural calamities to occur and enforcing birth control measures
(d) can be decreased by following the national programme of family planning

25. Niche of a species in an ecosystem refers to its : **(CBSE 1996)**
(a) function at its place of occurrence
(b) place of its occurrence
(c) competitive ability
(d) centre of origin

26. Nature of climax community depends upon : **(CBSE 1996)**
(a) water (b) temperature
(c) soil fertility (d) climate

27. The human population can be truly called civilized in a large measure on : **(CBSE 1996)**
(a) their ability of humanity to moderate its fecundity
(b) in increasing the food population
(c) colonization of under populated areas
(d) combating human diseases

28. The abundance of a species population within its habitat is called : **(CBSE 1995)**
(a) absolute density (b) regional density
(c) relative density (d) niche density

29. If decomposers are cleaned totally from any environment which of the following effect will occur ? **(CBSE 1995)**
(a) Blockage of mineral flow (b) Blockage of solar energy
(c) Light energy flow blockage (d) Rate of decomposition will increase

30. If the age pyramid of a population is bell shaped, it will show : **(CBSE 1995)**
(a) high percentage of old (b) high percentage of children
(c) low percentage of children (d) none of the above

31. Which of the following is applicable to human races ? **(CBSE 1994)**
(a) All races can interbreed but most will produce sterile young ones
(b) All races can interbreed and produce fertile offspring
(c) Different races can not interbrred
(d) Only few human races can interbreed

32. They are referred to as endemic organisms which : **(CBSE 1994)**
(a) grow near oceans
(b) grow in deserts
(c) are restricted to a particular geographical area
(d) are grown through genetic engineering

33. Killing an organism for feeding is : **(CBSE 1993)**
(a) predation (b) parasitism
(c) symbiosis (d) exploitation

34. A population with equal number of births and deaths will show : **(CBSE 1992)**
(a) acceleration phase of growth (b) plateau phase
(c) exponential growth phase (d) initial phase of growth

35. One who gets food at the cost of other is : **(CBSE 1992)**
(a) parasite (b) commensal
(c) saprophyte (d) insectivore

36. In commensalisms : **(AFMC 2000)**
(a) both partners are benefitted
(b) both partners are harmed
(c) weaker is benefited while stronger is unharmed
(d) none of the above

37. Dominant organisms found on earth now a days are : **(AFMC 2000)**
(a) herivores (b) carnivores
(c) rodents (d) marine mammals

38. A successful parasite is the one which : **(AFMC 2000)**
(a) grows rapidly (b) reproduces fast
(c) sticks to host for long (d) makes minimum demands from its host

39. All living organisms in an ecosystem interacting together constilute : **(AFMC 1998)**

(a) community (b) population
(c) planktons (d) ecology

40. The tendency of biological systems to resists changes so as to remain in a state of equilibrium is called : **(AFMC 1997)**

(a) homeostasis (b) epistasis
(c) bioenergetics (d) holocoenosis

41. A parasite living within tissue of host is : **(AFMC 1996)**

(a) epiphyte (b) endophyte
(c) ectophyte (d) none of these

42. **Assertion :** The sex ratio of Kerala is highest in India.
Reason : In countries like India the population is increasing at a rapid rate. **(AIIMS 2005)**

(a) Both A and R are true and R is the correct explanation of A
(b) Both A and R are true but R is not a correct explanation of A
(c) A is true but R is false
(d) A is false but R is true

43. Which one of the following is a matching pair of certain organism (s) and the kind of association ? **(AIIMS 2003)**

(a) Shark and sucker fish-commensalism
(b) Algae and fungi in lichens-mutualism
(c) Orchids growing on trees-parasitism
(d) Cuscuta (dodder) growing on other flowering plants-epiphytism

44. Mutualism occurs between : **(AIIMS 2001)**

(a) butterfly and flower (b) *Escherichia coli* and man
(c) *zoochlorellae* and *Hydra* (d) hermit crab and sea anemone

45. Scholars caught, marked and released 80 fishes in a pond. Later 100 fishes were caught at random. 40 of them were marked. The number of fishes in the pond is: **(AIIMS 2001)**

(a) 400 (b) 200
(c) 100 (d) 50

46. Most of the plants cultivated in agriculture are actually belonging to the succession level : **(AIIMS 2000)**

(a) early 1^0 succession (b) late 1^0 succession
(c) early 2^0 succession (d) late 2^0 succession

47. If the rate of addition of new members increases with respect to the individual lost of the same population, then the graph obtained has : **(AIIMS 2005)**

(a) declined growth (b) exponential growth
(c) zero population growth (d) none of these

48. Small fish get stuck near the bottom of a shark and derives its nutrition from it. This kind of association is called as : **(BHU 2005)**

(a) symbiosis (b) commensalism
(c) predation (d) parasitism

49. 11th July is : **(BHU 2002)**

(a) World Population Day (b) World Food Day
(c) World Environment Day (d) World Health Day

50. A bird enters the mouth of crocodile and feed on parasitic leeches. The bird gets food and crocodile gets ribs of blood sucking leeches. Both the partners can live independently. Such an association is : **(BHU 2001)**

(a) mutualism (b) amensalism
(c) commensalism (d) proto co-operation

51. The genepool of a population tends to remain stable, if certain conditions are met. These conditions include large populations, no mutation, no migration and : **(BHU 2000)**

(a) random mating (b) natural selection
(c) reduction of predators (d) moderate environmental changes

52. Population I and II are growing in the same habitat and are morphologically similar but these are intersterile. In terms of 'Biological Species Concept' they may be considered : **(BHU 2000)**

(a) one species (b) two distinct species
(c) two sibling species (d) none of these

53. The concept of life from classes and biological spectrum was proposed by : **(BHU 1999)**

(a) Raunkiaer (b) Clements
(c) Weaver (d) Turesson

54. Hermit crab and sea-anemone relationship is : **(BHU 1997)**

(a) mutualism (b) commensalism
(c) parasitism (d) proto co-operation

55. Actively moving animal organisms in any aquatic systems are : **(BHU 1996)**
(a) benthic (b) phytoplanktons
(c) nektons (d) zooplanktons

56. Camels and llamas are examples of dispersed individuals evolving separately; this is considered as a consequence of : **(BHU 1995)**
(a) ecological barriers (b) climatic barriers
(c) continental drift (d) topographical factors

57. Pick up the correct statement : **(BHU 1994)**
(a) Hone range of members of a species has larger area than that of territory
(b) Many insects mainly honey bees communicate among themselves by secreting pheromones
(c) The last succession in series of biotic succession is called a sere
(d) Number of individuals of particular species present in a unit area at a given time is known as population size

58. Biotic potential refers to : **(BHU 1994)**
(a) increase of population under optimum conditions
(b) increase of population under given conditions
(c) increase of population under natural conditions
(d) increase of population under climatic conditions

59. Proportion of reproductive age group is higher in : **(Manipal 2006)**
(a) triangular age pyramid (b) urn-shaped pyramid
(c) bell-shaped age pyramid (d) pyramid of biomass

60. Which of the following causes a species to be different from other and mainatain it's own specificity ? **(Manipal 2003)**
(a) Micro-evolution (b) Macro-evolution
(c) Genetic adaptation (d) Climatic variations

61. Lichens are : **(Manipal 2000)**
(a) commercial (b) parasites
(c) composite organisms (d) saprophytes

62. Hyperparasite is parasite which : **(Manipal 1998)**
(a) kills its host
(b) completes life cycle in one host
(c) uses host machinery for reproduction
(d) none of the above

63. Insectivorous plants are adopted to soils : **(Manipal 1995)**
(a) rich in water
(b) deficient in water
(c) deficient in nitrogenous compounds
(d) deficient in trace elements

64. Cuscuta is a : **(DPMT 2005)**
(a) parasitic plant (b) symbiotic plant
(c) predator (d) decomposer

65. Types of interaction that occur in predation and parasitism are : **(DPMT 2001)**
(a) +, + (b) –, –
(c) +, 0 (d) +, –

66. In any growing population the most contribution is of : **(DPMT 1997)**
(a) post-reproductive members (b) reproductive members
(c) pre-reproductive members (d) all of above

67. In increasing order of organizational complexity, which one of the following is in the correct sequence? **(DPMT 1997)**
(a) Population, variety, community, ecosystem
(b) Population, species, community, ecosystem
(c) Species, population, community, ecosystem
(d) Species, variety, ecosystem, population

68. A bird introduced from another country became a serious pest due to : **(DPMT 1996)**
(a) better adaptation to new area (b) more sexual reproduction
(c) better nesting habitats (d) absence of natural competitions

69. The graphic representation of age groups of a population given here, shows : **(DPMT 1996)**
(a) declining population (b) increasing population
(c) stable population (d) fluctuating population

70. When population reaches carrying capacity ? **(DPMT 1996)**
(a) Mortality rate = Birth rate (b) Mortality rate > Birth rate
(c) Mortality rate < Birth rate (d) None of the above

71. Living in same habitat organisms of same species form : **(CPMT 2005)**
(a) biosphere (b) community
(c) population (d) niche

72. Lichens show : **(CPMT 2005)**

(a) mutualism (b) commensalisms

(c) parasitism (d) saprophytism

73. One of the following ecosystem has the highest primary productivity :

(a) pond ecosystem (b) lake ecosystem **(CPMT 2000)**

(c) grassland ecosystem (d) forest ecosystem

74. Feeding on dead and decaying organisms represent what type of nutrition :

(a) Holozoic (b) Saprophytic **(CPMT 2000)**

(c) Parasitic (d) Autotrophic

75. Pants which prefer extreme sunlight are called : **(CPMT 1998)**

(a) heliophytes (b) halophytes

(c) sciophytes (d) xerophytes

76. One of them is a submerged hydrophyte : **(CPMT 1998)**

(a) *Ceratophyllum* (b) *Utricularia*

(c) *Vallisneria and Hydrilla* (d) *Lemna*

77. Which take part in symbiosis of lichen ? **(CPMT 1996)**

(a) Alga-fungus (b) Alga-alga

(c) Fungus-fungus (d) fungus-gymnosperm

78. If natality rate is parallel to mortality rate then population is : **(MH CET 2005)**

(a) slowly increases (b) remain stationary

(c) show J-shaped curve (d) slowly decreases

79. Which of the following is wrong ? **(Karnataka 1998)**

(a) Lichen, an association of fungus and alga, is an example of mutualism

(b) Epiphytes, using other plants for support only and not for water or food, show commensalism

(c) Sea anemone and hermit crab association is an example of proto co-operation

(d) Mutualism, proto co-operation and commensalism can not be included under symbiosis

80. In India, human population is heavily weighed towards the younger age group as a result of: **(Pb. PMT 2005)**

(a) short life span of many individuals and low birth rate

(b) short life span of many individuals and high birth rate

(c) long life span of many individuals and high birth rate

(d) long life span of many individuals and low birth rate

81. Territoriality occurs as a result of: **(Pb. PMT 2004)**
(a) parasitism (b) predation
(c) co-operation (d) competition

82. Proportion of young individuals is highest in case of: **(Pb. PMT 2002)**
(a) declining population (b) stable population
(c) both (a) and (b) (d) expanding population

83. After exponential increase, population becomes stagnant. The growth curve is: **(Pb. PMT 1998)**
(a) J-shaped (b) S-shaped
(c) fluctuating (d) circular

84. The association of roots of higher plants with fungi is: **(Pb. PMT 1998)**
(a) symbiosis (b) mutualism
(c) parasitism (d) hyperparasitism

85. Some animals do not like others to enter their habitat. This habit is: **(Haryana PMT 2003)**
(a) territorial (b) parasitism
(c) predation (d) symbiosis

86. In a population, if number of immigrated people are more than number of emigrated and death of people is low then in growth curve the phase is: **(Haryana PMT 2000)**
(a) decline phase (b) exponential phase
(c) steady phase (d) lag phase

87. Plants obtaining nourishment from other plants by haustoria are: **(Haryana PMT 1998)**
(a) epiphytes (b) parasites
(c) xerophytes (d) halophytes

88. The existence of sea anemone on the back of hermit crab is an example of: **(AMU 2006)**
(a) proto co-operation (b) parasitism
(c) antagonism (d) commensalism

89. The organisms inhabiting a common environment belong to the same: **(AMU 2005)**
(a) species (b) genus
(c) population (d) community

90. Population was termed as self perpetuating unit by: **(AMU 2003)**
(a) Malthus (b) Spencer
(c) Mobius (d) Odum

91. An association of individuals of different species living in the same habitat and having functional interactions is: **(AMU 2002)**
(a) population (b) ecological niche
(c) biotic community (d) ecosystem

92. Population of an area tends to increase when: **(AMU 1997)**
(a) reproductivity decreases (b) immigration increases
(c) emigration occurs (d) predation increases

93. If decomposers stop functioning in an ecosystem the functioning of ecosystem will be affected due to: **(AMU 1997)**
(a) the energy flow in ecosystem will be blocked
(b) cycling of material and minerals will be stoped
(c) both (a) and (b)
(d) none of the above

94. Ratio of natality and mortality of a population expressed in percentage is: **(AMU 1997)**
(a) vital index (b) growth rate
(c) survival rate (d) biotic potential

95. Term used to describe non-dominant species that dictates community structures is: **(Kerala PMT 2004)**
(a) pioneer species (b) keystone species
(c) transitional species (d) exogenous species

96. *Rhizobium* bacteria and root nodules of pea have an association called: **(Kerala PMT 2002)**
(a) predation (b) parasitism
(c) symbiosis (d) commensalism

97. An interaction between two individuals where one is benefitted while the other is neither benefitted nor harmed is called as: **(Kerala PMT 2002)**
(a) predation (b) symbiosis
(c) scavenging (d) commensalism

98. Relationship between two organisms in which one obtain some benefit at the cost of other is: **(Kerala PMT 2001)**

(a) parasitism (b) predation
(c) symbiosis (d) scavenging

99. A protective device in a community is: **(Kerala PMT 2001)**

(a) commensalism (b) mimicry
(c) parasitism (d) competition

100. The permanent decrease in population number occurs due to: **(Kerala PMT 2001)**

(a) mortality (b) natality and migration
(c) territorality (d) emigration

101. Density of a population (D) is: **(Kerala PMT 2000)**

(a) $D = \frac{S\,(\text{size})}{W\,(\text{weight})}$ (b) $D = \frac{S\,(\text{space})}{N\,(\text{number})}$

(c) $D = \frac{N\,(\text{number})}{S\,(\text{space})}$ (d) None of these

102. Severe competition for food, shelter and space is maximum is: **(CET Chd. 2002)**

(a) closely related species occupying the same niche
(b) closely related species occupying the different niche
(c) distantly related species occupying the same habitat
(d) distantly related species occupying the different habitat

103. Rhinoceros and pecker bird show: **(CET Chd. 1996)**

(a) proto co-operation (b) commensalism
(c) ammensalism (d) mutualism

104. Who said that population would exceed the food resources? **(CET Chd. 1996)**

(a) Oparin theory (b) Malthus theory
(c) Darwin theory (d) Theory of Lamarck

105. In commensalism: **(MP PMT 2004)**

(a) both partners are benefitted
(b) both partners are harmed
(c) weaker is benefitted while stronger is unharmed
(d) none of these

106. At asymptome stage, the population is: **(MP PMT 2002)**
(a) stabilised (b) increasing
(c) decreasing (d) changing

107. Human population shows: **(MP PMT 2000)**
(a) J-shaped growth curve (b) Z-shaped growth curve
(c) S-shaped growth curve (d) All of these

108. The percentage ratio of natality over mortality is defined as: **(MP PMT 2000)**
(a) population density (b) biotic potential
(c) vital index (d) population dynamics

109. Swarming is characteristic of one of the following populations: **(MP PMT 1998)**
(a) mosquitoes (b) houseflies
(c) locusts (d) *Pyrilla*

110. The aggregate of processes that determine the size and compositions of any population is known as: **(MP PMT 1995)**
(a) population dynamics (b) population dispersal
(c) population explosion (d) population density

111. Competition will be severe in a population with ……. **(MP PMT 1994)**
(a) non-random (b) irregular
(c) uniform (d) random

112. The information about birth rate, death rate, sex ratio, age distribution etc., of a population is obtained from : **(MP PMT 1994)**
(a) mortality table (b) survival table only
(c) natality table (d) both (a) and (b)

113. Study of population is: **(MGIMS Wardha 2002)**
(a) demography (b) seismonastry
(c) population graphics (d) desmography

114. Sigmoid curve is: **(MGIMS Wardha 2002)**
(a) rate of transpiration (b) rate of respiration
(c) rate of photosynthesis (d) growth of population

115. Man and *Ascaris* show: **(DY Patil Mumbai 2003)**
(a) predation (b) parasitism
(c) commensalism (d) symbiosis

116. Plants growing in dry and saline soil are called: **(RPMT 2005)**
(a) xerophyte (b) hydrophyte
(c) halophyte (d) heliophyte

117. Plants growing on sandstone are: **(RPMT 2000)**
(a) psammophytes (b) oxylophytes
(c) lithophytes (d) phanerophytes

118. Indian population contain more of younger people because of: **(RPMT 1995)**
(a) short life-span and low birth-rate
(b) long life-span with low birth-rate
(c) short life-span and high birth rate
(d) none of the above

119. Sigmoid curve represents: **(ASCC J.K. PMT 1998)**
(a) slow initial phase, rapid growth in logarithmic phase and steady final stage
(b) rapid growth in initial phase, slow in second and steady in the last stage
(c) no growth initially as well as in final stage but rapid in lag phase
(d) none of the above

120. The population density can be calculated by: **(JIPMER 1997)**
(a) D = S/N (b) D = N/S
(c) D = S/W (d) D = W/S

121. A force which checks the achievement of the highest possible level of population of world growth is: **(JIPMER 1997)**
(a) population pressure (b) saturation level
(c) immigration (d) none of these

122. Aspects such as sedentary habit, territoriality etc., not aiding diversification are classified as: **(JIPMER 1995)**
(a) reproductive isolation (b) mutational effects
(c) biological barriers (d) genetic incompatibility

123. An association of two species where both the partners derive mutual benefit from each other is: **(HP PMT 2005)**
(a) parasitism (b) symbiosis
(c) commensalism (d) predation

ANSWER SHEET

1. (d)	**2.** (d)	**3.** (b)	**4.** (a)	**5.** (d)	**6.** (a)	**7.** (a)	**8.** (a)	**9.** (b)	**10.** (c)
11. (c)	**12.** (b)	**13.** (a)	**14.** (b)	**15.** (d)	**16.** (b)	**17.** (b)	**18.** (a)	**19.** (d)	**20.** (c)
21. (c)	**22.** (c)	**23.** (c)	**24.** (d)	**25.** (a)	**26.** (d)	**27.** (a)	**28.** (d)	**29.** (a)	**30.** (a)
31. (b)	**32.** (c)	**33.** (a)	**34.** (b)	**35.** (a)	**36.** (c)	**37.** (a)	**38.** (d)	**39.** (d)	**40.** (a)
41. (b)	**42.** (b)	**43.** (a)	**44.** (c)	**45.** (b)	**46.** (d)	**47.** (b)	**48.** (b)	**49.** (a)	**50.** (d)
51. (a)	**52.** (c)	**53.** (c)	**54.** (d)	**55.** (c)	**56.** (d)	**57.** (a)	**58.** (a)	**59.** (b)	**60.** (a)
61. (c)	**62.** (d)	**63.** (c)	**64.** (a)	**65.** (d)	**66.** (b)	**67.** (c)	**68.** (d)	**69.** (b)	**70.** (a)
71. (c)	**72.** (a)	**73.** (d)	**74.** (b)	**75.** (a)	**76.** (c)	**77.** (a)	**78.** (b)	**79.** (d)	**80.** (b)
81. (d)	**82.** (d)	**83.** (b)	**84.** (b)	**85.** (a)	**86.** (b)	**87.** (b)	**88.** (a)	**89.** (d)	**90.** (a)
91. (c)	**92.** (b)	**93.** (c)	**94.** (a)	**95.** (b)	**96.** (c)	**97.** (d)	**98.** (a)	**99.** (b)	**100.**(a)
101.(c)	**102.**(a)	**103.**(a)	**104.**(b)	**105.**(c)	**106.**(a)	**107.**(c)	**108.**(c)	**109.**(c)	**110.**(a)
111.(b)	**112.**(d)	**113.**(a)	**114.**(d)	**115.**(b)	**116.**(c)	**117.**(c)	**118.**(c)	**119.**(a)	**120.**(b)
121.(d)	**122.**(c)	**123.**(b)							

Chapter 4
Ecological Succession

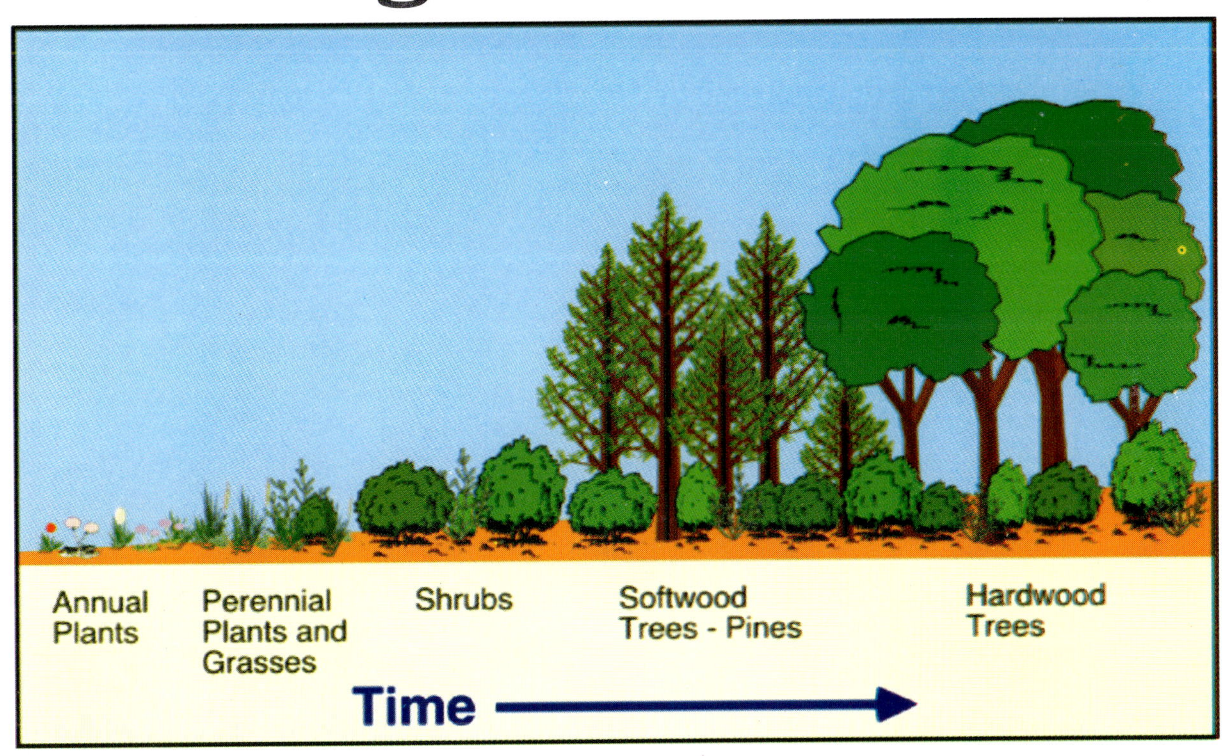

ECOLOGICAL SUCCESSION

Succession was first of all studied by King 1685, while term Succession was coined by Hult (1885). Succession may be defined as –

"Act of repeating following up of one another in order of time at a given space".

The ecological succession refers to orderly sequence of communities of plants as well as animals which occur at period of time at a given space.

In any of the basic environment (terrestrial, fresh water, marine), where there was no previously any sort of living matter, the first group of organisms establishing there are known as pioneers or primary community. The particular community around pioneer community is called **seral community** and where the maximum growth of particular species is established, called the **climax community**.

The succession in the early stage, of the life cycle of plants when they are short is called r-selection and in the later phase, it is called as k-selection.

Productivity of community: The productivity of community increases gradually because at the early stage, the rate of P/R is >1 and in case of a climax community, the ratio of P/R = 1. It is called dynamic equilibrium.

Types of succession

1) **Autogenic** or **Self generated succession**: In this type of succession, the change of communities is due to plant themselves i.e. forces or factors derive from plants.
2) **Allogenic succession**: In this type of succession, the change in habitat and change in community structure is due to the factor like change in climate, deposition of salt or sand.
3) **Primary succession**: It takes place in bare areas, which are not previously occupied by plants and such soils are called as Biosterile soil because they are rich in inorganic substances but generally totally lacking in organic matters. In this, the rate of succession is very slow.
4) **Secondary succession**: Succession takes place on areas which are Biofertile, rich in organic matter and are previously occupied by vegetation. The rate of succession is much faster than primary succession.
5) **Autotrophic succession**: This succession is dominated by green plants or Autotrophs. In this case, organic matter may or may not represent but generally

rich in inorganic substances. In the advanced stages of succession gradually organic matter is added to the soil.

6) **Heterotrophic succession**: This succession is dominated by heterotrophs like-Bacteria, fungi, actinomycetes and generally occurs in areas which are rich in organic matter.

General Process of Succession

The whole process of a primary autotrophic succession is actually completed through a number of sequential steps, which follow one another. These steps in sequence are as follows:

(i) Nudation

This is the development of a bare area without any form of life. The area may develop due to several causes such as landslide, erosion, deposition, or other catastrophic agency. The cause of nudation may be:

1. **Topographic**: Due to soil erosion by gravity, water or wind, the existing community may disappear. Other causes may be deposition of sand, landslide, volcanic activity and other factors.
2. **Climatic**: Glaciers, dry period, hails and storm, frost, fire etc. may also destroy the community.
3. **Biotic**: Man has very important role and is responsible for destruction of forests, grasslands for industry, agriculture, housing etc. Other factors are disease epidemics due to fungi, viruses etc. which destroy the whole population.

(ii) Invasion

This is the successful establishment of a species in a bare area. The species actually reaches this new site from any other area. This whole process is completed in the following three successive stages:

1. **Migration (dispersal)**: The seeds, spores, or other propagules of the species reach the bare area. This process, known as **migration**, is generally brought about by air, water etc.
2. **Ecesis (establishment)**: After reaching the new area, the process of successful establishment of the species, as a result of adjustment with the conditions prevailing there, is known as ecesis. In plants, after migration, seeds or propagules germinate, seedlings grow, and adults start to reproduce. Only a few of them are capable of doing this under primitive harsh conditions, and thus most of them disappear. Thus as a result of ecesis, the individuals of species become established in the area.

3. **Aggregation**: After ecesis, as a result of reproduction, the individuals of the species increase in number, and they come close to each other. This process is known as **aggregation**.

(iii) Competition and coaction

After aggregation of a large number of individuals of the species at the limited place, there develops **competition** (inter- as well as intraspecific) mainly for space and nutrition. Individuals of a species affect other's life in various ways and this is called **coaction**. The species, if unable to compete with other species, would be discarded. To withstand competition, reproductive capacity, wide ecological amplitude etc. are of much help to the species.

(iv) Reaction

This is the most important driving force for succession. The mechanism of the modification of the environment through the influence of living organisms on it, is known as **reaction**. As a result of reactions, changes take place in soil, water, light conditions, temperature etc. of the environment. Due to all these, the environment is modified, becoming unsuitable for the existing community which sooner or later is replaced by another community (seral community). The whole sequence of communities that replaces one another in the given area is called **a sere**, and various communities constituting the sere, as **seral communities, seral stages** or **developmental stages**. The pioneers are likely to have low-nutrient requirements, more dynamic and able to take minerals in comparatively more complex forms. They are small-sized and make less demand from environment.

(v) Stabilization (Climax)

Finally, there occurs a stage in the process, when the final terminal community becomes more or less stabilised for a longer period of time and it can maintain itself in equilibrium with the climate of the area. This final community is not replaced, and is known as **climax community** and the stage as **climax stage**.

Generalised process of succession and different plant communities appearing in the process with developing environmental complex are shown in figure. General process of succession as outlined below will show that the whole process involves several stages, each stage having characteristic organisms together with their environment. Each such developmental stage is called a **seral stage**. These stages are infact continuous with each other and the whole sequence from beginning till the climax stage is known as a **sere**. The species which colonise the bare area in the beginning of succession are called **pioneers**.

several stages, each stage having characteristic organisms together with their environment. Each such developmental stage is called a **seral stage**. These stages are infact continuous with each other and the whole sequence from beginning till the climax stage is known as a **sere**. The species which colonise the bare area in the beginning of succession are called **pioneers**.

Due to destructive effects of organisms, sometimes the development of disturbed communities does not occur and the process of succession instead of progressive becomes retrogressive, as for example- forest may change to shrubby or grassland community. This is called **retrogressive succession (Gleason, 1929)**.

Sometimes due to changes in local conditions as soil characteristics or microclimate, the process of succession becomes deflected in a different direction than that presumed under climatic condition of the area. Thus the climax communities are likely to be different from the presumed climatic climax community. This type of succession is called **deflected succession**. In India, with a monsoon type of climate, in some habitats like temporary ponds, pools etc., it is common to observe each year the development of different kinds of communities in different seasons of the year. Dudgeon (1921) called them as **seasonal succession**. However, according to Misra (1946, 1955), such changes are simply recurrent and not developmental and thus should not be designated as succession.

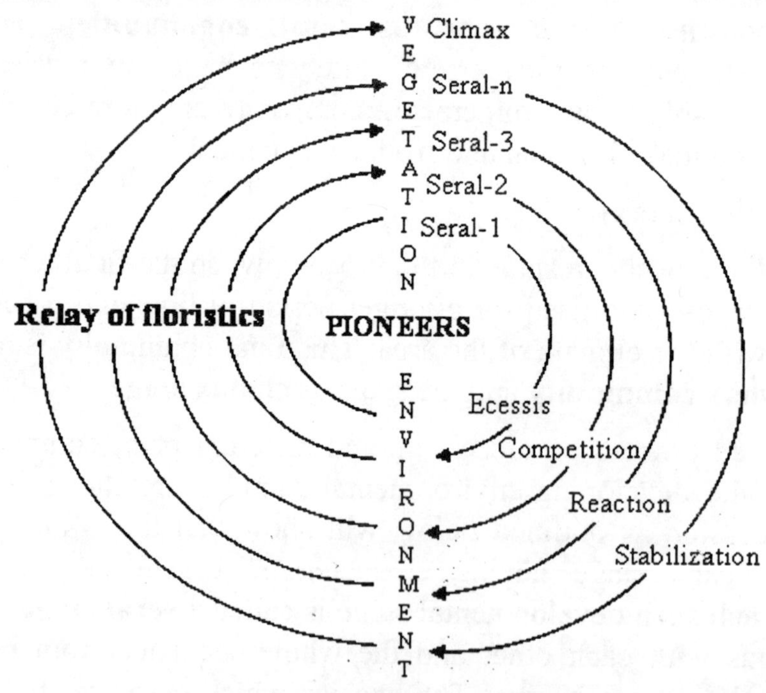

Hydrosere or Hydrarch

A good example of succession is the hydrarch succession or hydrosere succession, in which a pond and its community are converted into a land community.

(a) Phytoplankton stage: It is the pioneer stage of hydrosere. Different spore reach the water body through wind or animals. The first organisms appear are minute and autotrophic, called phytoplankton, e.g., diatoms, green flagellates, single-celled colonial or filamentous green algae. Death and decomposition of plankton produce organic matter. The litter mixed up with clay and silt at the bottom to form soft mud, favourable for growth of next seral stage.

(b) Rooted submerged stage: The bottom lined by soft mud having organic matter is favourable for growth of submerged plants *Hydrilla, Potamogeton, Vallisnaria* and *Utricularia* etc. They are rooted in the mud and form dense growth. As a result sand and silt get deposited around the plants. The bottom level rises slowly. The older plants and buried parts of other plants form humus on their death and decay. This enriches the newly built up bottom and makes it favourable for growth of next stage.

(c) Rooted floating stage: Floating leaved anchored plants (*Nymphaea*) appear where water becomes shallow. These plants have sub-terranean stems like rhizome and tuber. The plants make the water rich in mineral and organic matter. It becomes suitable for growth of free floating plants like *Wolffia, Azolla* etc. They cover the water quickly. Rapid growth of floating stage further built up bottom so that water becomes shallow on periphery.

(d) Reed-swamp stage: Amphibious plants grow where the water body becomes shallow (0.3-1.0 m) e.g., *Sagittaria*. The plants of swamp stage transpire huge quantities of water. They also produce abundant organic matter. Their tangled growth accumulates silt.

(e) Sedge or Marsh Meadow stage: The shores built up by reed swamp stage are invaded by *Carex* (sedge), *Juncus*, grasses like *Themeda* and *Dichanthium* and herbs like *Caltha, Polygonum* etc. The plants transpire rapidly and add abundant humus, therefore, soil is build up to invite next stage.

(f) Woodland stage: The periphery of sedge meadow stage is invaded by some rhizome bearing shrubby plants which can tolerate bright sunlight as well as water logged conditions, e.g., bogwood, *Cephalanthus* (Bustton Brush). The shrubs shade away the plants of sedge medow stage. They invite invasion by trees capable of bearing sunlight and water logging e.g., *Populus* (cotton wood); *Alnus* (alder). The plants of woodland stage lower the water table by their transpiration. They also built up more soil. Shade loving plants come to grow below them.

(f) Woodland stage: The periphery of sedge meadow stage is invaded by some rhizome bearing shrubby plants which can tolerate bright sunlight as well as water logged conditions, e.g., bogwood, *Cephalanthus* (Bustton Brush). The shrubs shade away the plants of sedge medow stage. They invite invasion by trees capable of bearing sunlight and water logging e.g., *Populus* (cotton wood); *Alnus* (alder). The plants of woodland stage lower the water table by their transpiration. They also built up more soil. Shade loving plants come to grow below them.

(g) Climax forest: New trees invade the area. They have shade loving seedlings. These trees grow to greater heights. The trees and shrubs of woodland stage disappear. The climax forest depends upon the climate-rainforest in moist tropical area and mixed coniferous in temperate area. A mixed temperature forest includes broad leaved trees like oak, elm etc.

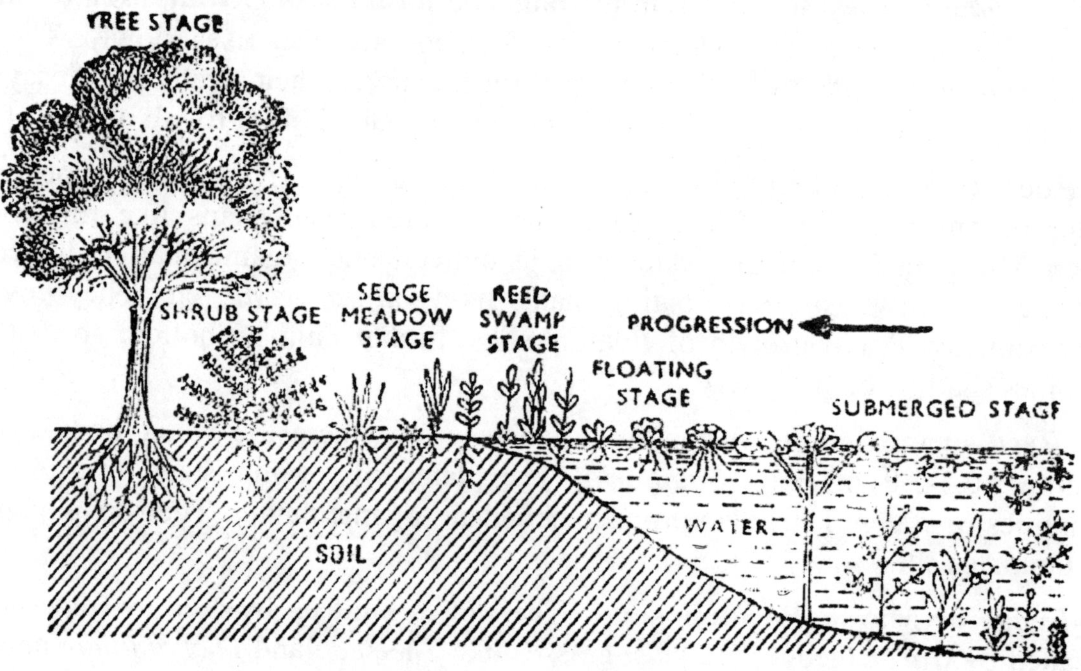

Fig. Stages of plant succession in a pond.

A possible trend of succession in the aquatic environment is as follows-

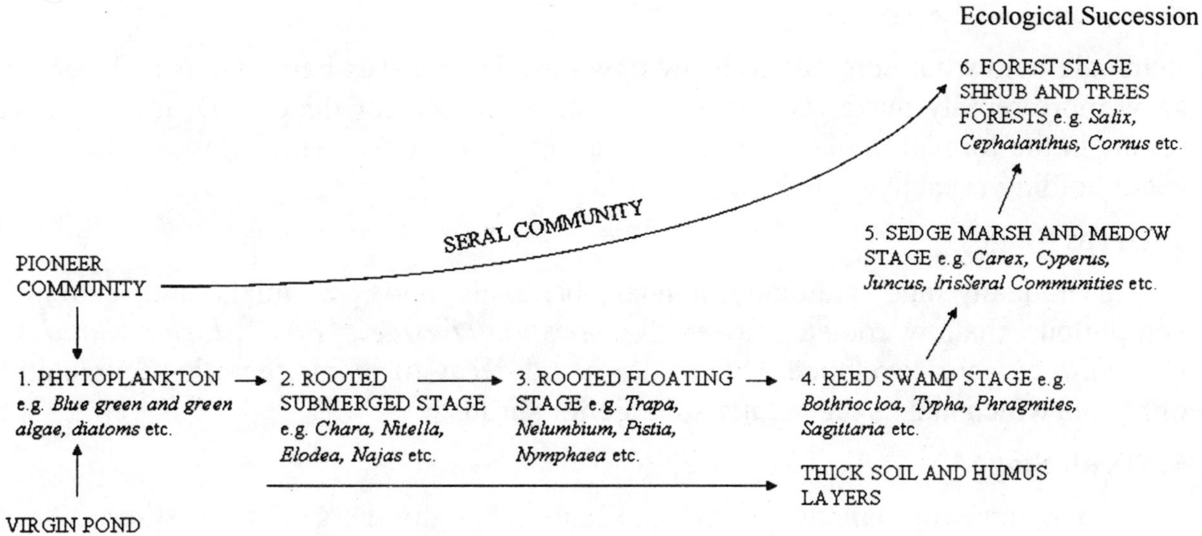

Fig. The Trend of Succession in Hydrosere

XEROSERE (A SUCCESSION ON BARE ROCK SURFACE)

Xerosere is the series of changes in the vegetation of bare rocky beds, rock, hill slopes, sand bades and other areas with extreme scarcity of water. Succession of plants occur on bare rocks in the following stages-

1. Crustose Lichen Stage

The bare rocks show completely xerophytic conditions like total absence of water, high temperature, direct exposure to sun light etc. On these rocks, the development of rooted plant is not possible. Under such conditions only *crustose lichens* can grow on these bare rocks. These crustose lichens gradually spread and secrete some sort of acids with decomposes the superficial layers of rock and make them rough enough for the growth of foliose lichen.

2. Foliage Lichen Stage

Crustose lichen makes the substratum suitable for **foliose lichen**. These lichens include *Parmelia, Dermatocarpon, Umbilicaria* etc. which have large leaf like thalli, and overlap the crustose lichens. The crustose lichens thus cut off from direct sun light and this results in death and decay of it. The foliage lichen absorb and retain more water and make suitable environment for next community.

3. Moss Stage

The thin layer of soil, formed by the death and decay of crustose and foliose lichen, favours the growth of moss and hence different types of mosses migrate and develop on the rock. The rhizoids of mosses penetrate into the substratum and form a mass with the foliose lichens. Accumulation of soil takes place rapidly as the erect stems die below but continue to grow upwards. The mosses being taller and stable to grow more rapidly make the competition unfavourable for lichens. Death of mosses results in the formation of a matting on the surface of rocks, which possess the higher water holding capacity.

4. Herbs Stage

Gradually and gradually annual, biennials and perennials and extremely xerophilous shallow rooted grasses like *Aristida, Festuca, Poa, Elusine, Solidago, Potentilla* etc. are developed. Due to death and decay of plants there is accumulation of humus which makes the habitat suitable for shrubs.

5. Shrub Stage

The increased amount of soil and humus favours the growth of shrubs. These plants are taller and their leafy shoot cover up underlaying herbs. When the shrubs have become quite dense the herbs find the growth conditions almost impossible and they disappear. The fallen leaves, the thick net work of roots, dense stems produce a good condition for quick accumulation of soil and humus. The roots of the shrubs further corrode the rock which is now more or less completely covered with soil. Thus soil is further enriched by humus formed from fallen leaves and twigs. The soil is shaded and the evaporation forms it dry. Wind movement is retarded and humidity increased. All these conditions are favourable for the growth of trees.

6. Forest Stage

Some xerophytic tree species invade the habitat. They are sparsely distributed and are stunted because the environment are still not so favourable for them. Further weathering of rocks and increasing humus content of the soil favour vigorous growth of a much larger number of trees. Finally there develops a stable forest commuity which is **Climax Community**.

Retrogressive Succession: In this type of succession, continuous biotic influences have some degenerating influence on the process. As for example, forest may change to shrubby or grassland community.

Deflected Succession: Due to changes in local conditions, the process of succession becomes deflected in a different direction than that presumed under climatic conditions of the area. Thus the climax communities are likely to be different from the presumed climatic climax community.

Seasonal Succession: It is common to observe each year, the development of different kinds of communities in different seasons of the year. This is called seasonal succession.

The various stages of a xerosere (bare rock) and the chief component plant species appearing at each stage are as shown below:

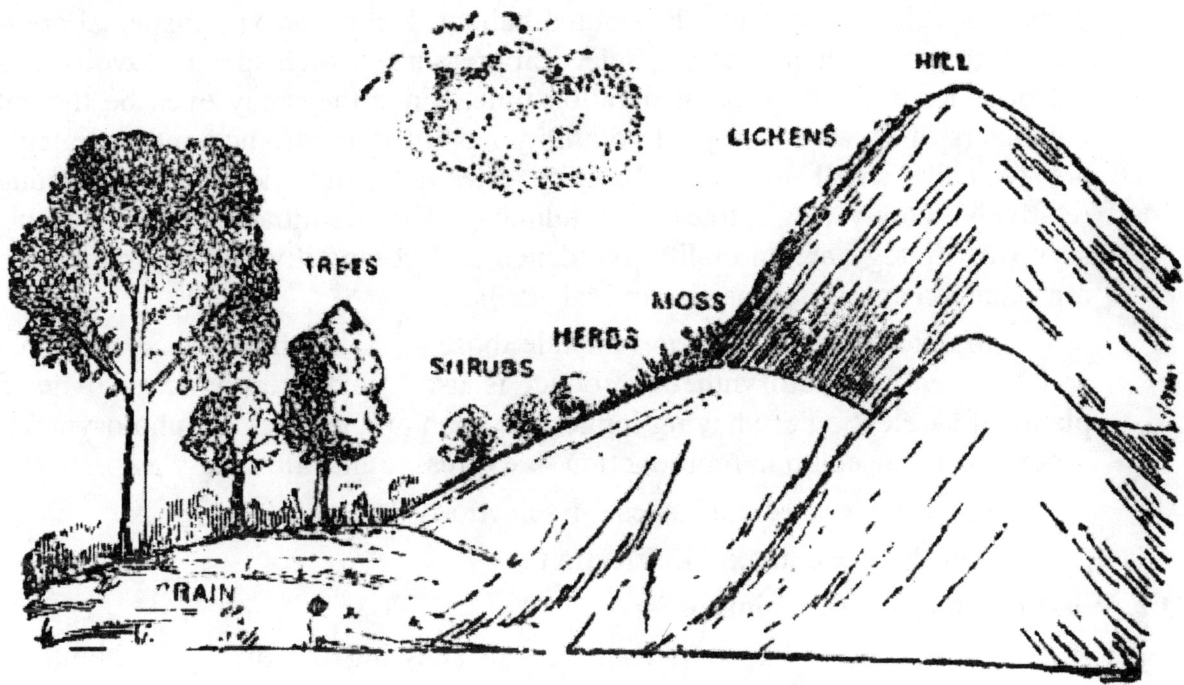

Fig. The stages of plant succession on a rock

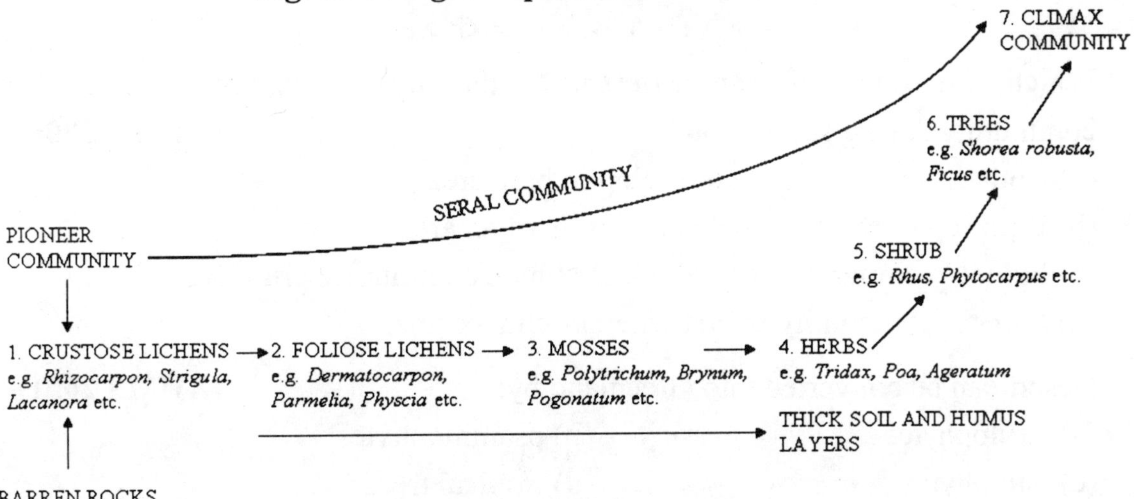

Fig. The Trend of Succession on Bare Rock

MASTER STROKES

Q. 1. What are r and k-species ?

Ans **r and k-species :** Species that reproduce rapidly and have a high value of r (intrinsic rate of increase or biotic potential) are termed r-species or r-strategies. These are generally opportunist species and represent the typical pioneer species of new and distributed habitat. Migration and dispersal are an important part of their strategy. Selection pressure in such species favours high reproduction rate and short generation time. Since they may even be the sole colonizers of an area, competitive ability has little importance and r-strategies are typically small in size. Mortality rates are high. k-species reproduce relatively slowly, e.g., trees and humans. The dominant thrusts of their survival, strategy are mortality avoidance and competitive ability rather then reproduction speed is a major survival attribute.

In other words r-selection include those species whose life span is short and their entire photosynthetic product is used up in reproduction whereas plants of k-selection are having longer life span and their entire photosynthetic product is not used up in reproduction. It expressed in following way-

Pioneer + early seral stages = r-selection

Late seral + climax = k-selection

Q.2. What do you mean by biotope?

A clearly demarcated unit of environment showing uniformity of habitat of conditions is known as a **biotope**. A sand desert, a sandy or rocky beach, a marshy land or a unit of ocean are examples of biotypes.

ENTRANCE CORNER

1. Which one statement can differentiate the pioneer community and climax community during succession ? **(AFMC 2006)**

(a) Both these communities develop in bare area

(b) Both represents the final community of an area

(c) Pioneer community is final while climax community primitive

(d) Pioneer community is first whereas climax final

2. Desert can be converted into Greenland by: **(AIIMS 2001)**

(a) oxylophytes (b) psammophytes

(c) halophytes (d) tropical trees

3. Last stabilized community in a plant succession is known as: **(DPMT 2004)**

(a) seral community (b) pioneer community

(c) ecosere (d) climax community

4. Succession is: **(Haryana PMT 2006)**

(a) series of physical changes that occur in an area

(b) development of biotic communities on a bare area

(c) series of biotic communities that appear in a previously bare area
(d) both (a) and (b)

5. In succession complexicities in structure: **(Haryana PMT 2005)**
(a) drastically increasing (b) slowly increasing
(c) does not increasing (d) constant

6. A community which starts succession in a habitat is: **(Manipal 1997)**
(a) pioneer community (b) seral community
(c) biotic community (d) ecosere

7. A rocky area, once bare, is now a green forest. The sequence of succession has been: **(CPMT 2002)**
(a) woody shrubs, annual herbs, mosses and lichens
(b) lichens, mosses, woody shrubs and annual herbs
(c) mosses, lichens, annual herbs and woody shrubs
(d) lichens, mosses, herbs and woody shrubs

8. If green plants are removed from an aquarium, the fishes will : **(Karnataka 1998)**
(a) die (b) increase in number
(c) decrease in size (d) not be affected

9. Which is not a primary colonizer ? **(Pb. PMT 1998)**
(a) Viruses (b) Lichens
(c) Blue-green algae (d) Mosses

10. Succession on secondary bare area is: **(BHU 2006)**
(a) primosere (b) xerosere
(c) subsere (d) none of these

11. The primary succession in a sere is: **(AMU 1998)**
(a) subsere (b) mesosere
(c) pioneer sere (d) none of these

12. Succession showing changes in communities at a place is called: **(AMU 1997)**
(a) geographical succession (b) biotic succession
(c) physiographic succession (d) climatic succession

13. Ecological succession on sand is: **(CET Chd. 2002)**

(a) psammosere (b) xerosere

(c) halosere (d) hydrosere

14. Last stage of plant succession is: **(RPMT 2000)**

(a) ecotype (b) seral community

(c) climax community (d) ecotone

15. Biotic succession is caused by: **(AIEEE 2003)**

(a) competition amongst species

(b) occurrence of diseases

(c) changes is grazing habits

(d) adaptive ability to environmental changes

ANSWER SHEET

1. (d) **2.** (b) **3.** (d) **4.** (c) **5.** (b) **6.** (a) **7.** (d) **8.** (a) **9.** (a) **10.** (c)
11. (c) **12.** (b) **13.** (a) **14.** (c) **15.** (d)

Chapter 5
Energy Resources

ENERGY RESOURCES

The capacity to do work is called **energy**. Without energy no work can be done and all life processes would stop. Hence, energy is the society's fundamental need and is the basic need to national economy. Like other natural resources, energy resources are also **inexhaustible** as well as **exhaustible**.

(A) Inexhaustible resources (= unlimited)

These are the natural resources which are unlimited. The quality of these natural resources may be degraded but not the quantity. These include rainfall, solar energy, wind power, tidal power, hydro power, air, etc.

(i) Solar Energy: Sun is the source of energies on the planet earth. It is a large nuclear reactor where hydrogen is continuously burning at high temperature. Solar energy originates from the thermonuclear fusion reactions occurring in sun. The energy generated by sun into the space is received on the earth as radiant energy.

India is located between 7°N and 37°N latitude and prospects of using solar energy is very bright indeed. If India can trap 1% of the incidental radiation, it can generate many tonnes the energy of its actual requirement at present. But it utilizes only 25×10^{-7}% (13×10^{7} kWH/year) of the incident solar radiation (5×10^{15} kWH/year).

Sun's heat energy is being used directly in solar cookers to cook food, or in solar heaters to heat water. The sun's energy can be converted by big concave reflectors and used to boil water to obtain steam and produce electricity. Solar cells can convert the sun's energy directly into electricity. The biggest advantage of solar energy is that it does not bring about any trace of pollution.

Large energy demand of domestic heating, water supply etc. can easily be met by using solar energy. Solar photo-voltaic (SPV) converts solar radiation directly into electricity through silicon solar cells, thus provides instant electricity in a non-polluting manner, in remote and unelectrified areas for pumping water for irrigation and drinking, street light, domestic lighting, community T.V., wrist watches, calculators, railways and road signals and heating of swimming pools etc. In September 1992, the Prime Minister of India dedicated to Nation the amorphous silicon cell plant set up at Gurgaon by BHEL. The development of solar energy was taken up as a mission in 1986 and the country has made advances in the field by developing amorphous silicon cells. It is proposed to electrify 1 lakh villages through solar energy during 8th Plan.

A solar energy centre has been set up at Gurgaon, Haryana. Rural electrification through solar photovoltaics continued in different parts of our country.

More than 320 villages in Uttar Pradesh, Rajasthan, Gujrat and Haryana are being supplied with street lighting now. Solar energy can also be used for manufacturing microwater plants. The Department of Metallurgy College of Engineering Pune, designed a solar furnace that can generate heat upto 2000°C. This could be used for various heat treatments. This is claimed to be the first kind in India. The first multipurpose solar dryer cum ware house is currently under construction at Ganaur (Haryana) to be used to dry chillies, potato chips and other food stuffs. A solar power station has also been launched in space by French Technology Survey.

National Physical Laboratory (NPL), Delhi has developed a solar water pump of 1 KW rating called Abhimanyu. Solar refrigerators and solar milk cookers have been designed by the REC (Regional Engineering College), Kurukshetra. The first solar water heating system was installed in 1982 at Haryana Breveries Ltd., at Murthal. The system is capable of heating 15,000 litres of water per day at an average temperature of 65°C. Plants are also being made to heat with solar energy, the water of Olympic size swimming pool of the Motilal Nehru School of Sports at Rajasthan during the winter months. The Government of India has introduced box type solar cookers with heavy subsidies. A solar cooler consisting of an aluminium reflector of 10 sq.feet area has been tried in India. Now India is undergoing a major social and economic revolution as a result of indigenously developed solar energy consuming and converting technologies.

(ii) Nuclear Power: Nuclear power or atomic energy will probably become the principal source of energy as the fossil fuels are used up. The most important advantage of this power is the generation of enormous amount of energy from a small amount of radioactive material. A tonne of U^{235} would provide as much energy as 3 million tonnes of coal or 12 million barrels of oil. Nuclear power is not only a means of producing electricity, but also has been used as a fuel for marine vessels, in continuous generation of heat for chemical and food processing plants.

At present, in India, only 2.58% of the total power generating capacity is obtained by nuclear reactors. Atomic Energy Commission of India will install 12 more natural uranium fueled reactors of 235 MW capacity each by 1995. Ten unit of such power plants of 500 MW will be installed upto 2000 A.D., which will be based on plutonium fueled prototype fast breeder. India, today has five nuclear reactors in operation, two each at Tarapur (Maharashtra) and Dota (Rajasthan) and one at Kalpakkam (Tamil Nadu) with a combined installed capacity of about 1,095 MW, representing about 2.5% of the total power capacity in the country.

(iii) Wind Energy: Moving air is called **wind**. The wind has energy, the energy possessed by wind is due to its high speed. When the average wind speed exceeds seven meters per second, wind power can be cost effective today. This energy is useful in remote areas, helps in saving fossil fuels, and is free from any kind of pollutants. This energy can be converted into mechanical and electrical energies.

Wind energy is fast emerging as the most cost effective source of power as it combines the abundance of a natural element with modern technology. A wind driven power station consumes no raw materials and nor it gives off waste gases or other waste materials. Its technology is very simple to manage and presenting no problems for environment.

In Peninsular India and coastline of Gujrat, Western Ghats and certain parts of central India, there are 10 KW/m/day and 4 KM/m/day wind densities during winter and other months, respectively. These areas would become a good source of wind energy if utilized and managed properly. In India, efforts to use the latent power in wind began in 1985, when the joint sector Gujrat Wind Farms Ltd., was set up. In the state of Tamil Nadu, the State Government has decided to set up 100 MW capacity wind farm during the 8th Plan. The current estimate for wind power generation in the country, in terms of potential, is placed around 20,000 MW. Of this Tamil Nadu alone is expected to account 5000 MW. But wind power generation received first impetus by the Gujrat state in the country. The wind farms with a total capacity of 3.3 MW have already been set up at Mandvi, Okha, Deogarh, Tamil Nadu and Orissa states (DNES). DNES has installed 924 wind pumps with pumping capacity of about 20 m throughout the country. In India, the exercise to harness wind power includes wind pumps, wind battery chargers, wind electric generators and grid connected wind farms. Wind heating, wind-ice making, and wind fertilizer production are all technically possible and may prove to be useful to some locations.

(iv) Geothermal Energy: It is the exploitation of heat energy within 10 km of earth crust. It can generate power where the geothermal fluid has about 130°C temperature. Geothermal energy utilizes the great heat source of steam, or super heated water, from 1,000 to 2,000 meters deep to power an electric generator or used directly as primary heat. Geothermal energy is quite common in volcanic regions and where hot springs occur.

(B) Exhaustible resources (= Limited)

These are the natural resources with definite stock or supply, they are vulnerable to both qualitative and quantitative degradation. These includes minerals, fossil fuels, forests etc. Exhaustible are mainly of two types-

(a) Renewable resources

These are living or biotic resources. They are able to reproduce or replace themselves and to increase. The renewable resources get replenished, recycled or reproduced and can lost for ever, provided they are not used beyond their renewability e.g. A forest is a biotic resource as are wildlife and fisheries. Water, is also a renewable resource, e.g., water, soil fertility, forests, range lands, wild life, agriculture system , aquatic plants and animals. It is also termed as **non-conventional** or **non-commercial** sources of energy.

(i) Forest Resources: Forest is an important natural resource. One third of the land surface is covered with forests with about 50% as tropical forest. Of the total geographical area of India, 22.74% is covered by forest. It is too less than 33% forest cover prescribed under National Forest Policy 1952. The forests are storehouse of biodiversity and provide important environmental services to main kind.

Functions of forests: Forests has mainly three types of functions i.e., **Productive functions, Protective functions** and **Regulative functions.**

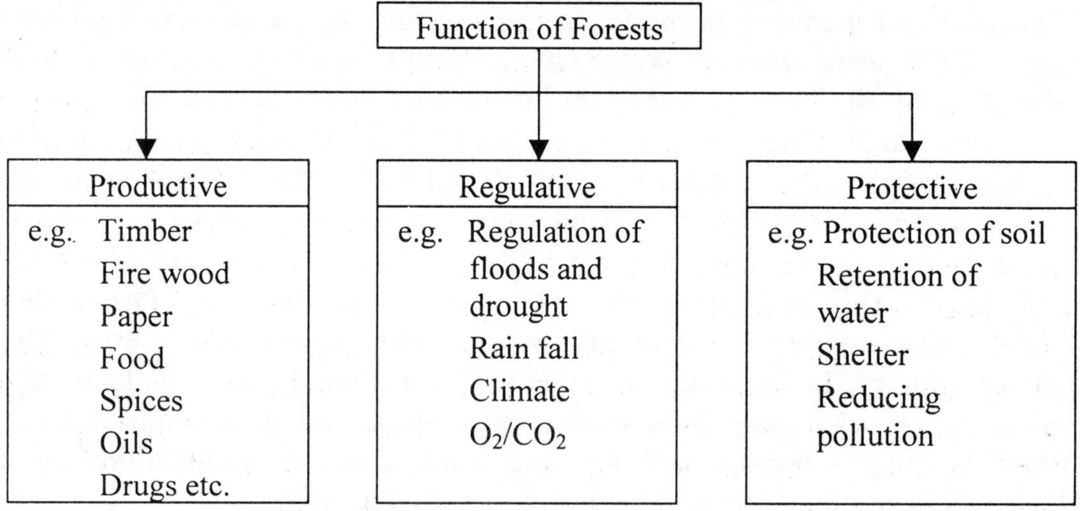

Forest area in India: At the beginning of the twentieth century, about 30 percent of land in India was covered with forest. But by the end of the twentieth century, the forest cover was reduced to 19.4 per cent.

Table : Forest Cover in India (Estimated in 1999)

Class	Area (sq.km.)	% Geographic Area
Dense forest[1]	3,77,358	11.5
Open forest[2]	2,55,064	7.8
Mangrove[3]	4,871	0.1
Sub-total	6,37,293	19.4
Scrub[3]	5,896	1.6
Non-forest (other land use)	25,98,074	79.0
Total	32,41,263	100.0

[1]Canopy cover > 40 per cent of land
[2]Canopy cover 10-40 per cent of land
[3]Canopy < 10 per cent of land.

Social forestry is one of the main key behind the conservation of forest.

Social Forestry

The term social forestry was for the first time coined by Westoby in his lecture, delivered at the Ninth Commonwealth Forestry Programme at New Delhi in 1968 and he defined social forestry as, "A forestry that aims at the flow of protection and recreation of benefits for the community."

Generally social forestry may be defined as, "forestry for the people, by the people and of the people" and for this multipurpose or 5F plants are grown i.e.

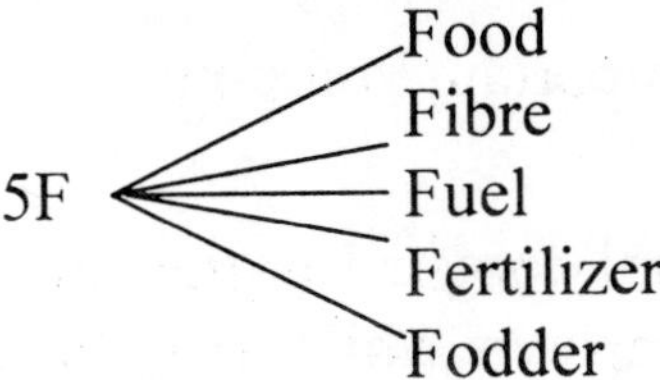

Now concept of 5F plant change into 7F plants-

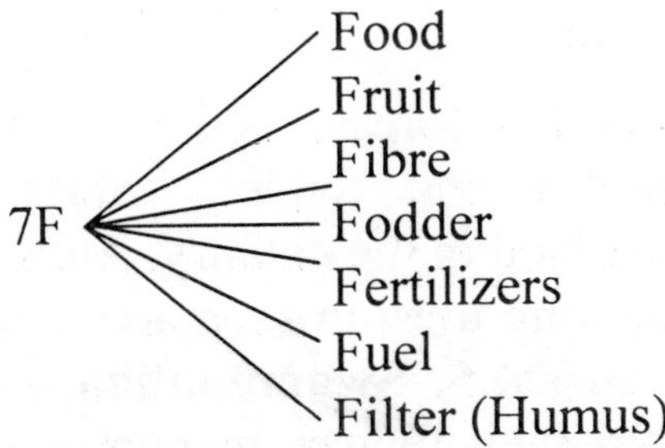

Following are the objectives of social forestry:

Van Mahotsava: The annual festival of trees was for the first time inaugurated in 1963 and actually the concept of Van-mahotsava was given by K.M. Munshi when he was the Agriculture Minister in Government of India. He for the first time in this country recognized the interrelationship between agriculture and forestry and this was the main theme behind the social forestry.

Aims of Social Forestry

The revised National Forest Policy 1952 have implesized the **Aforestation** programme under social forestry. National Commission of Agriculture recognized in 1976 have following policies-

- To provide fuel as well as manure for the rural people.
- To increase production of fruits and thus to increase the potential food resource of this country to help conservation of soil and stop further deterioration of soil fertility.

- To create shelterbelts around agriculture fields to increase fertility.
- To provide fodder for cattles and thus to minimize the pressure on grazing land.
- To provide shade and ornamental trees.
- To provide small holes and timber for agricultural implements, house construction and finishing.
- To increase knowledge of trees, tree plantation among the people.
- To popularize planting of trees in villager's municipal and public lands for their aesthetic, economic and protective value.

Components of social forestry: Following are the 3 important components of social forestry:

i) Agroforestry: The technology and concept of social forestry is the growing of multipurpose trees (MPT) along with crop. Trees are grown which may be used as fodder, fuel and to increase the fertility of soil. They also provide a fence between two plants, as well as they behave like- **shelter belts**. Agroforestry is more useful in desert areas where high velocity of wind too of the gear.

In India, agroforestry of tree planting by individual forester is the only solution to meet the challenge by scarcity of fuel, timber and fodder. "**State forest departments**" are generally not sufficient for vacant land in the country. Agroforestry may be the substitute for government agency in planting trees in such area where the green revolution in agriculture can not be achieved (**M.S. Swaminathan gave the term green revolution**) without self sufficiency in fuel, fodder, manure and small constructional timber. Therefore, agroforestry should be the objective of green revolution.

ii) Rural forestry: This is also called as extinction forestry, in which trees are planted on community and **Panchayat land**, degraded forest, road and railway side canal banks etc. for meeting the needs of rural people. It also includes the restoration of degraded land, mining areas, road construction areas etc. The objectives of rural forestry are similar to those agroforestry, meeting the rules of rural people. The most important difference from agroforestry is the ownership of land which is mostly individual in agroforestry and communal in rural forestry. This difference in ownership often possess peculiar protection problem and thus need the environment. The community as a whole number of cottage industry may develop with the help of rural forestry, tree resources. This may include silk worm, sports wood, honey bee forming, oil from seeds of trees, household furniture and dairy industry etc.

iii) Urban forestry: The main aim of urban forestry is to bringing trees to the door of urban people. It lays emphasis on the aesthetic development of the urban areas.

Flowers and fruits trees of ornamental variety planted along road side, canal banks, near town and cities. It also includes beautification of houses, roads and vacant lands.

(ii) Animal Power: Over half of energy consumed on India's farming, comes from the muscles of hard-worked animals. India has about 80 million working animals, out of which 70 million bullocks, 8 million buffaloes and one million, donkeys, elephant and yaks are also used as work animals. If each of these animals generates 0.5 Horse Power, the installed capacity of the animal labour force comes to 40,000 HP or about 30,000 MW, which is equal to the installed capacity of electric power generation in the country. The main uses of this power are in operation of bullock carts. About 20 million people are involved in part-time or full-time in the bullock cart business.

(iii) Ocean Energy: Basically there are three ways of generating power from the ocean. They are taming the waves (wave energy), harness the tidal power (tidal energy) and conversion of ocean thermal energy into electrical energy (OTEC).

(b) Non-renewable energy sources: The sources of energy which have accumulated in nature over a very-very long time and can not be quickly replaced when exhausted are termed as **non-renewable** sources of energy. These resources are also known as **conventional sources of energy** or **commercial sources of energy**. Minerals, metals, salt, coal, oil, deposits, natural gas etc. are such type of resources. These are available in limited amounts and in no manner can be rebuilt or increased. For instance, fossil fuel is non-renewable source of energy because fossil fuel has accumulated in the earth over a very very long time, and if all the fossil fuel gets exhausted, it can not get reproduced quickly in nature. Coal, oil and natural gas are the common non-renewable source of energy. These are of organic origin and called of **fossil fuels**.

Region	Coal	Gas	Oil
America	30%	15%	10%
Africa	Small	10%	10%
Australia	5%	Small	Small
E. Europe	5%	Small	Small
W. Europe	15%	5%	5%
China	15%	Small	5%
Russia	20%	40%	10%
Rest of Asia	Small	5%	5%
Total reserves	40,000	2,500	4,000

Table showing world scenario of fossil fuel.

(i) Coal: Coal is a dark brown combustible mineral substance, composed of C, H, O and small amounts of N_2 and S. It is found in deep coal mines under the surface of earth and largely mined for use as fuel. Coal was known to the natives from the time immemorial, but the first licence to dig coal in England was granted in 1239. In 18th Century, coal mining started in India. Most of the coal raised in India has been obtained from the Gondwana system of strata of peninsular India. By 1906, main coalfields were in Raniganj and Jharia. After 1906, main coal mining started in new fields like- Giridih in Bengal and Pench valley, Narsinghpur district, Chanda district and Umaria. The younger coals are nearly all of cretaceous and tertiary age, although something and poor seems of upper Jurassic age. Coal of tertiary age occurs in foot hills of Himalayas, eastern Assam and Andaman and Nicobar. Deposits in north-east Assam are youngest, probably of Miocene age.

Gondwana coal has excess of ash and moisture, while in the tertiary coal percentage of ash is less but that of combustible volatile matter is high, so it is a lighter fuel. Coal gas, coaltar and coke are related products of coal. Best types of fossils are obtained from coal balls which are irregularly rounded masses ranging in diameter from a few millimeters to a meter. Each ball is a mass of calcium and magnesium carbonate, sometimes with iron sulphide. These show petrified remains of great number of plant fragments, even delicate parts remain intact in the coal balls.

Coal is important because it might be used as a source of energy as such, or it can be converted into other forms of energy like coal gas, electricity and synthetic petrol. The major drawback from coal use is environmental pollution. It is a great contributor to the 'global warming' and 'green house effect'.

(ii) Natural Gas: Natural gas consists mainly of methane with small quantities of ethane and propane. It occurs deep under the crust of the earth either alone or along with oil above the petroleum deposits. Thus, some well dug into the earth produce only natural gas, whereas others produce natural gas as well as petroleum.

Natural gas serves as the most desirable fuel for domestic use. It is said that it supplies about 60% of the world's energy today. Oil and Natural Gas Commission (ONGC, 1978) has reported that the gas reserves are equivalent to about two billion tonnes of oil. At present 40% of daily output of 20 million cubic tonnes of gas is being flared away. In our country, many new natural gas fields have been discovered recently. These new gas fields have been found in Tripura, Jaisalmer, off shore area of Bombay and the Krishna-Godawari delta. The ONGC estimated that natural gas fulfill 90.6% of global energy consumption. It is used up at this incredible rate that its supply may last only a few more decades unless abundant new resources are sought.

(iii) Oil: Crude oil occurs in the form of petroleum. It is a mixture of mainly various saturated aliphatic hydrocarbons. It occurs naturally inside the earth and is obtained commonly by drilling oil wells. It is used as a fuel and can also be separated into various other forms such as kerosene oil, neptha, gasolines etc. These products are used as by products or petrochemicals in the preparation of synthetic rubber, fertilizers, pesticides, plastic, explosive etc.

Originally, it was estimated that close to 1,250 billion barrels of oil were deposited in the earth, and out of this, about 100 billion barrels were extracted by 1960s. ONGC has estimated that total oil reserves in India is very least and India is still 90% dependent upon imported oil today. In the country, 27 localities are marked for oil extraction.

The use of petroleum has increased as such a tremendous rate during last few years that it's supply is likely to exhausted in less than a hundred years. This source of energy is already severely short in supply.

Tehri Dam conflict		
1. Location of river	:	Tehri town of Uttaranchal at the confluence of Bhagirathi and Bhilganga
2. Height of Dam	:	260.5 metre
3. Electricity generation capacity	:	1,000 megawatt
4. Irrigation facility available	:	9,10,000 hectare
5. Drinking water facilities available	:	500 cusec
6. Water cover area of reservoir	:	4,250 hectare
7. Expected area under inundation	:	1,600 hectare cultivated land, 2,000 hectare forest and 23 villages with over 50,000 peoples
8. Tehri dispute onset (by Tehri Dam Opposition Struggle Committee)	:	1977

(iv) Biogas: **Italian physicist Volta** has firstly described the presence of biogas in the bottom of ponds, lakes or marshy places. Biogas is formed by the partial decomposition of organic matter.

Raw materials for biogas are-

i) Raw material of animal husbandry: e.g. cattle dung (cow dung or "Gobar").

ii) Raw materials of agricultural sources: e.g. straw.

iii) Raw materials of domestic sources: e.g. sewage.

Biogas has -

CH_4 = 60-63%

CO_2 = 30-40%

H_2 = 2.4%

N_2 = 2.7%

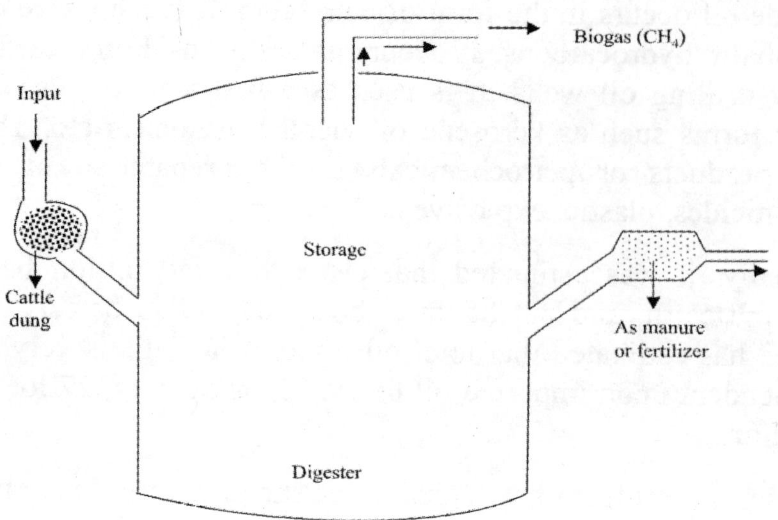

Fig. Biogas plant

The production of biogas comprises following three steps:
i) Hydrolysis
ii) Acidification/Acidogenesis
iii) Methanogenesis

The anaerobic or partial decomposition of organic matter, cattle dung is responsible for production of biogas. The methanogenic bacteria wuch as *Methanococcus, Methanothrix, Methanobcterium* and *Methanosarcina* involves in the production of CH_4 from cow dung.

Use of Biogas: It can be used as fuel for the purpose of cooking and lightening.

(v) Hydrocarbons: **Kelvin 1979** has described that, higher plants and certain algae may contain hydrocarbons after the act of photosynthesis, such hydrocarbons having the carbon atom C_{18} upto C_{40} would be used as a source of fuel. Such crops/algae which can be used as a source of fuel are called as **petrocrop** e.g.

Name of plant families	**Name of plants**
1. Algae	*Chlorella* sp.
2. Euphorbiaceae	*Euphorbia hirta*
3. Asclepidaceae	*Calotropis*
4. Moraceae	*Ficus ilatica*

(vi) Hydrogen: All photosynthetic organisms produce hydrogen by photolysis of water. This production of H_2 from water by biological organisms is called **Biophotolysis**. Microorganisms, such as blue-green algae and photosynthetic bacteria can be employed for commercial production of H_2. The hydrogen production involve the use of two different enzyme-

1) Nitrogenase and
2) Hydrogenase

1) Production of Hydrogenage and Nitrogenase: All nitrogenase fixing propagates such as, Blue-green algae, for e.g., *Nostoc, Anabaena* or *Bacteria, Chlorobium, Rhodopsuedomonas* and *Rhodospirullum* have nitrogenase during nitrogen fixation and there is occur production of H_2-gas. This gas would be isolated, collected and used as a source of fuel for automobiles etc.

$$N_2 \xrightarrow[\text{ATP} \quad \text{NADPH}]{\text{Nitrogenase}} NH_3 + H_2 \uparrow$$

2) Hydrogen production by use of Hydrogenase: Blue green algae as well as nitrogen fixing bacteria possess hydrogenase activity. This enzyme has also been exploited for production of hydrogen. Following organisms have been used world wide for the production of hydrogen gas-

i) *Chlorella* (A green algae)
ii) *Anabeana* and *Nostoc* (Blue-green algae)
iii) *Chlorobium-chromaceum* (Photosynthetic bacteria)

(vi) Biomass: The biomass of plant system can be used as a source of energy. Plant utilizes solar radiation and fixes CO_2 in the form of organic substances. The biomass of higher plants comprises organic materials such as carbohydrates, proteins, lipids, fats etc. Certain crop plants such as – *Eucalyptus, Ficus bengalensis, Ficus religiosa, Lucina leucocephala* (Sabababul), *Butea monosperma* (Palas or Dhak) etc. can be used in energy translation.

(vii) Alcohols: In America, **Gasohol** (comprised of 20% alcohol + petrol). Alcohol can be produced by the formation of sugars.

The raw materials for the productions of alcohols is cellulose, starches, lignocellulose and simple sugars such as – Glucose and Fructose. Cellulose is most abundant product on this earth. The cell wall of higher plants comprises cellulose in the form of microfibrils or as fibrillar component.

Cellulose can be hydrolysed with enzyme cellulose. The cellulose can be obtained from certain microbes such as *Cellulomonas* and *Trichoderma recsei* and *Trichoderma viridae*.

MASTER STROKES

Q. What is silent valley? Where it is situated in India?

Ans. Silent Valley occupies an area of 8950 hectares at an altitude of 3000 ft. in **Palaghat district, Kerala**. It is surrounded by the **Nilgiri forests** to the north and **Attappadi forests** to the east, together they comprise 40,000 hectares of pristine (i.e. primitive) forest. This **tropical rain forest in the Western Ghat** is a precious reservoir of genetic diversity which has not fully exploited here plant species and other forms of life have survived for centuries in the forest. It is this gene pool to which man has to turn in future for new materials for agricultural and life-saving drugs, etc.

The Prime Minister (Indira Gandhi) in 1983 had accepted the recommendation of top scientists and environmentalists and declared the Silent Valley as the **Biosphere Reserve** by cancelling the hydel project proposal of the Kerala State Government.

ENTRANCE CORNER

1. More than 70% of world's freshwater is contained in : **(CBSE 2005)**
   (a) polar ice (b) glaciers and mountains
   (c) antarctica (d) greenland
2. Amount of alcohol allowed to be mixed with petrol in India is: **(CBSE 2004)**
   (a) 2.5% (b) 5%
   (c) 10% (d) 10-15%
3. Major fuel component of biogas is: **(CBSE 2004)**
   (a) CO (b) CO_2
   (c) CH_4 (d) CH_3
4. Major reason for extinction of animals is: **(CBSE 2002)**
   (a) predation (b) pollution
   (c) habitat destruction (d) felling of trees
5. Who will maintain the organic matter of the soil ? **(CBSE 2001)**
   (a) Fungi and semiparasites (b) Cyanobacteria and parasitic animals
   (c) Protozoa and slime moulds (d) Bacteria and fungi
6. *Idri idri* is found in : **(CBSE 2000)**
   (a) Australia (b) Madagascar
   (c) Mauritius (d) Tasmania
7. Now-a-days biological reserves are commonly destroyed by : **(CBSE 2000)**
   (a) pollution (b) population
   (c) rains (d) none of these

8. Maximum productivity is shown by : **(CBSE 1999)**
   (a) tropical rain forest (b) temperature forest
   (c) savanna (d) all of these

9. Phytotron is : **(CBSE 1999)**
   (a) the technique of growing plants in controlled environment
   (b) device to grow cactus
   (c) device to induce mutations
   (d) none of the above

10. Land mass occupied by forest is about : **(CBSE 1999)**
    (a) 11% (b) 22%
    (c) 30% (d) 60%

11. Worker bee lives for : **(CBSE 1999)**
    (a) 30 days (b) 15 weeks
    (c) 10 weeks (d) one hour

12. Ecofriendly method is : **(CBSE 1999)**
    (a) plantation of C_3 plants (b) plantation of sugarcane
    (c) energy plantation (d) none of these

13. Land mass occupied by forest is about or according to Indian forest policy what percentage of the land area should be under forest cover : **(CBSE 1999)**
    (a) 11% (b) 22%
    (c) 30% (d) 60%

14. Which communities are more vulnerable to invasion by outside plants and animals? **(CBSE 1998)**
    (a) Tropical evergreen forests (b) Temperate forests
    (c) Mangroves (d) Oceanic island communities

15. The response of different organisms to environment rhythms of light and darkness is called : **(CBSE 1999)**
    (a) vernalization (b) phototaxis
    (c) photoperiodism (d) phototropism

16. What is the major cause of diminishing wildlife numbers ? **(CBSE 1998)**
    (a) Cannibalism (b) Habitat destruction
    (c) Intra-specific competition (d) None of these

17. If we remove half of the forest cover of earth, the crisis that will occur: **(CBSE 1996)**
    (a) many species would become extinct
    (b) population, pollution and ecological imbalance will rise
    (c) energy crisis will commence
    (d) the remaining forest will correct the imbalance

18. Bulk fixation of carbon through photosynthesis takes place in: **(CBSE 1994)**
    (a) tropical rain forests
    (b) tropical rain forest and crop plants
    (c) crop plants
    (d) oceans

19. Deforestation does not lead to: **(CBSE 1994)**
    (a) quick nutrient cycling
    (b) soil erosion
    (c) alteration of local weather conditions
    (d) destruction of natural habitat of wild animals

20. With the exception of water, which is most important accessory chemical used in industry ? **(CBSE 1994)**
    (a) Ethanol (b) Petroleum
    (c) Nitrogen (d) Rubber

21. Largest amount of freshwater is found in: **(CBSE 1994)**
    (a) lakes and streams (b) underground
    (c) polar ice and glaciers (d) rivers

22. American water plant that has become a troublesome water weed in India is: **(AMFC 2000)**
    (a) *Cyperus rotundus* (b) *Eichhornia crassipes*
    (c) *Trapa latifolia* (d) *Trapa bispinosa*

23. Poikilotherm animal amongst the following is: **(AMFC 1999)**
    (a) tortoise (b) otter
    (c) whale (d) penguin

24. Characteristic of mangroves of mashy Sunderbans is : **(AMFC 1999)**
    (a) vivipary (b) breathing roots
    (c) pneumatophores (d) all of these

25. Wild life of nature is renamed as : **(AMFC 1999)**
(a) World Wild Life conservation of nature
(b) World Wild Life of nature
(c) World Wild Fauna
(d) World Wild Fund

26. The map given below indicates the former and the present distribution of an animal:

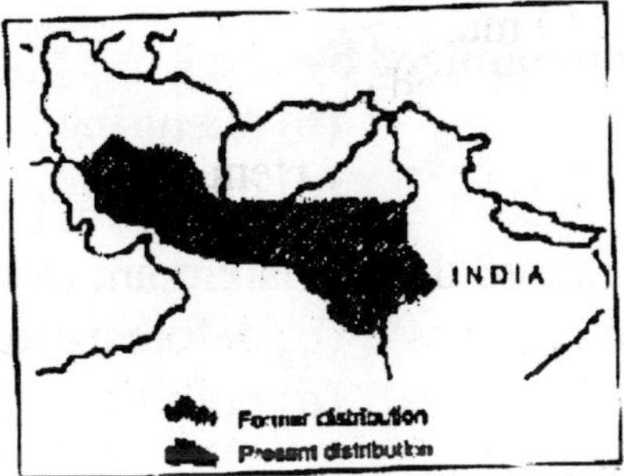

Which animal could it be ? **(AIIMS 2003)**
(a) Wild ass (b) Nilgai
(c) Black buck (d) Lion

27. **Assertion**: Tropical rain forests are disappearing fast from developing countries such as India.
Reason: No value is attached to these forests because these are poor in biodiversity. **(AIIMS 2003)**
(a) A is correct and R is explanation of A
(b) A is correct and R is not explanation of A
(c) A is correct but R is wrong
(d) A is wrong but R is correct

28. Social forestry is : **(AIIMS 1996)**
(a) management of forests by co-operative societies
(b) growing and management of useful plants on common lands
(c) growing one type of trees
(d) growing different types of plants together

29. Which of the following is a renewable resources ? **(BHU 2006)**
(a) Coal (b) Petroleum
(c) Natural gas (d) Wind energy

30. Uniform soil erosion by running water is: **(BHU 2001)**
(a) gully erosion (b) rill erosion
(c) riparian erosion (d) sheet erosion

31. Biogas can help to decrease the load on forest reserves as it is good substitute for: **(BHU 2000)**
    (a) fuel wood (b) petroleum and oil
    (c) coal (d) charcoal

32. Soil salinity is measured by : **(BHU 2000)**
    (a) porometer (b) potometer
    (c) calorimeter (d) conductivity meter

33. In hilly areas, erosion is minimized by : **(BHU 1999)**
    (a) terracing (b) manuring
    (c) ploughing (d) mixed cropping

34. Wild life destruction is caused due to maximum extent by : **(BHU 1998)**
    (a) less production of young ones (b) deforestation
    (c) hunting (d) pollution

35. Chipko movement is connected with : **(BHU 1994)**
    (a) conservation of natural resources (b) plant/forest conservation
    (c) plant breeding (d) project tiger

36. A non-renewable resource is : **(Manipal 2006)**
    (a) wind energy (b) water energy
    (c) coal (d) geothermal energy

37. A renewable resource is : **(Manipal 2001)**
    (a) fossil fuel (b) metals
    (c) sunlight (d) both (a) and (b)

38. A recent controversy broke out in which there was a demand to change the site of the dam. The proposed demand was in favour of one of the following: **(Manipal 2000)**
    (a) to get more water
    (b) to make irrigation facilities available to a larger area
    (c) to make the surrounding beautiful
    (d) to generate hydel power slowly and steadily

39. Crop rotation is employed for : **(CPMT 2003)**
    (a) increasing nitrogen content of soil (b) checking soil erosion
    (c) community development (d) enhancing soil fertility

40. 'Chipko movement' is associated with : **(CPMT 1998)**
    (a) soil conservation (b) forest conservation
    (c) pollution (d) green revolution

41. Which of the following is non-renewable resource of energy ? **(CPMT 1997)**
(a) Solar energy (b) Petroleum
(c) Biomass (d) Hydropower

42. Forests participate in : **(CPMT 1994)**
(a) controlling pollution (b) prevention of soil erosion
(c) maintenance of ecological balance (d) all of the above

43. Soil erosion is prevented by : **(DPMT 2005)**
(a) deforestation (b) afforestation
(c) reduction of CFCs production (d) use of CNG in all transports

44. Biogas is combustive as it contains : **(DPMT 2003)**
(a) CO_2 (b) CH_4
(c) H_2 (d) H_2S

45. Stratification can be observed in : **(DPMT 2001)**
(a) tundra (b) tropical forest
(c) temperate (d) desert

46. Soil erosion can be prevented by: **(DPMT 1997)**
(a) restricted human activity (b) good plant cover
(c) checking movement of animals (d) wind screen alone

47. Decreased rainfall in northern areas is due to : **(DPMT 1996)**
(a) changes in sun
(b) global phenomenon
(c) deforestation in catchment areas
(d) deforestation in evaporating area near equator

48. Which one is a non-renewable resource ? **(DPMT 1996)**
(a) Coal and minerals (b) Plants and coal
(c) Water and natural gas (d) Energy and water

49. Soil conservation is a practice in which : **(Karnataka 2005)**
(a) soil is protected from being carried away by wind and water
(b) soil is well aerated
(c) soil fertility is enhanced
(d) soil erosion is allowed

50. The organization which has been published the 'Red Data Book' is : **(Karnataka 2003)**
(a) National Environment Engineering Research Institute
(b) International Union for Conservation of Nature and Natural Resources
(c) Conservation of International Trade in Endangered species of Wild Fauna and Flora
(d) National Wildlife Action Plan

51. NEERI is : **(Karnataka 2000)**
(a) National Ethological and Ecological Research Institute
(b) National Eugenics and Ecological Research Institute
(c) National Ecological and Environment Research Institute
(d) National Environment Engineering Research Institute

52. The Siberian crane from Russia is a regular visitor to the bird sanctuary in one of the following place in India : **(Karnataka 1996)**
(a) Lallbagh, Bangalore
(b) Vendanthgol Sanctuary, Tamil Nadu
(c) Ranganathittis Sanctuary, Karnataka
(d) Bharatpur Sanctuary, Rajasthan

53. Which one is the inexhaustible resources? **(Pb. PMT 2006)**
(a) Fossil fuels (b) Solar energy
(c) Coal (d) Petroleum

54. Social forestry is useful in yielding : **(Pb. PMT 2003)**
(a) floriculture (b) timber
(c) medicines (d) multipurpose uses

55. An antiforest measure is : **(Pb. PMT 2003)**
(a) afforestation (b) selective grazing
(c) clearing forest (d) selective felling

56. The type of forest mainly found in India is: **(Pb. PMT 2003)**
(a) sub-tropical deciduous (b) tropical moist deciduous
(c) tropical deciduous (d) temperate deciduous

57. Afforestation with regard to urban development is called : **(Pb. PMT 2000)**
(a) social forestry (b) integrated forest management
(c) both (a) and (b) (d) none of the above

58. Multinational source of energy is : **(Pb. PMT 1999)**
(a) soil (b) water
(c) sun (d) all of these

59. Sal forest is : **(Pb. PMT 1998)**
(a) tropical rain forest
(b) tropical evergreen forest
(c) temperate deciduous
(d) deciduous

60. Geothermal energy is : **(Pb. PMT 1998)**
(a) non-renewable non-conventional energy source
(b) non-renewable conventional energy source
(c) renewable non-conventional energy source
(d) renewable conventional energy source

61. Jhoom cultivation refers to : **(Pb. PMT 1998)**
(a) cultivation of Jamun trees
(b) cultivation of medicinal plants by tribes
(c) tribal method of shifting cultivation
(d) cultivation of terraces

62. Which of the following is non-renewable energy source? **(Haryana PMT 2000)**
(a) Hydropower
(b) Tidal power
(c) Geothermal energy
(d) Nuclear energy

63. Which of the following can be traced on hilltops ? **(Haryana PMT 1998)**
(a) Alpines
(b) Conifers
(c) Prairies
(d) Savannas

64. MAB stands for : **(Haryana PMT 1998)**
(a) Man Antigens Biology
(b) Man and Biosphere
(c) Man and Biology
(d) Man and Biotic Community

65. Environmental planning organization is : **(Haryana PMT 1997)**
(a) CSIR
(b) NEERI
(c) CPHERI
(d) ICAR

66. Inexhaustible, non-conventional energy source is : **(Haryana PMT 1994)**
(a) solar
(b) tidal
(c) wind
(d) all of these

67. Forest destruction results in : **(Haryana PMT 1994)**
(a) loss of wild life
(b) famine, floods and drought
(c) soil erosion
(d) all of the above

68. Percentage of water present in oceans and seas is : **(Haryana PMT 1994)**
(a) 60%
(b) 70%
(c) 80%
(d) 95%

69. Afforestation should be with : **(Haryana PMT 1994)**
(a) exotic species (b) indigenous species
(c) bamboos (d) *Eucalyptus*

70. Air, clay, sand, solar energy are ..... resources of energy. **(AMU 2006)**
(a) inexhaustible (b) exhaustible
(c) renewable (d) non-renewable

71. Minerals, metals and fossil fuels are which type of resources of energy ? **(AMU 2005)**
(a) Renewable (b) Non-renewable
(c) Biodegradable (d) Degradable

72. After Na, K and Cl, sea water has higher concentration of : **(AMU 2001)**
(a) Mg (b) Zn
(c) Ca (d) Pb

73. LPG is : **(AMU 1998)**
(a) low price gas (b) fossil fuel
(c) low pressure gas (d) biogas

74. Deforestation leads to : **(AMU 1998)**
(a) soil erosion (b) global warming
(c) soil protection (d) both (a) and (b)

75. Chipko Andolan (Movement) which was started in 1970 in Garhwal/Himalayas (Gopeshwar) near Alaknanda river was for the first time initiated by : **(AMU 1997)**
(a) Chander Prasad Bhatt (b) Sunder Lal Bahuguna
(c) Baba Amte (d) Vinoba Bhave

76. National Environment Engineering Research Institute (NEERI) is in : **(Kerala PMT 2003)**
(a) New Delhi (b) Nagpur
(c) Kolkata (d) Chennai

77. What is the animal symbol of WWF (World Wild-life-Fund)? **(Kerala PMT 2002)**
(a) Dolphin (b) Kangaroo
(c) Tiger (d) Giant Panda

78. Animals like mice, lizards and cockroach share their habitat with human dwellings. They are : **(Kerala PMT 2001)**
(a) inquilines (b) cultigens
(c) transplants (d) humus

79. "Chipko movement" by Sunderlal Bahuguna was started in : **(CET Chd. 2003)**
   (a) 1965 (b) 1970
   (c) 1983 (d) 1947

80. In India highest amount of coal is present in : **(MP PMT 2003)**
   (a) West Bengal (b) Maharashtra
   (c) Jharkhand (d) Assam

81. Savannas are : **(MP PMT 2002)**
   (a) tropical rain forest
   (b) desert
   (c) grassland with scattered trees
   (d) dense forest with close canopy

82. Domestic cooking gas cylinder is filled with : **(MP PMT 1998)**
   (a) alcohol (b) diesel oil
   (c) liquid petroleum gas (d) coal gas

83. Possibility of coal in an area is guessed through : **(MP PMT 1998)**
   (a) mining the area (b) economic survey
   (c) pollen grain analysis (d) ecology of the area

84. Which one is not dangerous for life ? **(MP PMT 1998)**
   (a) Biopollutants (b) Ozone layer
   (c) Nuclear blast (d) Deforestation

85. Which one is a renewable source of energy ? **(MP PMT 1995)**
   (a) Nuclear fuel (b) Trees
   (c) Coal (d) Petroleum

86. Universal non-polluting source of energy is : **(MGIMS Wardha 2003)**
   (a) fossil fuel (b) sun
   (c) nuclear (d) wind and water

87. Kew, London is famous for : **(MGIMS Wardha 2001)**
   (a) being the largest biological reserve
   (b) herbarium
   (c) being the largest botanical garden
   (d) diverse flora and fauna

88. The most diverse forest is : **(CMC Ludhiana 2003)**
   (a) sub-tropical (b) coniferous
   (c) tropical rain forest (d) deciduous

89. *Rhizophora* occurs in : **(Orissa JEE 2004)**
    (a) tropical rain forest
    (b) tropical deciduous forest
    (c) mangrove
    (d) both (a) and (b)

90. Mangrove forests occur in : **(Orissa JEE 2003)**
    (a) laterite soil
    (b) estuaries
    (c) aried areas
    (d) hilly areas

91. Which are the richest sources of fossils ? **(DY PATIL, Mumbai 2003)**
    (a) Lava
    (b) Sedimentary rocks
    (c) Volcanic rocks
    (d) Oceans

92. Source of energy which does not evolve CO_2 is : **(RPMT 1996)**
    (a) coal
    (b) oil
    (c) organic compounds
    (d) nuclear energy

93. Which will contribute to water rise in sea after few years of global warming? **(JIPMER 1994)**
    (a) Excessive rain
    (b) Melting of polar ice caps
    (c) Both (a) and (b)
    (d) None of these

ANSWER SHEET

1. (a)	**2.** (b)	**3.** (c)	**4.** (c)	**5.** (d)	**6.** (b)	**7.** (b)	**8.** (a)	**9.** (a)	**10.** (c)
11. (c)	**12.** (c)	**13.** (c)	**14.** (d)	**15.** (c)	**16.** (b)	**17.** (b)	**18.** (d)	**19.** (a)	**20.** (b)
21. (c)	**22.** (b)	**23.** (a)	**24.** (d)	**25.** (d)	**26.** (a)	**27.** (c)	**28.** (b)	**29.** (d)	**30.** (d)
31. (a)	**32.** (d)	**33.** (a)	**34.** (b)	**35.** (b)	**36.** (c)	**37.** (c)	**38.** (d)	**39.** (d)	**40.** (b)
41. (b)	**42.** (d)	**43.** (b)	**44.** (b)	**45.** (b)	**46.** (b)	**47.** (c)	**48.** (a)	**49.** (a)	**50.** (b)
51. (d)	**52.** (d)	**53.** (b)	**54.** (d)	**55.** (c)	**56.** (b)	**57.** (b)	**58.** (c)	**59.** (d)	**60.** (a)
61. (c)	**62.** (d)	**63.** (a)	**64.** (b)	**65.** (b)	**66.** (d)	**67.** (d)	**68.** (d)	**69.** (b)	**70.** (a)
71. (b)	**72.** (a)	**73.** (b)	**74.** (d)	**75.** (b)	**76.** (b)	**77.** (d)	**78.** (a)	**79.** (b)	**80.** (c)
81. (c)	**82.** (c)	**83.** (c)	**84.** (b)	**85.** (b)	**86.** (b)	**87.** (b)	**88.** (c)	**89.** (c)	**90.** (b)
91. (b)	**92.** (d)	**93.** (b)							

Chapter 6
Plant Adaptations

PLANT ADAPTATIONS

Any feature of an organism that may be morphological, anatomical, physiological or reproductive, which enables it to exist under conditions of its habitat is called **adaptation**. Warming 1895 had realised for the first time, the influence of controlling or limiting factors upon the vegetation in ecology. He classified plants into several ecological groups on the basis of their requirements of water and their substratum (soil) on which they grow. Some of them are as follows-

A) On the basis of substratum (soil)

- **Oxylophytes**- Plants growing in acidic soil
- **Halophytes**- Plants growing in saline soil
- **Psammophytes**- Plants growing on sandy soil
- **Lithophytes**- Plants grow on the surface of rocks
- **Chasmophytes**- Plants growing in the crevices of rocks

B) On the basis of water requirements

- **Hydrophytes**- Plants growing in water
  - Submerged hydrophytes
  - Floating hydrophytes
  - Amphibious hydrophytes
- **Xerophytes**- Plants grow in xeric condition i.e. in poor supply of water
  - **Oxylophytes**- Plants growing in acidic soils
  - **Halophytes**- Plants growing in saline soils
  - **Lithophytes**-Plants growing on rocks
  - **Psammophytes**- Plants growing on sand and gravels
  - **Chersophytes**- Plants growing on waste land
  - **Eremophytes**- Plants growing in deserts and steppes
  - **Psychrophytes**- Plants growing in cold soil
  - **Psilophytes**- Plants growing in Savannah
  - **Sclerophytes**- Plants growing in forest and bushland
- **Mesophytes**- Plants growing in an environment which is neither very dry nor very wet

HYDROPHYTES

(Greek, Hudor = water and phyton = plant; water plant)

Plants which grow in wet places or in water, either partly or wholly submerged are called hydrophytes or aquatic plants. Examples- *Hydrilla, Chara, Trapa, Pistia* etc. According to their relation to water and air, the hydrophytes are classified into the following categories:

1. Submerged hydrophytes
(grow below the water surface)

Submerged floating hydrophytes
Plants are completely submerged but not rooted in the mud. Occur in floating condition e.g. *Naja, Utricularia, Ceratophyllum* etc.

Submerged rooted hydrophytes
Plants are completely submerged and are rooted in the mud. Only their shoots floats on or above the surface of water e.g. *Potamogeton, Vallisneria* etc.

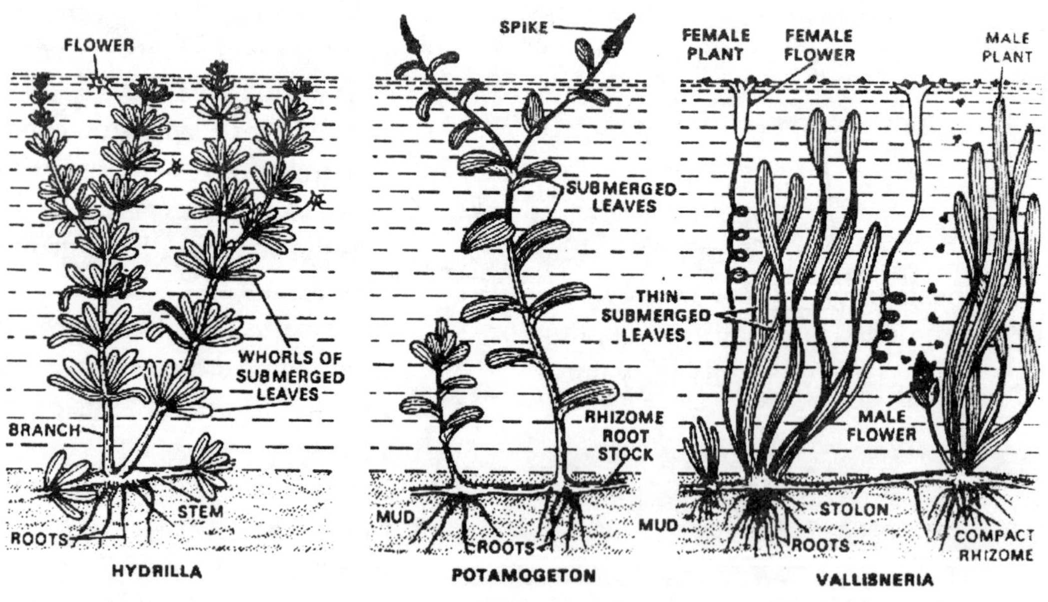

Fig. Submerged plants.

2. Floating hydrophytes
(plants float on the surface or slightly below the surface)

- **Free floating hydrophytes**
  Plants float freely on the water surface but are not rooted in the mud e.g. *Eichhornia, Trapa, Azolla, Lemna, Salvinia* etc.

- **Floating but rooted hydrophytes**
  Plants are rooted in the mud but their leaves and flowering shoots float on or above the surface e.g. *Nymphaea, Nelumbium, Marsilea* etc.

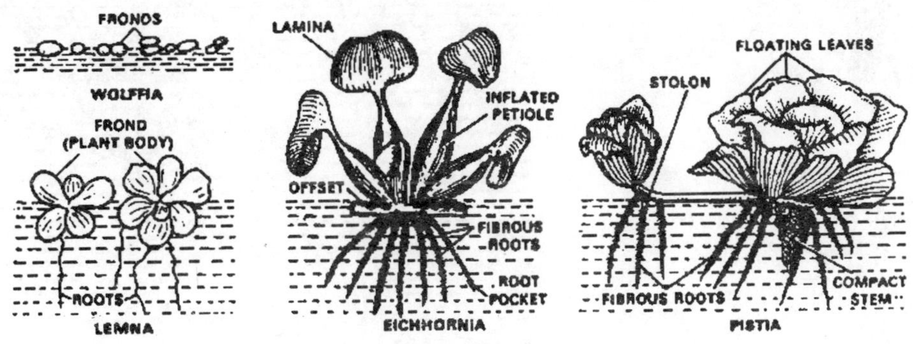

Fig. Free floating plants

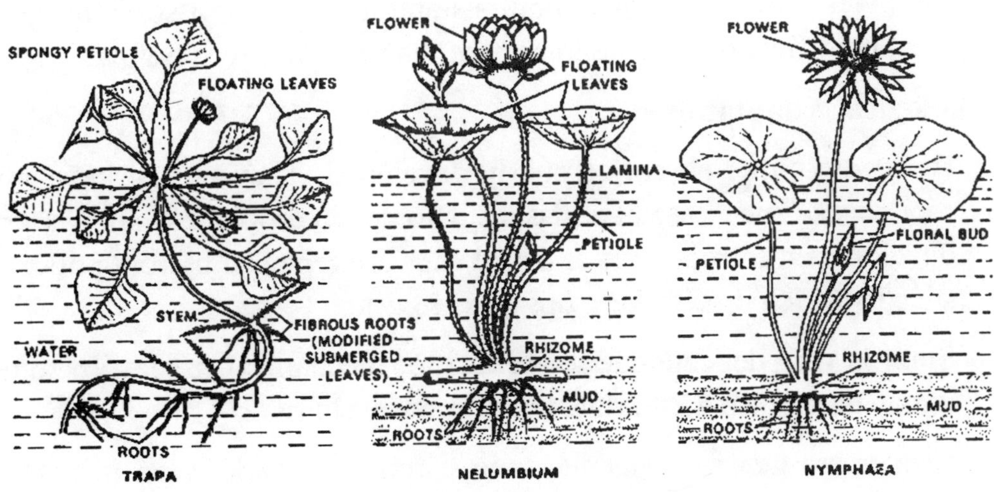

Fig. Fixed floating plants.

3. Ambhibious hydrophytes : Plants are adapted for both aquatic and terrestrial modes of life e.g. *Saggittaria, Ranunculus* etc.

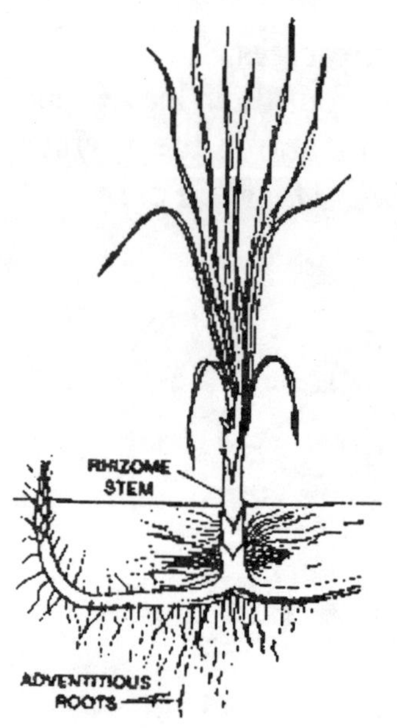

Fig. *Typha latifolia*

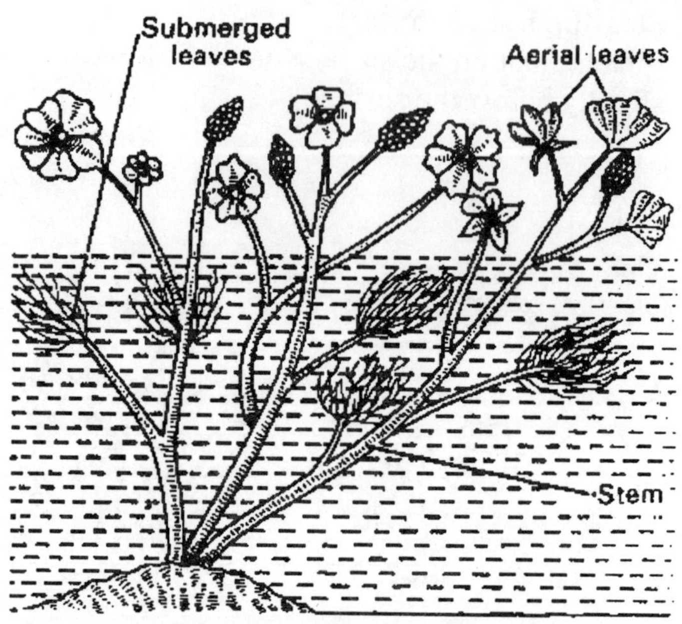

Fig. ***Ranunculus*** **showing heterophilly (Submerged leaves are very much dissected.)**

HYDROPHYTIC ADAPTATIONS

Hydrophytes grow in hydric conditions and hence exhibit some morphological and anatomical modifications. Some important features are given below:

A) Morphological Adaptations

i) **Roots** are either absent or poorly developed. They do not have root hairs, root caps and vascular tissues. Some hydrophytes show very poor development of root hairs and root caps are replaced by root pockets or root sheaths which protect their tips from injuries.

ii) **Stem** is very delicate due to the lack of mechanical tissues. In some cases, it may be modified into rhizome or runner, etc.

iii) **Leaves** are generally peltate, long, circular or dark green in colour. Their upper surface are exposed to the air but lower surface are generally in touch of water. Leaves are generally covered with wax to prevent the wetting of upper surface.

Some aquatic plants develop two different kinds of leaves called **Heterophylly** e.g. *Saggitaria, Salvinia, Azolla* etc.

iv) **Reproduction** is vegetative, asexual and sexual. Vegetative method takes place by propagation or fragmentation. In sexual reproduction, pollination is **Hydrophyllus** i.e. pollination takes place by water. When pollination occurs below water level called **hypo-hydrophily** e.g. *Naja, Ceratophyllum* and when pollination takes place at the surface of water, called **epihydrophily** e.g. *Vallisneria.*

B) Anatomical Adaptations: The anatomical adaptations in hydrophytes occur due to-

i) Reduction in protecting structures
ii) Well developed air spaces
iii) Poorly developed mechanical tissues
iv) Poorly developed vascular tissues.

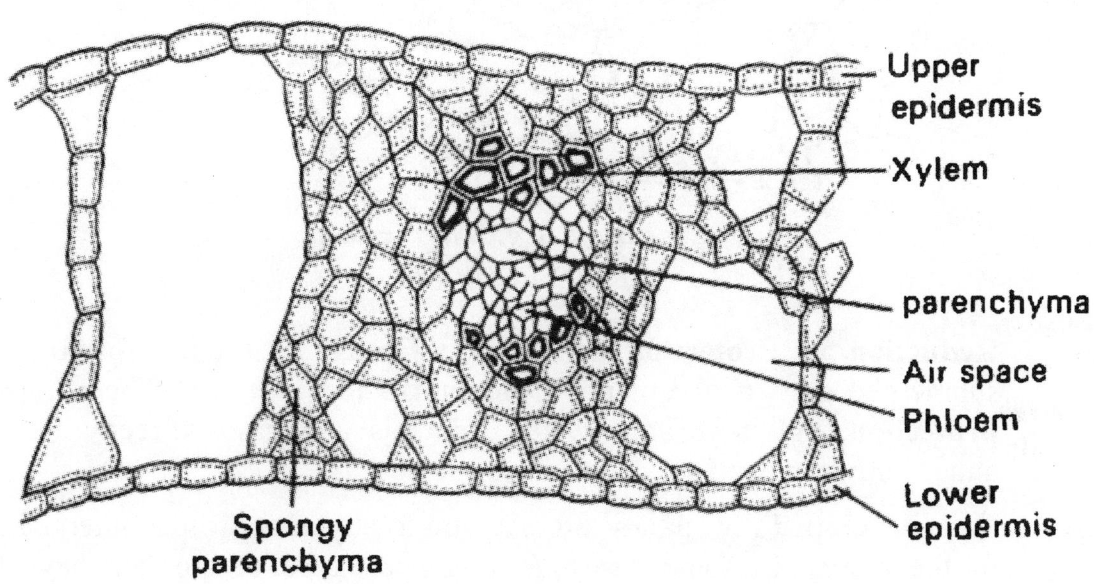

Fig. Part of T.S. of *Hydrilla* leaf

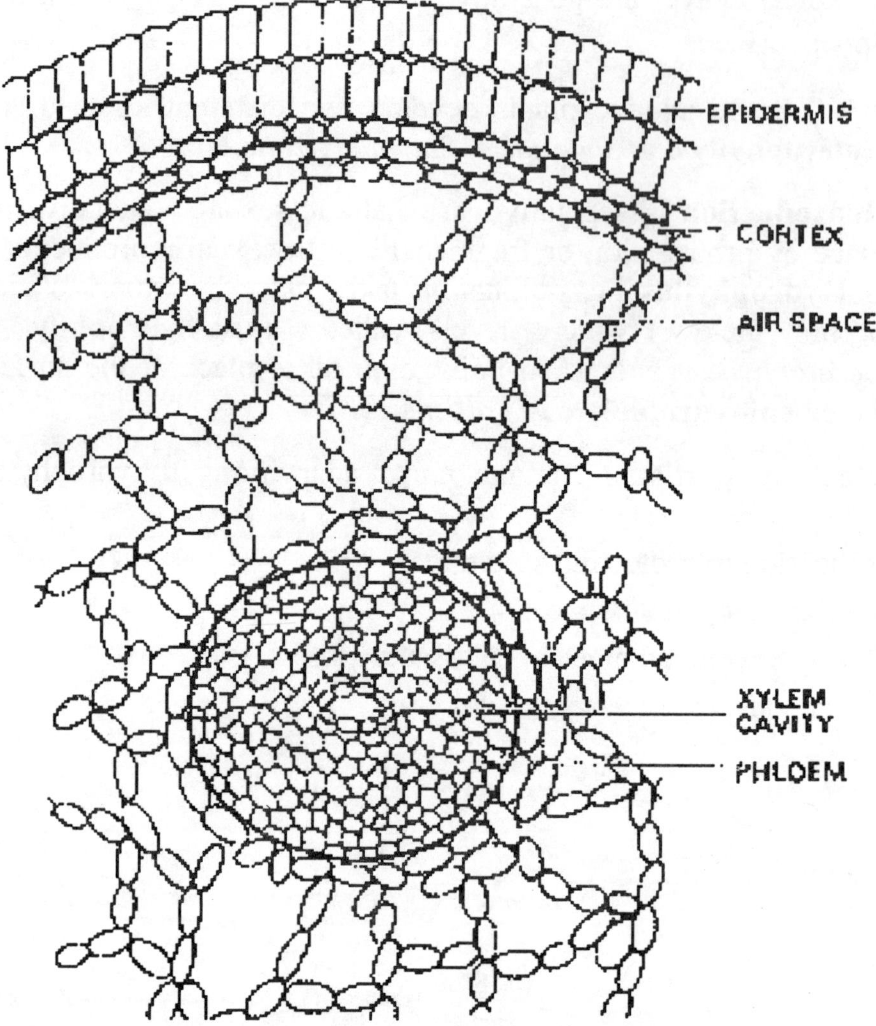

Fig. T.S. of *Hydrilla* stem.

i) **Reduction in Protecting Structures**: **Cuticle** is totally **absent** in the submerged parts of the plants. **Epidermis** does **not** serve the purpose of **protection**, but it **absorbs** water, minerals and gases directly from the aquatic environment.

ii) **Well developed air spaces**: Stomata are totally **absent** in submerged parts of the plants. In some cases, *Potamogeton* type of stomata have been noticed. In these cases, exchange of gases directly occurs through the cell walls. They have well developed **air spaces** which act as aerating system and also provide **buoyancy** to the plants.

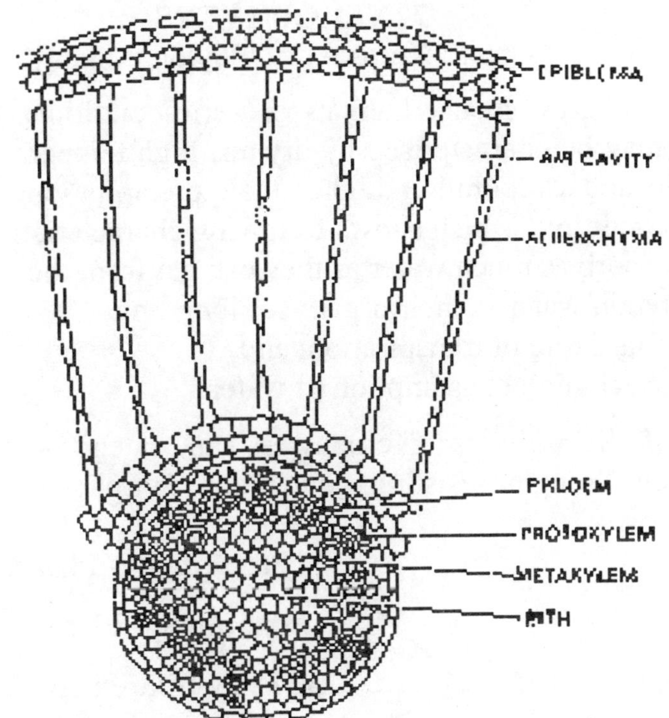

Fig. T.S. root of *Eichhornia*.

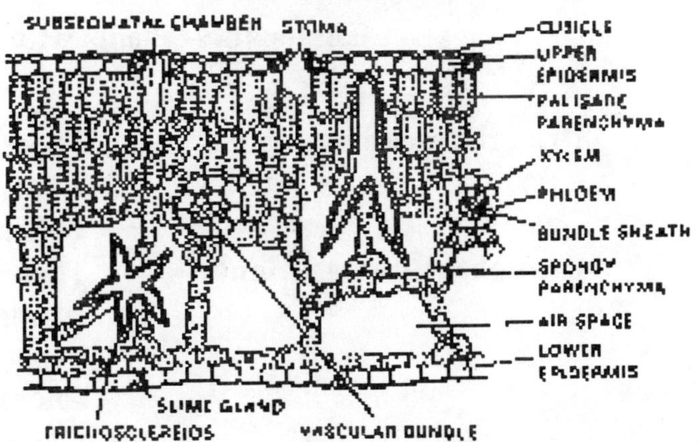

Fig. Part of T.S. of *Nymphea* leaf-lamina.

iii) **Poorly developed mechanical tissues**: The well developed air spaces, which provide **buoyancy** to plants, **save** them from **physical injuries**, hence they do not need mechanical tissues to strengthen their body.

iv) **Poorly developed vascular tissues**: Conducting tissues (vascular system) are poorly developed in hydrophytes because the absorption of water and nutrients takes place through their entire body surfaces, hence they do not need well developed vascular tissues.

2. XEROPHYTES

(Greek: Xeros = dry, Phyton = plant)

Plants which grow in dry habitats or xeric conditions are called xerophytes. Xeric condition may be characterised by dry air, high temperature, strong winds, hgh transpiration rate and evaporation higher than precipitation. So, only those plants survive in xeric conditions which shows following characteristics-

i) They absorb as much water as they can get from the surroundings.
ii) They retain water in their organs for long time.
iii) They retard rate of transpiration; and
iv) They check high consumption of water.

Classification of Xerophytes: Xerophytes can categorised into several groups according to their drought resisting power and morphology of the plants. These groups are as follows-

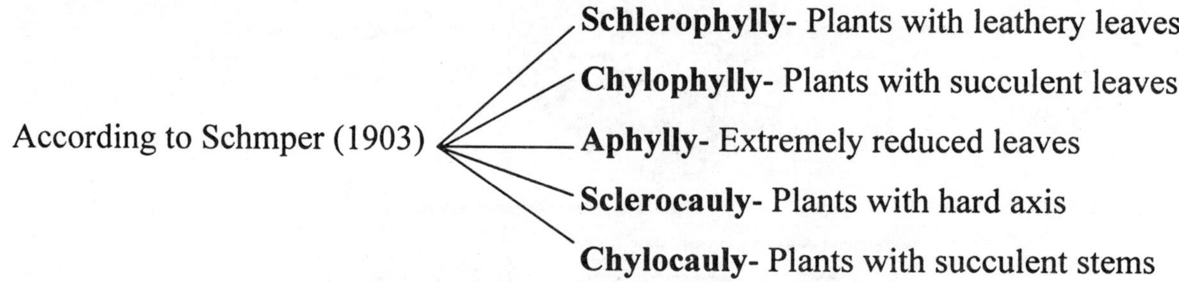

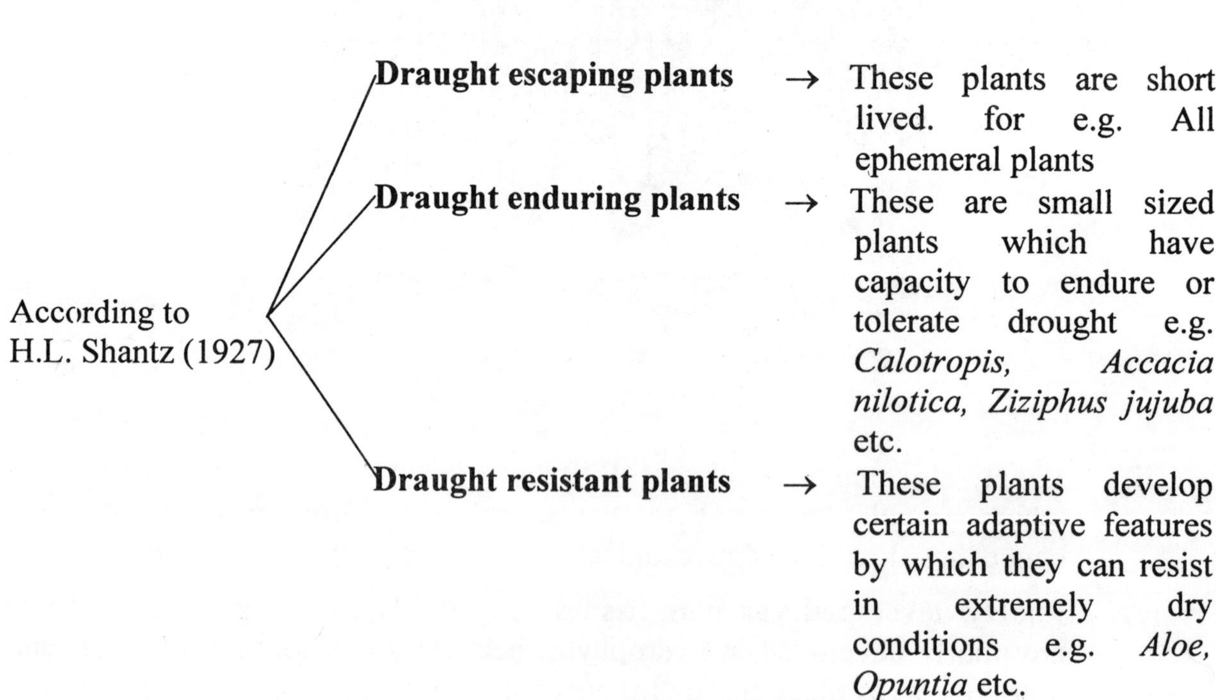

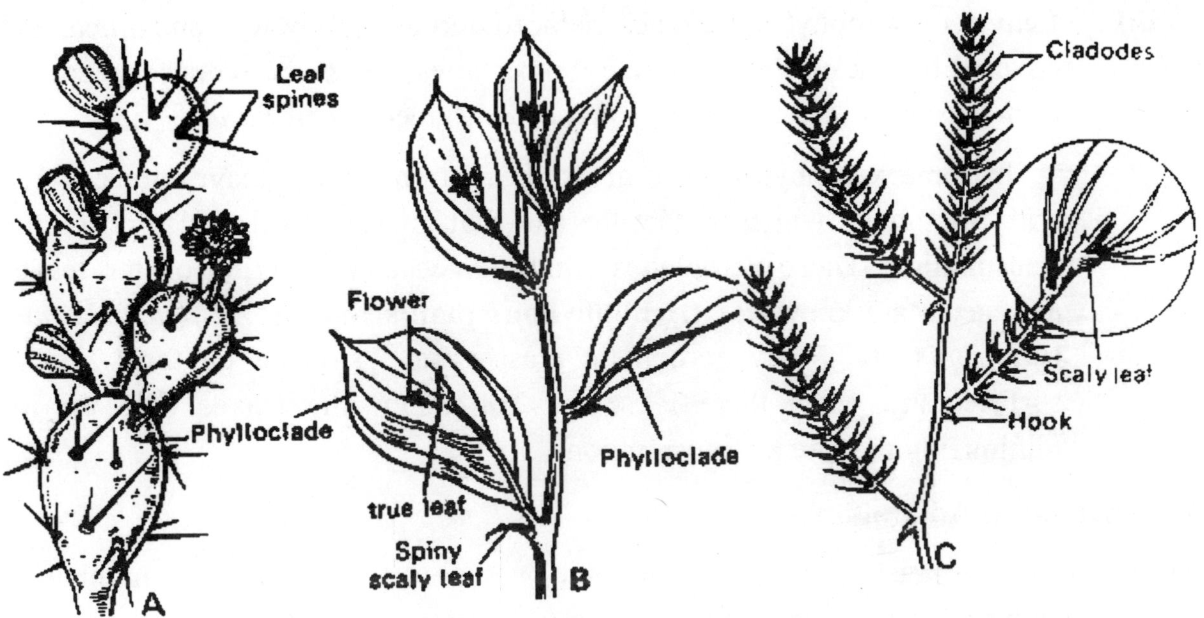

Fig. **A. Phylloclade of *Opuntia*, B. Phylloclade of *Ruscus* C. Cladode of *Asparagus*.**

Xerophytic Adaptations

Xerophytic adaptations can be described under following heads-

(A) Morphological Adaptations
(B) Anatomical Adaptations
(C) Physiological Adaptations

(A) Morphological Adaptations

i) **Roots** are well developed i.e. tap root; which is profusely branched and may go as deep as 20-40 feet e.g. *Calotropis, Acacia* etc. and may reach to the water table.

ii) **Stems** of some xerophytes become very hard and woody. They are covered with wax and silica as in *Equisetum*, while some may be covered with dense hairs as in *Calotropis*. In some xerophytes stems may be modified into thorn e.g. *Duranta*.

Some times stem modifies into leaf like structure called **phylloclade** or **cladodes** e.g. *Opuntia, Ruscus* etc.

iii) **Leaves** of xerophytes are much reduced and coated by wax and silica, and covered by thick layer of cuticle. Some times leaves are reduced to scales, as in *Casurina, Asparagus* etc. or "needle shaped" as in *Pinus*.

In some xerophytes those growing in strong wind, leaves are covered with thick hairs which protect the stomatal guard cells and also check the transpiration. Those xerophytes which have hairy covering on the leaves and stems are known as **trichophyllous plants** for e.g. *Ziziphus, Nerium, Calotropis* etc. Many xerophytic grasses have stomata on their adaxial surface, and when there is scarcity of water, the leaves **role up** for minimizing the rate by transpiration.

(B) Anatomical Adaptations

Due to the need of water conservation, xerophytic plants show a number of anatomical modifications. Some of them are given below-

i) **Epidermis** is well developed with, heavily thickened cell wall. Some times it becomes multilayered as in *Nerium* leaf.

ii) Many xerophytes, posses simple or compound **hairs** which protect the stomata and prevent excessive water loss. In some plants, surfaces of stems and leaves develop characteristic ridges and furrows or pit as in *Casurina* which also prevent from excessive loss of water.

iii) **Stomata** of xerophytes is **succulent** type which shows **diurnal** pattern. It is **Potato** type (i.e. Number of stomata is more in lower epidermis than upper epidermis) and located in furrows or pits.

iv) **Hypodermis** is sclerenchymatous, thick walled, compactly arranged in single or multilayered and present just below to the epidermis.

v) **Conducting tissue**, i.e. xylem and phloem, develop very well in the xerophytic plants.

vi) **Water storage tissue**, is well developed and it store water which is used in drought period. In some cases, latex and mucilage are also stored in parenchymatous tissues. Due to the viscocity of latex, the rate of transpiration is checked to some extent.

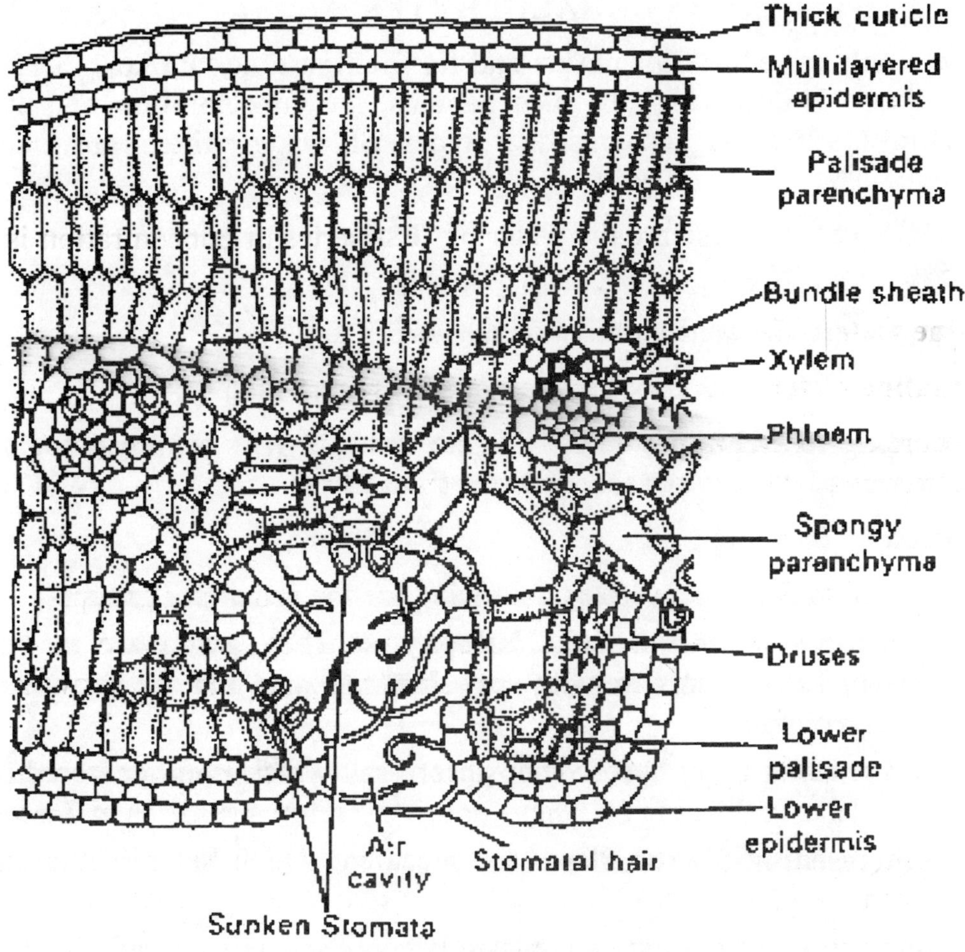

Fig. Part of leaf of *Nerium*.

(C) Physiological Adaptations

It was long assumed that the structural adaptation in the body of xerophytes were useful in reducing the transpiration but now a number of experiments related with the physiology of these plants reveal some facts which are contrary to the early assumptions. The structural modifications are directly governed by their physiology.

i) It contain polysaccharides, pentosans and a number of acids which resist them from drought because these chemicals have water binding property.

ii) Some enzymes, such as catalase, peroxidase are more active in xerophytes which hydrolyse starch very actively.

iii) High osmotic pressure in cells of xerophytes can increase the turgidity of cell which induce the water absorption.

HALOPHYTES

Commonly called as **Salt plants** and for the first time "**Pallas**" in 19th century coined the term "**Halophytes**". In such habitats, the concentration of salts is very high. **Brehm** (1926) has recognised 3 types of water which are rich in salt concentration.

1) Mixohaline water: These are the water of "**Estuaries**" where salt concentration is 0.05 to 3.5%.

2) Euhaline water: The concentration of salts is 3.5 to 4.5%.

3) Hyperhaline water: The concentration of salt is more than 4.5%.

According to **Schimper** (1898) saline habitats will be physiologically dry, because common plants can not absorb water from them.

Classification

According to Schimper (1898), following are the important groups-

i) **Pseudohalophytes** or **false halophytes**: They germinate in brief rainy season. These plants are with very short life span and can grow in salty soil e.g. *Agropyron*.

ii) **Salt erading halophytes**: They absorb salt which is either stored inside the vacuole as in the case of *Juncus* or it is excreted outside as in *Tamarix*.

iii) **Salt resisting plants**: Plants are succulent which helps in the dilution of salts for e.g. *Sueda*.

iv) **Salt enduring plants**: Protoplasm become specialized and able to resist in high salt cocentration.

Osmotic pressure of halophytes is very high i.e. around 150 atm. The minimum osmotic pressure is about 40 atm. According to **Harris** (1934), the highest osmotic pressure occurs in *Atripex confertifolia* i.e. 202.4 atm.

Tsopa (1939) classified halophytes into the following four groups according to their response to salinity:

i) **Obligate halophytes**: Plants requiring salinity throughout their life.

ii) **Preferential halophytes**: Also called as **facultative glycophytes**. They are found in normal soil but optimum activity is observed in saline soil.

iii) **Supporting halophytes**: They are normal plants but can grow in saline soil.

iv) **Occasional halophytes** found in normal soil but some times very rarely can grow in saline soil.

Iverson (1936) has divided saline habitats and halophytes into three categories, depending upon the salt concentration.

S.No.	Habitat	Salt content	Plants
1.	Oligohaline	0.01-0.1%	Oligohalophytes
2.	Mesohaline	0.1-1.0%	Mesohalophytes
3.	Polyhaline	More than 1%	Euhalophytes

According to **Chapman** (1942), halophytes have been classified into the following categories-

i) **Miohalophytes**: Plants growing in the habitats of low salinity i.e. below 0.5% NaCl.

ii) **Euhalophytes**: When the concentration of salt is more than 0.5%. They have been further sub-divided into the following groups.

a) **Mesohalophytes**: The concentration of salt is about 0.5 to 1%.
b) **Mesoeuhalophytes**: The concentration of salt is 1 to 5%.
c) **Eneuhalophytes**: When the concentration is more than 5%.

Warming (1909) has classified halophytes (plants growing on saline system) into following 5 groups depending upon the nature of the substratum:

i) **Lithiphilous halophytes**: Growing on rocks near the ocean.
ii) **Psammophilous halophytes**: Growing on sandy regions.
iii) **Pelophilous halophytes**: Growing in loam + clay.
iv) **Salt desert**: Growing in calm water and muddy spots.
v) **Littoral swamp forests or Mangroove vegetation**: Growing on sea-shores. They form special vegetation called **mangroves**. In India mangroves are very common on sea coasts of Bombay near "**Elephanta Cave**" and Kerala, Andaman and Nicobar Islands.

CHARACTERS OF HALOPHYTES

A) External morphology

i) **Root**: In halophytes, in addition to normal roots, many **stilt** or **prop** roots develop from the aerial branches of stem for efficient anchorage in muddy or loose sandy soil. The soil of coastal region is poorly aerated and it contains very little percentage of oxygen because of water logging. In order to compensate this lack of soil aeration, the hydrohalophytes develop special type of negatively geotropic roots, called **pneumatophores** or **breathing roots**. The pneumatophores usually develop from the underground roots and project in the air for efficient gaseous exchange.

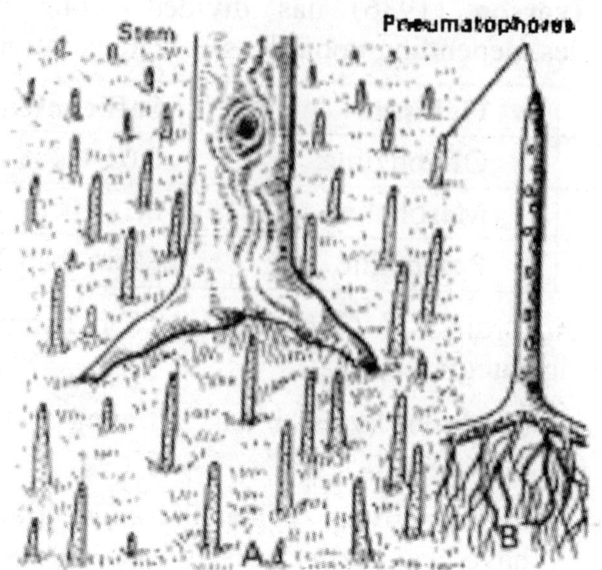

Fig. Pneumatophores of *Rhizophora*.

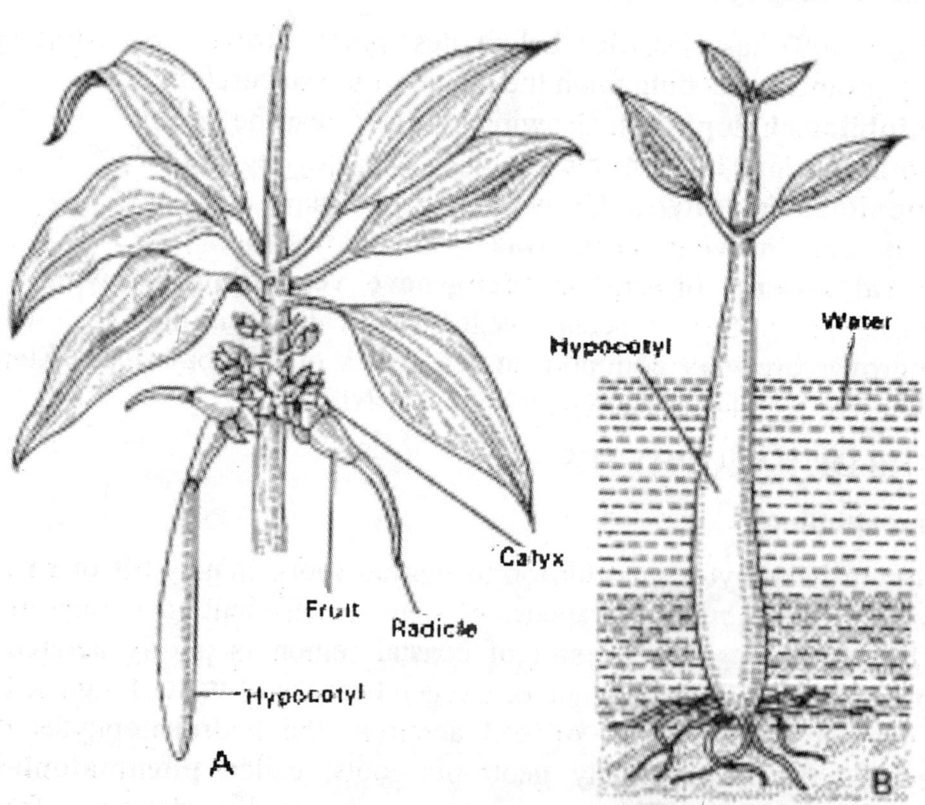

Fig. A. Vivipary, B. A young plant.

ii) **Stem**: Most of the halophytes have succulent stem for e.g. *Salicornia, herbacia, Suaeda* etc. The succulency of stem is due to decrease in number and increase in length of cells.

iii) **Leaves**: Leaves are small, succulent, thick and spiny to check transpiration.

iv) **Fruits, Seeds and their dispersal**: The fruits and seeds are generally light in weight can float freely on the water surface. The mangroves show the phenomenon of **Viviparous germination** i.e. seed start germinating while the fruit is still attached to the parent plant e.g. *Rhizophora, Cassula, Avicennia* etc.

(B) Anatomical features

The halophytes shows following anatomical features:

i) Large cells and small intercellular spaces.

ii) High elasticity of the cell wall.

iii) Extensive development of water storage tissues.

iv) Small relative surface area (surface/volume ratio).

v) Small and fewer stomata.

vi) Low chlorophyll content and

vii) Leaves of many species are dotted with local cork formation "Cork warts".

viii) Stem with epidermal layer which is covered by thick cuticle.

ix) Roots with well developed xylem and vascular bundle are conjoint, collateral, endarch and open type.

MASTER STROKES

Q. What do you mean by Teleology?

Teleology is a natural phenomenon which is explained without any scientific or experimental basis. It is the belief that events occur in response to a specific need. For example – stem bends towards light because they need light for growth.

ENTRANCE CORNER

1. The stem of submerged hydrophyte is soft and weak because- **(CPMT 2000)**
   (a) they totally lack phloem
   (b) they do not have stomata
   (c) they totally lack xylem
   (d) they have poorly developed mechanical and vascular tissue

2. Velamen tissue is found in- **(CPMT 1980)**
   (a) adventitious roots (b) aerial roots of orchids
   (c) hydrophytic roots (d) stilt roots

3. The characteristic plant of mangrove vegetation is-
   (a) *Rhizophora mangle* (b) *Ficus religiosa*
   (c) *Mangifera indica* (d) *Prosopis specigera*

4. Sunken stomata characterize the- **(CPMT 1992; RPMT 1997)**
   (a) hydrophytes (b) epiphytes
   (c) mesophytes (d) xerophytes

5. In hydrophytes following is very well developed- **(CPMT 2002)**
   (a) cuticle (b) aerenchyma
   (c) vascular tissue (d) sclerenchyma

6. Velamen tissue is present in roots of- **(CPMT 1989)**
   (a) orchids (b) parasites
   (c) all epiphytes (d) *Piper*

7. Sunken stomata are found in the leaves of- **(CBSE 2001)**
   (a) *Nelumbium* (b) neem
   (c) maize (d) *Nerium*

8. Among the plants listed, point out one that does not fit into ecological group represented by other plants- **(CPMT 1984)**
   (a) *Euphorbia* (b) *Acacia*
   (c) *Vallisnaria* (d) aloe

9. In submerged hydrophytes the functional stomata are found- **(CPMT 1990)**
   (a) on upper surface of the leaf (b) on the lower surface of the leaf
   (c) on both the surfaces of the leaf (d) no where on the plant

10. Extreme xerophytic characteristics are found in- **(CPMT 1991, 99)**
    (a) *Brassica* (b) *Nerium*
    (c) cactus (d) *Capparis aphylla*

11. Vegetation of Delhi is predominantly- **(CPMT 1991)**
    (a) hydrophytic (b) xerophytic
    (c) mesophytic (d) halophytic

12. Function of velamen is- **(CPMT 1991)**
    (a) absorption of water from soil (b) to help in transpiration
    (c) absorption of water from atmosphere (d) none of these

13. In a place there are spiny-thorny plants. Some plants have thick cuticle. Some plants are with small reduced leaves. Some plants are ephemeral-annuals and some groups of mosses are also there. This place is- **(CPMT 1989)**
(a) xerophytic and there is rainfall for a short period during the year
(b) mesophytic with heavy rain fall during the summers
(c) grass land with scattered trees
(d) mesophytic with small herbs and medium sized shrubs

14. Among the following characteristics which one does not characterize a hydrophyte- **(CPMT 1988)**
(a) abundant air spaces and air chambers
(b) plentiful xylem and sclerenchyma
(c) leaves having stomata only on the upper side or none
(d) poor development of roots

15. Thick cuticle on leaves are typical of plants growing in- **(CPMT 1984, 87)**
(a) wet habitats (b) warm habitats
(c) dry habitats (d) cool habitats

16. *Hydrilla* is a- **(CPMT 1983)**
(a) floating aquatic plant
(b) aquatic plant with floating leaves
(c) submerged aquatic plant
(d) amphibious plant with dimorphic leaves

17. Reduction in mechanical and vascular tissues and poor roots are found in- **(CPMT 1994)**
(a) halophytes (b) lithophytes
(c) hydrophytes (d) xerophytes

18. Heterophyllous plants are found as- **(CPMT 1986)**
(a) free-floating hydrophytes
(b) submerged hydrophytes
(c) rooted and floating hydrophytes
(d) halophytes

19. Plants growing near sea shore will usually have xerophytic characters like thick leaves because-
(a) there is plenty of water in the soil
(b) there is too high a concentration of salts for the plants to absorb sufficient water from soil
(c) light available to the plants is insuficient
(d) the soil is physically dry

20. Spongy tissue and less developed roots are present in- **(CPMT 1995)**
    (a) xerophytes (b) mesophytes
    (c) hydrophytes (d) halophytes

21. Which of the following is not the characteristic feature of the anatomy of xerophytes- **(CPMT 1990)**
    (a) spongy parenchyma
    (b) well-developed conducting tissue
    (c) well-developed mechanical tissue
    (d) thick cuticle

22. Succulents are likely to be found in- **(CPMT 1983)**
    (a) tropical rain forests (b) deciduous forests
    (c) deserts (d) tundras

23. Plants growing in xerophytic conditions have- **(CPMT 1994)**
    (a) extensively developed shoot system
    (b) extensively developed root system, small and thick leaves or stem performs photosynthesis
    (c) large number of stomata
    (d) extensive air spaces

'thought'

Hydrilla is a submerged aquatic plant. It shows larger spaces in the stem which are filled with air. This helps the plant to float. The amount of sclerenchym and xylem in the stem is highly reduced. **(CPMT 1983)**

24. The features in *Hydrilla* plant are-
    (a) xerophytic (b) hydrophytic
    (c) mesophytic (d) halophytic

25. When a *Hydrilla* plant is removed from water and placed in soil-
    (a) it starts producing collenchyma and sclerenchyma to get mechanical support
    (b) the plant will dry up and die
    (c) the plant will grow slowly
    (d) a thick cuticle will develop

26. Mechanical tissues and xylem are highly reduced in *Hydrilla* because-
    (a) water provides the support and there is no problem for the absorption and conduction of water
    (b) otherwise the weight of these tissues would have drown the plant
    (c) otherwise translocation of food would have been interrupted
    (d) of the anaerobic conditions created by submergence

27. Plants that grow in a marshy saline habitat are called- **(CPMT 1992)**
(a) hydrophytes (b) xerophytes
(c) mesophytes (d) mangroves

28. Aerenchyma helps plants to- **(CPMT 1992)**
(a) attachment (b) exchange of air
(c) float on water (d) mechanical support

29. One of the following does not have stomata- **(CPMT 1996)**
(a) hydrophytes (b) submerged hydrophytes
(c) xerophytes (d) mesophytes

30. The following is a amphibious plant- **(CPMT 2000)**
(a) *Hydrilla* (b) *Wolfia*
(c) *Casuarina* (d) *Polygonum*

31. An aquatic plant with floating leaves- **(AFMC 1999)**
(a) do not have stomata
(b) have stomata on both surfaces of leaves
(c) have stomata on upper surface of leaves
(d) have stomata on lower surface of leaves

32. Ephemerals are a type of xerophytes- **(AMU 1999; CPMT 2001)**
(a) drought resisting (b) drought escaping
(c) drought enduring (d) none of these

33. *Acacia arabica* (*A. nilotica*) is- **(AFMC 1997)**
(a) hydrophyte (b) xerophyte
(c) mesophyte (d) halophyte

34. Pneumatophores and viviparous germination are found in-
(CPMT 2001; RPMT 1999; CBSE 2000)
(a) *Jussiaea* (b) *Opuntia*
(c) *Eichhornia* (d) *Rhizophora*

35. A submerged rooted hydrophyte is-
(a) *Trapa* (b) *Vallisneria*
(c) *Utricularia* (d) all of these

36. Excessive aerenchyma is characteristic of- **(CPMT 1999)**
(a) heliophytes (b) xerophytes
(c) mesophytes (d) hydrophytes

37. Pneumatophores are meant for- **(CPMT 2005)**
(a) respiration (b) anchorage
(c) protection (d) nutrition

ANSWER SHEET

1. (d)	**2.** (b)	**3.** (a)	**4.** (d)	**5.** (b)	**6.** (a)	**7.** (d)	**8.** (c)	**9.** (d)	**10.** (d)
11. (c)	**12.** (c)	**13.** (a)	**14.** (b)	**15.** (c)	**16.** (c)	**17.** (c)	**18.** (c)	**19.** (b)	**20.** (c)
21. (a)	**22.** (c)	**23.** (b)	**24.** (b)	**25.** (b)	**26.** (a)	**27.** (d)	**28.** (c)	**29.** (b)	**30.** (d)
31. (c)	**32.** (b)	**33.** (b)	**34.** (d)	**35.** (d)	**36.** (d)	**37.** (a)			

Chapter 7
Soil Erosion and Its Conservation

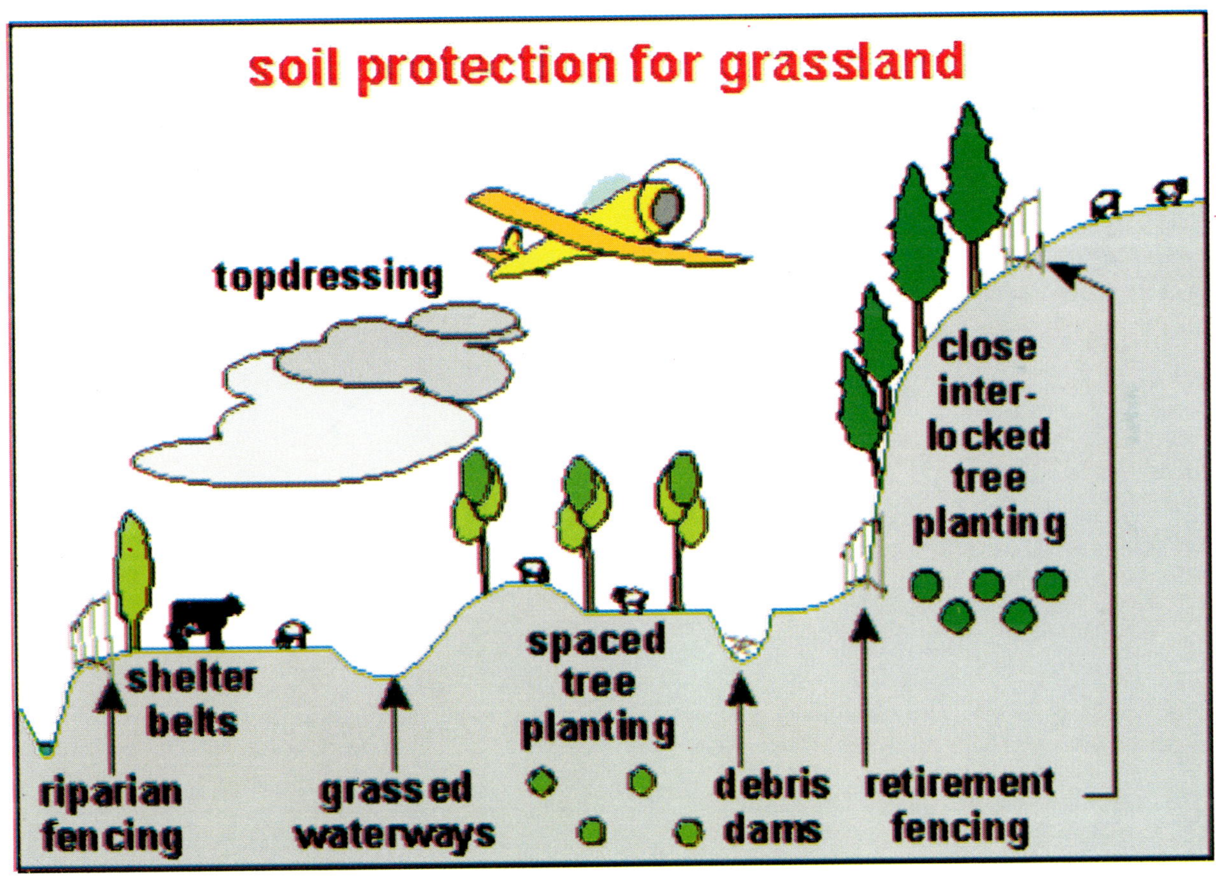

SOIL EROSION AND ITS CONSERVATION

The word "erosion" literally means to "wearing away". In soil erosion, fertile soil surfaces are detached and removed from their original places. Soil erosion is caused by the following two agencies –

A) Climatic
B) Biotic

(A) Climatic: These are **water** and **wind**. It is observed that in India near about 2,60,000 metric ton/year, soil is eroded by these two agencies.

1) Erosion due to water: The erosion of soil due to the agency of water is called water erosion. Water erosion occurs during the melting of snow and heavy rain fall, which can not be absorbed by the soil. The degree of water erosion determined by the soil cover and slope cover. In the absence of soil cover, rain drops bombard upon soil, due to this bombardment each rain drop bursts into the several parts and each part carrying soil particles in suspensions. Due to this phenomenon, erosion of soil takes place. Water erosion is of following types-

(i) **Sheet erosion**: This types of soil erosion occurs on the smooth and gentle slope; the top soil is removed as a thin layer of sheets, hence this is called sheet erosion. Due to this erosion the area becomes light in colour.

(ii) **Rill erosion**: As we know that the running water have great cutting and carrying capacity, hence whenever it meets to the loose or tilled soil and an uneven area, it causes cutting of soil. The cuts initially appears in the form of finger like or groove like narrow depressions and is called as rill erosion.

(iii) **Gully erosion**: It is more prominent type of erosion, in which heavy rainfall, rapidly running water and transporting water may result in deeper cavities or grooves called gullies.

Gullies may be "V" shaped or "U" shaped. When erosion occurs in a huge amount, "V" shaped gully is formed. It is the characteristic of coarse textured soil, and when erosion occurs in a small amount "U" shaped gully is formed and it is the characteristic of fine textured soil.

(iv) **Slip erosion**: It is also called as **Land slide erosion**. This type of erosion occurs in hilly areas where due to hydraulic and gravitational forces rock fragments come down.

(v) **Stream bank erosion or Riparian erosion**: It occurs due to the high velocity of river band. It cuts the soil horizontally called riparian erosion. It is very common at **Koshi river of Bihar**.

2) Erosion due to wind: Removal of soil by wind is called wind erosion. Stormy winds carry the soil particles to distant places and some times form sand-dunes. The wind erosion is based on-

i) Velocity of wind
ii) Dry and precipitated area
iii) It also depends on steepness

Wind erosion are of following types-

(i) **Suspension**: In this, fine soil particles (diameter less than 1 mm) are suspended in air. These suspended particles are deposited several kilometer away when the wind velocity decreases.

(ii) **Saltation**: Such kinds of wind erosion occurs when the size of soil particles is 1.0 to 1.5 mm in diameters. The particles are forced to roll along the ground for some distance and then suddenly kicked up into the air by the force of wind. Because due to their weight they can not remain in suspension for long time. This process occurs in several steps and called as saltation.

(iii) **Surface creep**: It occurs in case of heavier particles of soil, usually 5-10 mm in diameter. They can not easily be detached from the ground by the pressure of wind hence they creep on the surface of soil.

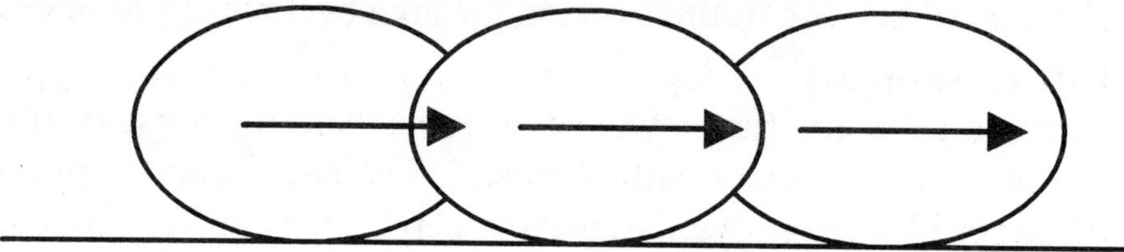

(B) Biotic: Excessive grazing, deforestation, undesirable forest biota and mechanical practices by man are important factors which cause soil erosion.

Some other causes of soil erosion

1. Uneven distribution of rain fall
2. **Shifting cultivations**: This practice of growing crop is very common in tribal people, particularly in a tropical country. It is also called as **slash** and **burn technique**, in which forest is cleared; it is burnt and organic matter becomes available. The land becomes fertile for the cultivation of crops. The crop is cultivated for 3 or 4 years and then they moved to the new areas and same practice is repeated. This is it termed as shifting cultivation.

In early days, the cycle used to get completed after 50 or 100 years, during this period natural regeneration of forest was possible and practice was according to the ecological principle. But today due to the scarcity of forest area, the cycle gets completed into 10-20 years, hence natural regeneration is not possible. Land is gradually degraded and converted into the waste land.

In Assam, it is called as **Jhuming** (Jhumias = forest farmer), in M.P. is called as **dahya** and in south it is called as **Podu**. It is estimated that about 5 hectare land degraded each year by shifting cultivation. Government of India in his **National Forest Policy 1952, Clause 4.7** has totally banned the shifting cultivation.

Effect of soil erosion: There are several serious effects of soil erosion, which are as follows-

i) Due to uprooting of trees, shortage of timber and fuel results.
ii) Loss of soil stability and fertility
iii) Shortage of fodder
iv) Destruction of land in plains
v) Higher temperature and lower rainfall.
vi) Greater frequency of floods and threat to communication channels.

SOIL CONSERVATION

- Term conservation is derived from two Greek words-

  Con = Together
  Servare = Guard

- According to IUCN (International Union for Conservation of Nature and Natural Resources), the conservation may be defined as "Management of biosphere for human use in such a way that it may yield greatest sustainable benefit to present generation and at the same time maintaining its potential to the needs and aspiration of future generation".

IUCN has given a new term, as WCU (World Conservation Union). In case of soil, conservation means, to protect soil from soil erosion and maintain the fertility of soil. In soil erosion, two main aspects are –

1) Detachment of soil particles
2) Transportation of soil particles

In India, **Central Soil Conservation Board (CSCB)** has been established having nine (9) important centres and its headquarter is situated at **Dehradun**.

* **Dr. H.H. Benett** is considered as the father of soil conservation. He was a soil conserver and great soil scientist.

Control of soil erosion by methods of soil conservation

According to CSCB, near about **23 million ton** soil is eroded every year. There are two types of particles involved in control of soil erosion-

A) Particles for control of water erosion

1) Biological methods

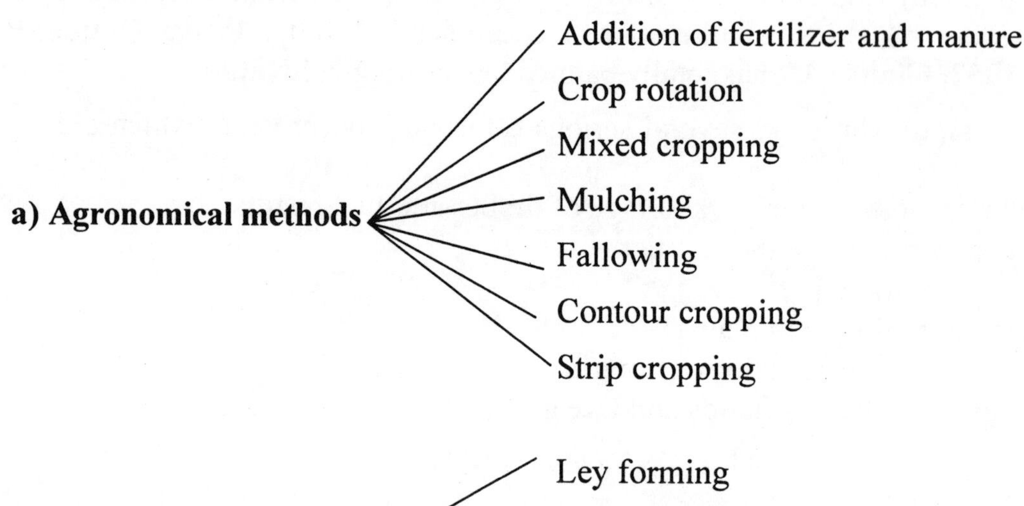

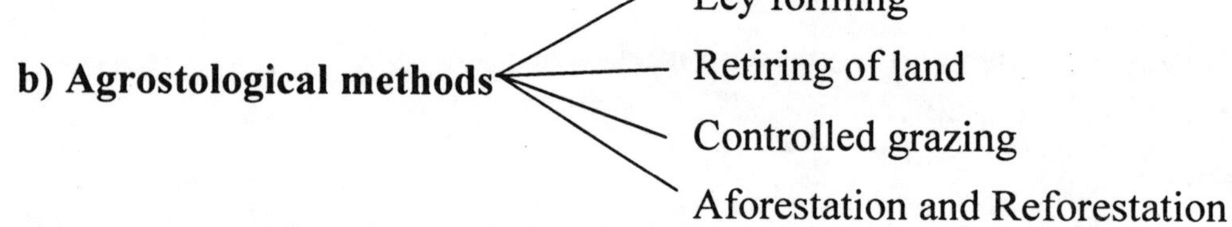

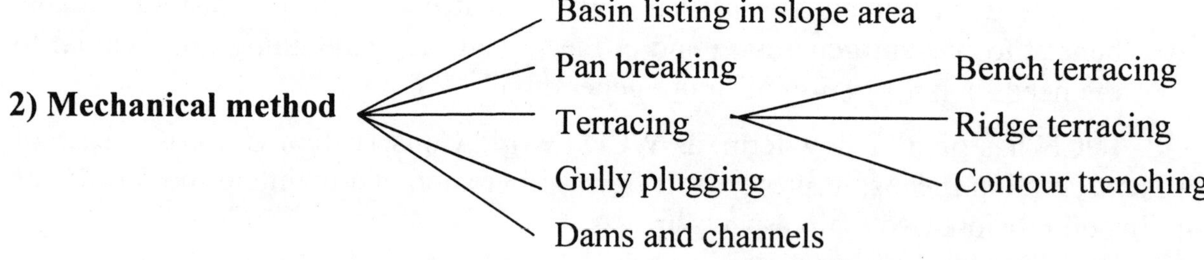

3) Dry forming practices

B) Control practices for wind erosion
- Wind breaks and Shelter belts
- Sand binders and soil binding

A) Practices for control of water erosion: Water erosion is controlled by 4 important principles-

i) To protect soil from direct rain beating
ii) To impede or check the rate of flow of water
iii) To check the movement of water from a narrow channel.
iv) To allow more water to percolate in the soil.

Methods

There are two methods to check the soil erosion:

1) **Biological methods**
2) **Mechanical methods**

1) Biological methods

The most important method, to check soil erosion is to grow vegetation (Mainly grasses). To calculate the conservation value (CV), there is a formula given by R.S. Ambasht-

$$CV = 100 - \frac{SW_P}{SW_D} \times 100$$

where – CV = Conservation value
SW_P = Dry weight of soil under vegetational cover
SW_D = Dry weight of soil which is unprotected.

Conservation value of some plants is given as under-

Saccharum munja, CV = 92-96%
Cyanodon dactylon, CV = 90-97%
Euphorbia hirta, CV = 10-12%

From the above values, it becomes clear that, the monocot plants are more efficient for the conservation of soil as compared to dicot plants, because monocots have adventitious roots and dicots have tap root.

There are two types of biological methods for the conservation of soil-

(a) Agronomical method
(b) Agrostological method

(a) Agronomical method: Study of crop is known as agronomy. Following are the agrononomical methods used for the conservation of soil-

i) **Addition of fertilizer and manure**: By addition of manure and fertilizers, the fertility of soil is increased and at the same time, water holding capacity of soil also increases and more water percolates into the soil. Thus, chances of erosion are decreased.

ii) **Crop rotation**: In this method, two different type of crops are grown in alternate season to maintain the fertility of soil for e.g. in most of the cases, cereals are alternated with pulses. Due to this, the competition decreases. Due to the differentiation in roots, they get water and nutritive materials from different zone of soil, hence there is less competition.

iii) **Mixed cropping**: In this method, following types of crops are grown-

i) Food crop	=	Those crops which are used as food e.g. wheat, paddy.
ii) Cash crop	=	Grown for monetary purposes e.g. potato, tobacco.
iii) Catch crop	=	To grow new crop at the place of original crop harvested by bacteria.
iv) Trap crop	=	To grow new crop around original crop.
v) Intensive crop	=	That crop, which is grown in large number but in a little area.
vi) Extensive crop	=	Grown in less number, but in large area e.g. Tea.
vii) Subsistence farming	=	The crop which is only available for food. No surplus production of crop is obtained in this crop.
viii) Interculture	=	When crops are grown in alternate row then called as interculture.

"The two different types of crop grown together are called mixed cropping" e.g. Wheat and Mustard, and Sugarcane and Arhar.

iv) **Mulching**: To check soil erosion, the surface of soil is covered with plant residue or dry plants like – dry leaves, twigs, which is known as mulch. Mulch protects the soil from direct rain beating as well as after decomposition it also increases the fertility of soil. In most of the crops harvesting is done above the soil surface, the lower portion of stem behaves like mulch and this mulch is known as **stubble mulch**.

v) **Fallowing**: Retiring of land known as fallowing. In this method, land is left as such, cultivation is not done for one or two seasons and during this period, cattles are also allowed to graze in the field. Grass grown in the field, helps to check the soil erosion and by adding organic matter, the fertility of soil is increased.

vi) **Contour cropping**: This method is mostly adopted in Hilly areas. Any line at the slopes, which is at right angle to the direction of slope is called **Contour line** and when cropping is done along contour line, it is called as **Contour cropping**. Due to ploughing along contour line, small depressions and elevations are resulted. Each elevation behaves like a small dam which protects or checks the flow of water, similarly each depression behaves like water reservoir and it allows more water to parcolate but in case of high intensity of rain fall, these deppressions or elevations get washed and it enhances the soil erosion effect, that is why contour cropping is generally done along with other methods to control the water erosion.

vii) **Strip cropping**: To check the soil erosion, crops of different habits are grown in alternate strip for e.g. long crops may be alternated with dwarf plants or legume may be alternated with cereals. The width of strip depends on the types of soil and slope percentage.

Land slope %	**Width of long crops or Erosion resistant crop**	**Width of short crops or Erosion permiting crop**
1. Less than 1%	9 meter	4-5 meter
2. Between 1-2%	6 meter	24 meter
3. Between 2-3%	4 meter	12 meter

The important types of strip cropping are as follows-

i) **Contour strip cropping**: In this, strip crops are grown on contour line.

ii) **Field strip cropping**: This practice is followed in areas, where slope is uneven. The crops of different heights are grown in different seasons.

iii) **Wind strip cropping**: Strips are grown at right angle to the direction of slope.

iv) **Buffer strip cropping**: In this, a permanent strip of crops (grasses) is alternated with strip of different types of crops in different seasons.

b) Agrostological methods

i) **Ley farming**: This is a crude method of farming, where check of erosion is not possible. In this method, crops are grown and at the same time grasses are also allowed to grow.

ii) **Retiring of land**: In this method, land is left as such without cultivating any crop for several years. During this period, grasses are allowed to grow, some times leguminous plants are also allowed to grow along with grasses.

iii) **Controlled grazing**: The grazing should not be allowed more than 60% of the carrying capacity of any grassland.

iv) **Aforestation and Reforestation**: Existment of new forest in bare area is called Aforestation. In other words, growing new forest on those areas where previously no forest exists. While existment of forest on those areas, where previously forest was destroyed due to any reason, is called reforestation.

Both check or decrease the rate of soil erosion.

2) Mechanical methods: Mechanical method is supplemented with biological method-

i) Slope is divided into flat segments

ii) Percolation of water should be more and erosion should be less.

i) Basin listing: With the help of instrument, basin-lister small hole are made on contour line to increase the amount of percolation, to decrease run-off and erosion.

ii) Pan-breaking: Sometimes, calcium carbonate ($CaCO_3$) or other inorganic compounds get deposited in the lower layer of soil and form a hard pan like structure, due to which percolation of water in deeper layers is not possible. With the help of instrument i.e. by pan breaker, we break the pan into several layers to increase the percolation of water inside the soil.

iii) Terracing

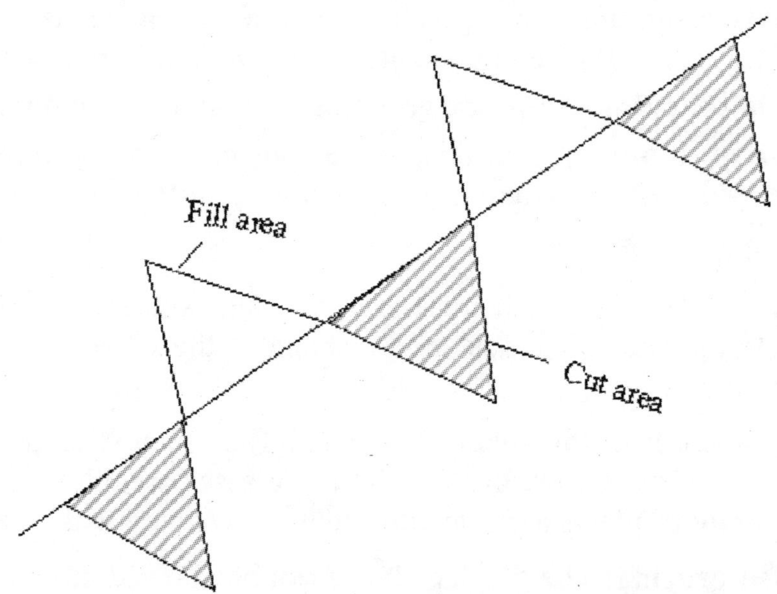

This practice is done in hilly areas where slope is divided into number of flat fields called as terraces. On these terraces, different crops are grown and this type of farming is called as Terracing cultivation or farming. It is of 3 types-

a) Bench terracing

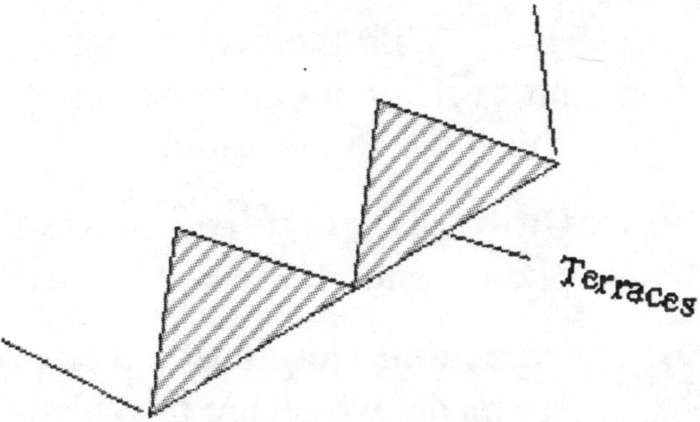

Terrace looks like bench on which crops are grown, hence this is known as bench terracing.

b) Ridge-terracing

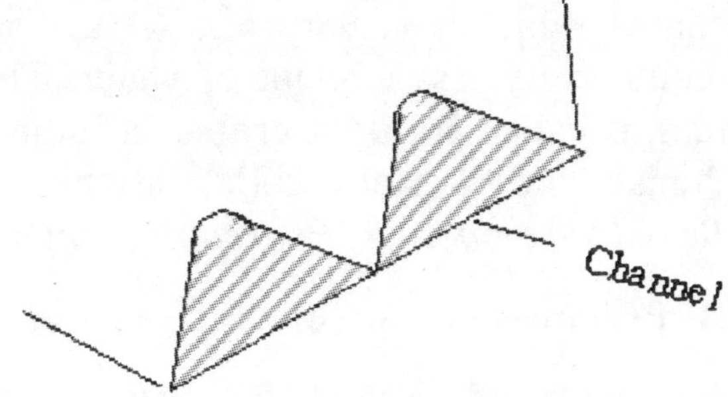

In this type, outer margin of benches is slightly raised, which allows more water to percolate and chances of erosion is decreased. It is known as ridge terracing or channel terracing. Sometimes complete channels are made at right angle to terraces which is known as channel terracing.

c) Contour-trenching

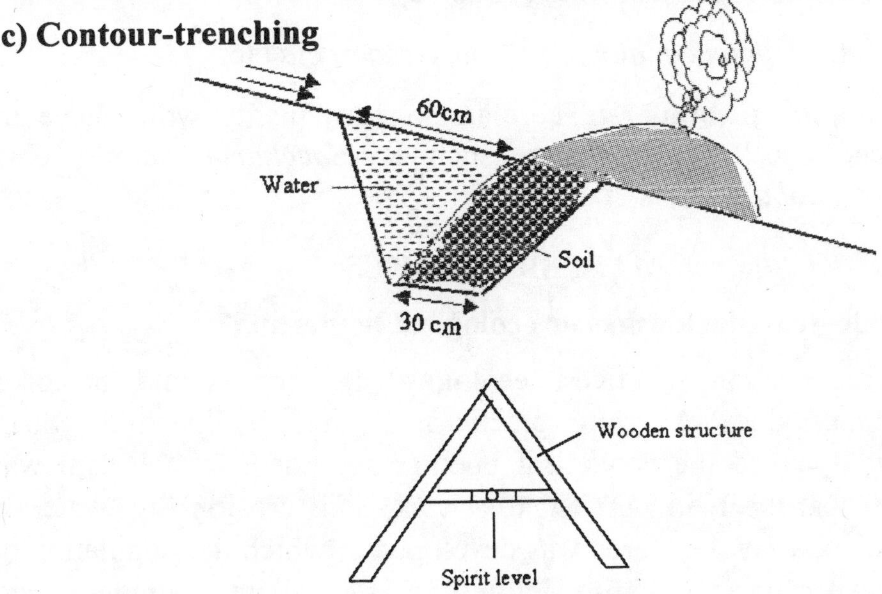

Fig. showing trenching device

With the help of contour trenching device, trenches are made at regular intervals or at contour lines. Trees are also planted on the margin of trenches to check the soil erosion.

iv) **Gully plugging**: To impede the flow of water through the gullies, stone and plants residues are kept in between gullies to check further damage to land.

v) **Dams and channels**: To check the soil erosion along the banks of river, dams are made, which are provided with concrete channels at regular intervals.

3) Dry farming practices: Dry farming practices are adopted in those areas, where annual rain fall is very less (less than 20 inches/annum), crops are grown which require very less amount of water. This crop, consistent with water available from rain, is called **Rainfed crops**. In India such areas are common in Rajasthan, parts of Punjab and Western U.P. and the examples of such crops are- *Cajanas cajan, Brassica compestris, Phaseolus munga, Phaseolous vulgaris* and *Pennicetum* etc.

B) Practices to control wind erosion

i) Wind breaks and shelter belts

In areas where wind velocity is very high, trees are planted at right angle to the direction of wind, when trees are planted in small blocks, it is known as **Wind break** and extensive plantation of tree is known as **Shelterbelt**. It has been calculated that, shelterbelt protects 5' area of tree height at wind ward side and conserve soil 30 times more than lee ward side. Important trees planted are-

Accacia nilotica, Accacia leucophloea, Albizia, Azaderecta etc.

ii) Sand binders and soil binders: There are so many plants, which have ability to bind the soil and check wind erosion, such as- *Saccharum munja, Cyanodon, Capparis, Ipomea biloba,* etc.

MASTER STROKES

Q.1. What is ecological blacklashes or ecological bomerang?

During conservation practices, ecological balance should be taken into account. Sometimes due to wrong practices, it may badly affect the human population, so it is called the ecological boomerang. For e.g. – A dam was made along the river **Zambezi** in **Africa** to check soil erosion by water. During construction of dam, a low land area was developed in which the population of **Tsisi-tsisi** fly was grown causing **sleeping sickness**, it is excellent example of ecological boomerang.

Q.2. What is green revolution?

In March 1968, the Director of the US-Agency for International Development (US-AID), **William Gaud**, first used the phrase "Green revolution" for the rapid increase in agricultural production (especially wheat and rice). It has been achieved through introduction of high yielding varieties (HYVs), increased irrigation facilities, fertilizer applications, weed pest and pathogen control, multiple cropping and better agricultural management. **Dr. N.E. Borlaug** recognized as the father of green revolution. He developed **semidwarf** varieties of wheat and received **Nobel Prize** for Peace in 1970. Borlaug used a Japanese variety **Norin 10** as a source of dwarfing gene. Semidwarf varieties of wheat are resistant against to the lodging, rust and other major diseases of wheat. ICAR (Indian Council of Agricultural Research) in 1963 introduced several dwarf varieties of wheat, from which **Sonalika** and **Kalyan Sona** varieties were selected by CIMMYT (International Centre for Wheat and Maize Improvement), Mexico. CIMMYT also directly introduced semidwarf varieties of wheat like- **Lerma Roja 64A** and **Sonora 64**.

A dwarfing gene **dee-geo-woo-gen** was noted in **Taiwan**. The scientists used these dwarfing genes and produced the first dwarf fertilizer responsive variety of rice- **Taichung Native-1** (TN-1) in 1956 at Taichung District Agricultural Improvement Station.

IRRI (Indian Rice Research Institute), Philippines developed a series of eight semidwarf varieties (IR-8) of rice in 1996 which become famous as miracle rice. It was short saturated, sturdy stemmed, photoperiod intensive, high yielding and fertilizer responsive variety adapted to a wide range of environment. IR-8 was followed by a series of several varieties (i.e. IR-20, IR-24, IR-26, IR-36, IR-54, IR-72, etc.). IR-36 is the most widely planted variety in history. IR-36 was produced by a team of IRRI scientists led by **Dr. Gurudev Singh Khush**.

ENTRANCE CORNER

1. The pH of a fertile soil is usually around -
   (a) 2-3 (b) 6-7
   (c) 8-10 (d) 11-12
2. Soil erosion can be prevented by-
   (a) deforestation (b) afforestation
   (c) overgrazing (d) vegetation removal

3. Uniform soil erosion by running water is -
   (a) gully erosion (b) rill erosion
   (c) riparian erosion (d) sheet erosion
4. In hilly areas, erosion is minimised by-
   (a) terracing (b) manuring
   (c) ploughing (d) mixed cropping
5. Chipko movement is connected with -
   (a) conservation is natural resources (b) plant/forest conservation
   (c) plant breeding (d) green revolution

ANSWER SHEET

1. (b) **2.** (b) **3.** (d) **4.** (a) **5.** (b)

Chapter 8
Plant Indicators

PLANT INDICATORS

The living beings are the measurements of the environment as they indicate environmental conditions of their respective habitats and thus serves as index of environmental conditions and commonly known as **Biological indicators**. The plant when indicates by virtue of their response to conditions about them they are said to be **plant indicators**.

Plants are very useful indicators and are of great value as they are able to indicate the uses for which land may be best employed such as agricultural forestation, grazing etc. Plants also indicate about the past climates. According to some scientists, plant also possess to clairvoyance to forecasting future changes and possibility of controlling them. Thus, an ecologist may know the present, past and future story of the landscape with the help of vegetation studies.

Many conditions of the environment are judged by the indicator species of communities. Some of them are as follows-

1. Agriculture Indicators. The most reliable indicators of the agricultural possibilities of a region are to be found in its native vegetation. It is of great importance as they indicate the type of farming and the kind of crop suited to a habitat. The natural plant cover indicates the crop producing capabilities of land better than any series of meteorological observations or soil analysis, e.g. the areas of maximum production of corn and winter wheat correspond to areas where the native grass communities are dominated by species of *Andropogon*, while spring wheat and durum wheat grow well where *Stipa* and *Andropogon glaucum* are abundant. Dhawan and Nanda (1949), Dhawan and Dhand (1950) have studied the soil types, calcium/boron content and plant communities of Punjab. They have discovered that the soils supporting *Salvadora oleoids, Zizyphus nummularia* and *Prosopis cineraria* are good soils, suitable for agriculture.

Table : Some Plant Indicators of Soil Types in Punjab (Dhawan and Nanda, 1949) are-

Plant Community	Soil Features
1. *Butea monosperma*	Heavy Alkaline
2. *Capparis deciduas*	Alkaline soil
3. *Pinus and Juniperus*	Uranium rich soil
4. *Tamarix articulata*	Saline soil
5. *Calotropis procera*	Light soil
6. *Salvadora oleoides*	Calcium and Boron rich soil, suitable for crop plants
7. *Zizyphus nummularia*	Good soil for agriculture
8. *Prosopis cineraria*	Good soil for crop cultivation if irrigation available
9. *Rumex acetosella*	Acid grassland soil

2. Overgrazing. Many plants are overgrazed which results in modification of grassland. Some plants show characteristic indication of overgrazing e.g., *Polygonum, Chenoppdium, Lepidium, Bromus, Sporobolus, Urochloa, Buchloe* and *Hilaria*. The presence of *Convolvus pluricaulis* and *Evolvulus alsinoides* also indicate overgrazing alongwith disturbed conditions of the soil.

3. Forest Indicators. Some plants indicate the characteristic type of forest and grow in an area which in not disturbed. *Narenga porphyrocoma* is a useful soil binding grass which indicates that Sal (*Shorea robusta*) can be profitably planted there. Similarly growth of *Viola* species in the western temperate Himalayas is a suitable indicator for plantation of *Cedrus deodara* and *Pinus wallichiana*.

4. Humus Indicators. Some plants act as humus indicator. Presence of *Monotropa, Neottia* and mushrooms indicated that soil is humus rich. Excessive humus prevents the regeneration of tree species, as indicated by *Strobilanthes* and *Impatiens*.

5. Soil Moisture Indicators. Some of the plants indicate moisture contents of soil. Some of them are given below:

Plant Community	**Soil Features**
1. *Citrullus colocynthis, Saccharum munja, Acacia nilotica, Calotropis, Agave, Opuntia, Argemone, Carlhamus oxycantha, Echinops echinatus, Cassia auriculata*	Indicate low moisture content of soil
2. *Acacia glandulifera, A. greggi, Prosopis chilensfs*	Indicate underground springs of the arid areas
3. *Typha, Phragmites, Vetiveria zizanioides*	Grow in water logged areas
4. *Phragmites, Juncus, Carex, Polygonum*	Grow in swampy soil
5. Mangrove plants	Indicate saline water logged soil

6. Indicators of Soil Types. Several plant species are helpful in indicating the soil type. The sandy loam soil is indicating by deeply rooted and taller species like *Andropogon scoparium* and *Ipomoea leptophylla*. Similarly *Woodfordia floribunda* and *Chloris virgata* indicate high lime content in the soil.

7. Soil Reaction Indicators. Slight acidity is useful to the growth of many forest trees which are calcifuge, e.g., *Shorea* (Sal), *Pinus roxburghii*. Calcicoles include *Tectona grandis* (Teak), *Cupressus torulosa*, *Ixora parviflora* and *Taxus baccata*. Mosses growing on limestone areas are *Tartella* and *Neckera*. *Rumex acetosa, Rhodadendron,*

Salicornia, Chenopodium, etc.

8. Soil Erosion Indicators. Plant like *Carrisa spinarum* and *Capparis sepiaria* are indicators of soil erosion where on *Zizyphus rotundifolia* is an indicator of soil formation.

9. Fire Indicators. Some plants are well adapted to grow in burnt and highly disturbed areas, for e.g., luxuriant growth of *Epilobium angustifolium, Paspalum notatum, Aristida stricto, Populus tremuloides, Quercus macrocarpa, Larix occidentalis* indicate the occurrence of fires in the area.

10. Altitude Indicators. Some of the plant species indicate altitude of a place. Some of them are listed below:

Plant community	**Range of altitude**
1. *Betula utilis, Rhododendron campanulatum* etc.	3200-3600 m
2. *Taxus baccata, Abies pindrow* and *Quercus* spp.	2200-3200 m
3. *Cedrus deodara* and *Pinus wallichiana*	1700-2800 m
4. *Pinus roxberghii*	900-1800 m

11. Indicators of Rock Beds. Number of plant species grow on rock beds. Some of them are as follows:

Plant community	**Rock beds**
1. *Calotropis procera* and *Aerua*	Gypsum bed (Rajasthan)
2. *Pterocarpus santalinus*	Quartzite bed (South India)
3. *Euphorbia antiquorum*	Granite bed (South India)
4. *Taxus baccata* and *Larix griffithiana*	Lime stone bed
5. *Betula, Salix* and *Sedum*	Igneous rock beds (USA)

12. Mineral Indicators. Those plant species which grow in mineral rich areas are called mineral indicators. They are also called as **metallicolus** plants or **metallicoles**. Metallicoles show various modifications or abnormalities in response to metal toxicity e.g., chlorosis double petal, petalless, dwarfism, gigantism, monstrosities, etc. Prospecting the minerals through plant indicators is called **geobotanical prospecting**. The first geobotanical linkage was found in Europe where it was found that *Viola calaminaria* was always associated with zinc containing soils. Metallicolous plants are of the following types:

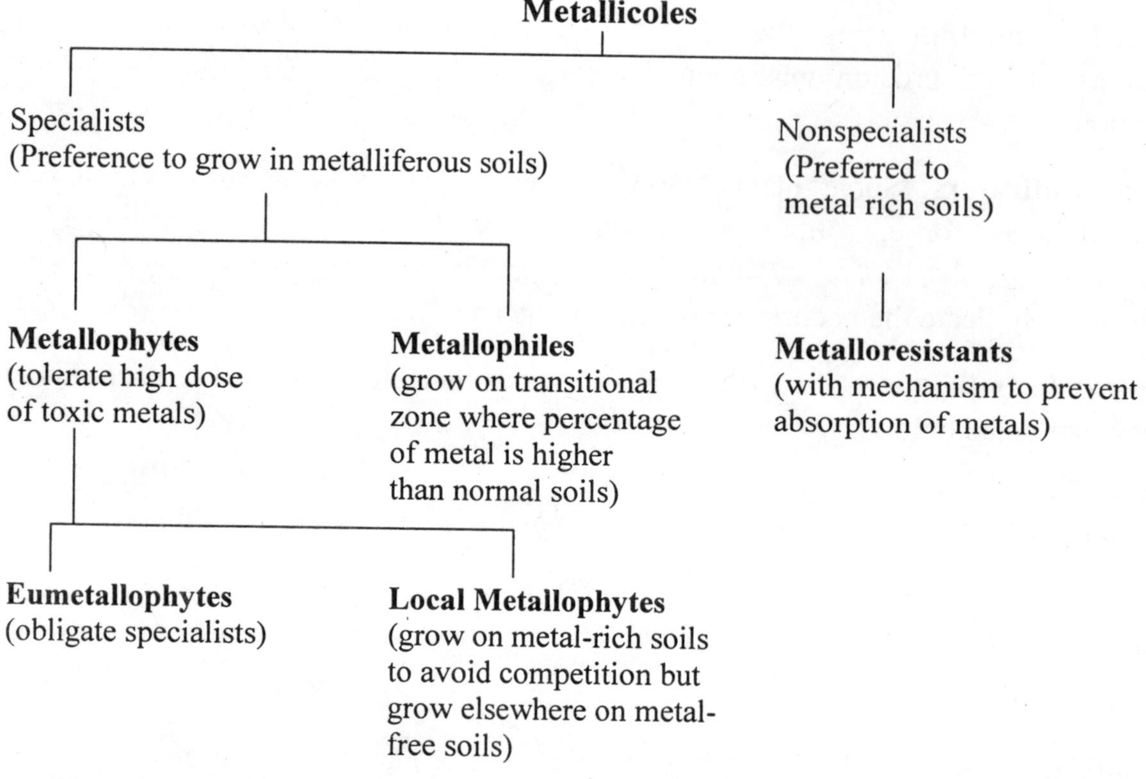

(i) **Aluminium**, e.g., *Ilex aquifolium* (in Italy).

(ii) **Bauxite**, e.g., *Memecylon* (in South India).

(iii) **Boron**, e.g., *Salsola nitrata, Eurotia cerutoides*.

(iv) **Cobalt**, e.g., *Silene cobalticola, Myssa sylvatica* var. *biflora* (in U.S.A.).

(v) **Copper**, e.g., *Viscaria alpina* (in Norway), *Polycarpea spirostylis* (in Australia), *Acrocephalus robertii* (in Katanga), *Becium homblei* (in Zaire and northern Zambia), *Becium ovovatum* (in Rhodesia), *Gymnocolea acutiloba* (in Ades, W.S. America), *Gypsophila patrini* (in U.S.S.R.), *Eschscholzia mexicana* (in America, U.S.A.). Mosses *Merceva* and *Midichhoferia* have been employed for copper prospecting in Scandinavia.

(vi) **Diamond**, e.g., *Vallozia candida* (in Brazil).

(vii) **Gold**, e.g., *Equisetum arvense* (in Czechoslovakia), *Lonicera confusa, Papaver libonoticum, Alpinia speciosa, Thuja* species (in Brazil).

(viii) **Iron**, e.g., *Damara ovata, Dacrydium caledonium* (in Scotland).

(ix) **Lithium**, e.g., *Lycium, Juncus, Thalictrum* species.

(x) **Mercury**, e.g., *Stellaria setacea* (in Spain).

(xi) **Nickel**, e.g., *Lychnis alpina* (in Sweden).

(xii) **Selenium**, e.g., *Astragalus* species, *Stanleya pinnata, Onopsis condensator, Neptunia amplexicaulis*, etc.

(xiii) **Silver**, e.g., *Eriogonium ovalifolium* (U.S.A.).

(xiv) **Sulphur**, e.g., *Allium, Arabis, Oenothera, Atriplex nuttalli*, etc.

(xv) **Uranium**, e.g., *Astragalus* species (in Colorado, U.S.A.).

(xvi) **Zinc**, e.g., *Viola calaminaria, V. lutea, Armeria valgaris, Alsine verna, Thalapsi alpestre* (in Europe).

MASTER STROKES

Q. What is Serendipity?

Serendipity is the phenomenon of unexpected discoveries by chance, accidently or intuition. However, according to Louis Pasteur "Chance favours the trained mind" i.e., only a person who is trained in scientific method is more likely to make a chance discovery. For example- Penicillin discovered by Alexander Flemming (1928) while working on the culture of bacterium *Staphylococcus*. Accidently, he found that *Staphylacoccus* infected with fungus *Penicillium*. Flemming got fungal extract and pouring it over bacteria. The extract killed the bacteria, then this extract was named as Penicillin.

ENTRANCE CORNER

1. The presence of gold is indicated by - **(CPMT 2006)**
   (a) *Equisetum* (b) *Viscaria*
   (c) *Viola* (d) *Salsola*
2. The presence of diamond is indicated by - **(AFMC 2005)**
   (a) *Astragalus* (b) *Vallozia candida*
   (c) *Lychnis* (d) *Memecylon*
3. *Capparis deciduas* commonly grows in - **(CBSE 2002)**
   (a) Acidic soil (b) Alkaline soil
   (c) Overgrazed soil (d) None
4. *Pinus roxburghii* grow at- **(AIIMS 2004)**
   (a) High altitude (b) Mineral rich soil
   (c) Fertile soil (d) None
5. Metallicoles are the plant which commonly grow in - **(PMT 2003)**
   (a) Mineral rich soil (b) Acidic soil

(c) Rock beds (d) Overgrazed soil

ANSWER SHEET

1. (a) **2.** (b) **3.** (b) **4.** (a) **5.** (a)

Chapter 9
Biodiversity and Its Conservation

BIODIVERSITY AND ITS CONSERVATION

The term biological diversity was first of all articulated by **Norse** and **Mc.Manus** (1980). Later on term 'Biodiversity' was coined by **W.G. Rosen** in 1985 in the first planning meeting of the national forum on Biodiversity held in Washington D.C. in 1986.

Before 1992, the term "Biodiversity" was used in technical sense, but after "Rio Summit", June 92 (Attended by 156 countries) termed as "United Nations Conference on Environment and Development" (UNCED), the term biodiversity is used in non-technical sense.

As we know that the number of different items and their relative frequency is called "**density**", whereas the variety and variability among living organisms and ecological complexes in which they occur is called "**biodiversity**". In biological diversity, these items are organized at various levels raising from complete ecosystem to chemical structure, which arc the molecular basis of heredity, thus the term includes various ecosystems, genus, species and relative abundance.

Magnitude of Diversity

It has been estimated that the total number of species on this planet earth is about 50 millions, out of which about 1.7 millions have been described by man.

Diversity of world (Plant)

Taxonomic group		**No. of spp.**
Bacteria	:	3600
Blue green algae	:	1700
Fungi	:	46983
Bryophytes	:	17000
Gymnosperms	:	750
Angiosperms	:	250000

Diversity of world (Animal)

Taxonomic group		**No. of spp.**
Insects	:	750000
Sponges	:	5000
Crustaceans	:	9000
Molluscs	:	38000
Star fishes	:	50000
Fishes	:	6100
Amphibians	:	19056
Reptiles	:	6300
Birds	:	9036
Mammals	:	40008

Diversity of India (Plant)

Taxonomic group		No. of spp.
Bacteria	:	850
Algae	:	2500
Fungi	:	23000
Lichens	:	1600
Bryophytes	:	2700
Pteridophytes	:	1022
Gymnosperms	:	64
Angiosperms	:	17000

Diversity of India (Animal)

Taxonomic group		No. of spp.
Protozoans	:	2517
Porifera	:	519
Cnidaria	:	237
Ctenophora	:	10
Platyhelminthes	:	1622
Nematoda	:	2350
Rotifera	:	310
Knoryncha	:	10
Acanthocephala	:	88
Sipuncula	:	110
Mollusca	:	38
Echiura	:	5042
Annelida	:	1093
Onychophora	:	1
Arthropoda	:	57525
Phoronida	:	3
Bryozoa	:	170
Entoprocta	:	10
Brachiopoda	:	3
Chaetognatha	:	30
Echinodermata	:	765
Hemichordata	:	12
Protochordata	:	116
Fishes	:	2546
Amphibians	:	204
Reptiles	:	428
Birds	:	1228
Mammals	:	372

* V.N. Gadgill and M.C. Mehta are the two living scientists of India who have worked on biodiversity. M.C. Mehta have been awarded by Megassase award in 1997.

Biodiversity of India:

Plants

Algae	= 5000 sp.
Lichens	= 1600 sp.
Fungi	= 20000 sp.
Bryophytes	= 2700 sp.
Pteridophytes	= 600 sp.
Flowering plants	= 15000 sp.

Animal

Insects	= 10000 sp.
Fishes	= 1193 sp.
Molluscs	= 4000 sp.
Reptiles and Amphibians	= 420 sp.
Birds	= 1200 sp.
Mammals	= 372 sp.

Out of total number of animal species, 33% reptiles and 62% amphibians are Endemic to India. There is very little knowledge about marine life, with regards to biodiversity of India. It is reported that about 42,000 plant species occur in India, which is 12% of total population.

India is treated as mega-diversity region and it is the meeting point of 3 great phytogeographical regions-

i) Endomalayan region
ii) Euratian region
iii) Afrotropical region

These regions are again divided into 16 Agroclimatic zones, 10 Vegetational zones and 25 Biotic provinces and 426 Biomes.

Why conservation is necessary?

As we know that all species has its own characteristic type of combination of gene pool. Once a species becomes extinct, that combination of gene pool is permanently lost and man can know more recover it for future use. Combination of gene pools of certain wild varieties has proved to be extremely useful for human society. Some examples are-

1. Before green revolution 60,000 varieties/species of paddy were reported from all over the world, out of which 30,000 were from India. But today we grow about 100 varieties/species all over the world which are high yielding varieties/species. In 1960s, viral disease attacked most of the high yielding varieties (**Grassy Stunt Viral disease**) by which vast area of paddy was badly

affected. Then resistant gene for this virus was discovered from a wild variety of paddy i.e. ***Oryza niviara*** from which this resistant gene was introduced in high yielding varieties and thus paddy yield was maintained.

2. A wild variety of melon collected from India provided genes for resistance to a mildew disease of melon crop as far away as in California (USA).

Causes of depletion of Biodiversity

A) According to IUCN (1978), there are 4 important causes behind the depletion of biodiversity. These are-

i) Modification of habitat
ii) Over-exploitation for commercial, scientific and educational purposes
iii) Diseases and
iv) Man made factor

B) Zeedyk (1978) again classified the causes of depletion of biodiversity, on the basis of IUCN. They are-

1. Natural threats

i) **Critically low population**- Some of the plant species have critically low population due that, they have maximum chances of extinction. Some of the examples of critically low population is-

Takakia – A rarest bryophyte, its only 100 (female) plants are found all over world.

Ginkgo biloba – A Gymnosperm, its only 50 species are found in India.

Taxus baccata – A bryophyte, found in Arunachal Pradesh. It is used in the treatment of cancer.

Ramosmania heterophylla – A tree of family Rubiaceae. It's only single tree is available in island of Indian ocean.

Diospyros hemiteles – Only one male plant surviving in Mauritus.

Paphiopedilum sp. – This is the orchid plant and its very few plants are available and its cost is 50,000 dollar/plant.

Sencio hadrosemus – Sencio is the largest genus, but its only 50 species are available all over world.

Dialiptera dodsonnii – Only single plant species is available in the world.

ii) **Lack of pollination**: The pollination takes place by water in some plants, for example- Coconut and mainly by winds e.g. flowering plants, while in some species it takes place by animals e.g. *Orchids* in which pollination takes place by *Ophyris*.

iii) **Invasion of Exotic species or Invasive or Alien species**: This is the second most important cause of depletion of biodiversity. Those species which come from outside are called as exotic species, for example-

Agrostemina githago (family- Caryophyllaceae) – Came from Europe.
Argemone maxicana (family- Papavaraceae) – Came from Mexico.
Lantana camara (family- Verbenaceae) – Came from Sri Lanka.
Parthenium hysterophorus (family- Asteraceae) – Came from America. Its first plant was grown in Pune in 1956 in India and its first plant was grown in U.P. in 1977.
Tribulus terrestris (family- Zygophyllaceae) – Came from Africa.
Verbena loraniencis (family- Asteraceae) – Came from America.
Xanthium strumarium (family- Asteraceae)- Came from America.
Alternethera philexeroides (family- Amaranthaceae) – Came from South America. First of all grown in Calcutta. It is dominated in West Bengal.
Eichhornea crassipes (family- Ponteridiriaceae) – Came from Brazil. Its first plant was grown in West Bengal (Narayangonj) in 1914.
Salvinia modleska (family- Salvinaceae) – Came from Africa.
Eucalyptus (family- Myrtaceae) – Came from Australia.
Casuarina (family- Casuarinaceae) – Came from Africa having single genera *Casurina*.

iv) **Air and water pollution**: Environmental pollution is one of the most important cause behind depletion of biodiversity. Air and water pollution takes place due to-

i) **Ozone depletion**– By CFCs and CO_2.
ii) **Acid rain**– Due to the oxides of sulphur and nitrogen.
iii) **Global warming**– Due to global warming, vegetation shift takes place towards the North pole. It has been estimated that with increase in 1°C temperature, the vegetation shift will be 100-150 km towards North. But in case of plants of high altitudes and Islands, where the chances of shifting is very less, plants gradually get extinct i.e. loss of biodiversity.

Water pollution causes Biomagnification or Bioamplification, which also depletes the diversity of organisms.

v) Flood, draught and earthquake
(For detail about ozone depletion, acid rain and global warming, see Chapter 10)

2. Man made threat:

i) **Destruction or modification of habitat**: This is most important cause of depletion of biodiversity. Habitat is modified by deforestation, urbanization, industrialization and sometime by this **habitat fragmentation** takes place which is not suited to most of the wildlife.

ii) **Over exploitation:** More use of plant and few animals is the main factor which depletes the diversity of organism. For e.g.-

Dipteraconpus terbinatus – found in Assam, used for commercial purposes.

Rauwolfia serpentina – This is a medicinal plant found in Meghalaya. 108 alkaloids have been reported from this plant.

Lycopodium sp.– A pteridophyte, its spores are used as gun powder. They are highly explosive. It is also used for medicinal purposes.

Nepenthes khasiayana – found in Meghalaya, used for educational purposes.

Panthera tigris (tiger) – It's skin is economically important. It is sold in the market @ 95,000 dollars/skin.

Rhinoceros – It's horn (Bunch of hair) used for medicinal purpose in China.

iii) Overgrazing

iv) Regeneration of scrub – Due to this, land becomes degraded.

v) Conversion into agriculture land –

Usable land	–	64,000 M. hectare
Forest area	–	31,000 M. hectare
Grass land area	–	1,800 M. hectare
Agricultural land	–	1,500 M hectare
Food crop	–	710 M. hectare

Due to monospecific monoculture of agriculture, other plants' species are gradually depleted, traditional rural practices and urbanization is other factor which depletes the biodiversity. It is observed that, 2 lac hectare land every year is depleted and converted into urban area and is responsible for depletion of biodiversity.

Red Data Book (RDB)

For wildlife preservation, the Red Data Book has been introduced. It gives a complete list of all endangered plant and animal species. In India about 253 species of wild fauna including mammals, birds and reptiles have so far been included in the Schedule I of Wildlife Protection Act. As regards flora, the subject has been under intensive study for the past few years and about 2000 species of flowering plants have been reported to be endangered. So far an inventory of 135 threatened species and subspecies of rare and endangered plants have been prepared by the Botanical Survey of India. The Red Data Book is expected to keep a full record of all endangered species.

Green Data Book (GDB)

It is also published by IUCN, include those species which are endangered but at present they are conserved.

vi) **Categories of threat**: Species Survival Committee of IUCN has classified depleted plants and animals in following categories on the following basis-

i) Present and past distribution
ii) Decline in number of population in course of time.
iii) Quality of natural habitat
iv) Biology and potential value of the species.

Following are the other categories –

i) **Extinct (Ex)**: Species which have no longer known to exist in living state, they are found only in the form of museum, photograph or herbarium e.g. Dodo.

ii) **Endangered (E)**: Plants and animals which have danger of extinction and whose survival is unlikely, if the causal factor continues operating. It's number is so less that any time they may move into extinct category or they are at the verge of the extinction.

In general, a species is considered as endangered, when it's natural regeneration is not possible due to exploitation or destruction or by natural or unnatural means. 19% mammals, 17% birds, 21% reptiles, 22% amphibians, 19% angiosperms are endemic to world. In India, their number is 54 animals and 113 plants, e.g., *Ailurus fulgens* (Red panda), *Bentinckia nicobarica*. Some other examples are Blue Whale, Largest Lemur Idri-idri of Madagascar, Asiatic Ass (*Asinus hemionus khur* now restricted to Rann of Kutch) and Lion tailed macaque etc.

Critically endangered: The taxon facing high risk of extinction and can become extinct any moment in the immediate future. Their number world wise is 425 animals and 1014 plants (10% mammals, 9% birds, 15% reptiles, 16% amphibians and 16% angiosperms). The number of critically endangered animals and plants in India is 18 and 14 respectively, e.g., *Sus salvinus* (Pigmy Hlog), *Berberis nilghiriensis* etc.

iii) **Vulnerable (Vu)**: The population is slightly more than the endangered and if the causal factor remains operating, first they will move in the category of endangered and then into the category of extinct. Out of the total threatened species, 34-51% are vulnerable (34% mammals, 36% birds, 43% reptiles, 48% amphibians and 51% angiosperms). In India, their number is 143 animals and 87 plants, e.g., *Antilope* (Black Buck), *Cupressus cashmeriana* etc.

iv) **Rare (R)**: Taxa with thin population are known as rare plants or animals. They are neither in the category of endangered nor vulnerable but they have the risk of thin population distributed in large or a limited area. When plants or animals are found in a localized area, they are known as Endemic, found in large or small area. For example- *Agedivecta indica, Ficus religiosa, Oscimum sanctum* are rare in India. Similarly, *Flacourtia jangemas* is endemic to **Gorakhpur**.

v) **Threatened (Th)**: When a taxa is known to be having thin population but it's exact category is not known, whether they are endangered, vulnerable or rare, then they are kept in the threatened category. Out of the total species, 37% mammals, 38% birds, 21% reptiles, 16% amphibians and 14% angiosperms come in this category. In India, their number is 109 animals and 73 plant species.

vi) **Out of danger (O)**: If they are having healthy population at present without any threat of extinction, they are called as, out of danger.

Conservation of Biodiversity: Out of millions of species, there is single species of **homosapiens** which is the main cause of depletion of biodiversity. We can conserve biodiversity by following ways-

i) Environmental education
ii) Environmental laws
iii) Creation of protected areas
iv) Conserving biodiversity in seed banks and germ plasm
v) Promoting green colour and wilderness
vi) Living with biodiversity.

i) Environmental Education:

The Environmental Education Conference was held at Belgrade in 1975 (called Belgrade charter) by UNESCO and then by UNEP (United Nation Education Programme) at Tblisi, USSR in 1977.

The most relevant to the subject is a Chinese perception about education, which says "If you plan for one year – plant rice; if you plan for 10 years – grow trees; and if you plan for 100 years – educate the people".

With acceptance of Tiwari Committee Report (1980), the country has realised the need for environmental education.

Objectives of Environmental Education

To educate children and general public towards:

1. **Awareness**: i.e. Acquire sensitivity to the total environment and its allied problems.

2. **Skill** : i.e. Acquire skills for identifying environmental problems.
3. **Knowledge**: To know conservation of natural resources.
4. **Evaluation ability**
5. **Altitude and participation**

Principal of environmental education:

- To consider environment in its totality (natural, artificial, technological, ecological, moral and aesthetic).
- To consider it continuous life process (from pre school to higher level).
- To be interdisciplinary in approach.
- To examine major environmental issues.
- To focus on current and potential environmental situations.
- To emphasise active participation in prevention and solution to problems.
- To develop critical thinking and problem solving skills.
- To promote cooperation in solving problems.
- To discover the symptoms and root cause of environmental degradation.
- To provide an opportunity for making decisions.

Environmental education among children:

A) Formal environmental education: The chief goal of environmental education in India must be-

- To improve the quality of environment
- To create awareness in people
- To create capabilities of decision making

The spectrum of EE (Environmental education) has 4 major interrelated components i.e. i) Awareness ii) Real life situation iii) Conservation and iv) Sustainable development.

Following are the main constituents of this education-

i) **Primary school stage**: Attempt should be made to sensitize children about environment. 75% emphasis should be given on awareness, followed by 20% on real life situation and 5% on conservation.

ii) **Lower secondary stage**: At this level objective must be, real life experience, awareness and problem identification. The quantum of awareness must decrease with the increase in real life situation. The contents should be supplemented with general science, teaching, practicals and field visits.

iii) **Higher secondary school stage**: The emphasis must be on conservation, assimilation of knowledge, problem's identification and action skills.

iv) **College stage**: Maximum emphasis should be given on knowledge regarding sustainable development based on experience with conservation. The content must be college based on Science and Technology. NCERT may play an important role at this stage.

v) **University education**: The university education has 3-major components- Teaching, Research and Extension. At postgraduate level, 4 major areas are recognized:-
   i) Environmental engineering
   ii) Conservation and management
   iii) Environmental health
   iv) Social ecology

B) Non formal environmental education: The education is designed for any age group, participating in cultural, social and economic development of the country. They form clubs, arrange exhibitions, public lectures and meetings etc. Following are the main constituents of this education:

i) **Adult education**

ii) **Rural youth and non-student youth**

iii) **Tribals and forest dwellers**

iv) **Children activities**: Department of Environment with the help of United school organisation of India, has organised essay competitions among different age groups of children. Short term courses are also oreganised by the **National Museum of Natural History (NMNH)**. NMNH conducts spot painting, modelling and poster design about environment for children.

v) **Ecodevelopment camps**: A set of guideline has been prepared by D.O.En (1984). The objectives are-
   - To create awareness in youth about basic ecological principles.
   - To find out root cause of ecological problems
   - To promote for solutions of ecological problems
   - To develop spirit of national integration

vi) **Non governmental organisations**: Near about 200 NGOs are involved.

vii) **Public Representatives**: For example- MPs and MLAs

viii) **Training executives**

ix) **Research and development programmes**

x) **Foundation courses**

xi) **Development of Educational materials and Teaching Aids**: Materials for media (TV, Radio, Films and News papers etc.) must be operated by competent manpower. India has a Centre for Environmental Education at Ahmedabad.

xii) **World Environment Day**: Celebrated on 5th June of each year.

C) Environmental Information: DOEn has set up a programme i.e. **Environmental Information System (ENVIS)** in 1982. It is a decentralised system, using distributed network of data bases for collection of environmental information.

Not only this but different leaders give different slogans to know about biodiversity. For example-

"This earth is enough for everyones need, but not for everyones greed".

Mahatma Gandhi

"Tree means water, water means bread, and bread means life".

K.M. Munshi
(First Agriculture Minister of India)

- Conserving indigenous knowledge, traditional culture, which are friendly to the environment. For example- *Ocimum sanctum* and **Neelkanth** (Magpie) etc. are saved because they are directly related from our tradition.

Chipko movement: The term 'Chipko movement' was first of all given by Chandi Prasad Bhatt and Vijay Datt Sharma on 27th March 1973.

The actual meaning of chipko is embressing or hugging the trees. The idea of chipko movement was actually adopted from "Khejari movement" occurred in 1730, in which about 363 people of Bishnoi community from 84 villages in the leadership of Amrita Devi laid down their life in effort to Khejari trees from men of Jodhpur maharaja.

Sir-Sundar Lal Bahuguna had presented a plan for this movement. The Chipko plan is infact a slogan of planting **5F-plant** i.e. food, fibre, fertilizer, fodder and fuel. In Karnataka, Chipko is known as **Apiko** and it is also known by other names in different states.

ENVIRONMENTAL LEGISLATION

Central and State governments implement different laws for the protection of environment:

A) Protection for specific animal and group of organisms

i) Madras Wild Elephant Protection Act 1873
ii) All India Wildlife Protection Act 1879
iii) Wild Bird and Animal Protection Act 1912
iv) Bengal Rhinoceros Protection Act 1932
v) Assam Rhinoceros Protection Act 1954
vi) Indian Wildlife Protection Act 1972 **Amended in 1983, 1986 and 1991**

B) Protection for wildlife including preservation of their natural habit

i) Bengal Smoke Nuisance Act 1905
ii) Indian Post Act 1907

iii) Motor Vehicle Act 1983
iv) Maharashtra Prevention of Water Pollution Act 1953
v) Orissa River Pollution and Prevention Act 1924
vi) Gujrat Nuisance Act 1963
vii) Water Prevention and Control of Pollution Act 1974 **Amendment in 1988**
viii) Air Prevention and Control of Pollution Act 1981
ix) Motor Vehicle Act 1988
x) Forest Conservation Act 1980 **Amendment in 1988**
xi) National Forest Act 1927
xii) National Forest Policy 1952
xiii) Environment Protection Act 1986
xiv) National Biodiversity Act 2002

C) Act regarding the right of people to healthy environment

i) Water Act 1977 **Amendment in 1991**
ii) Public Liability Insurance Act 1991

Creation of Protected Areas (PAs) for Wildlife Conservation

For the first time, the term wildlife was used in 1913 by **William Hornaday** (1913) in his book "**Our vanishing wildlife**". He was the Director of **New York Zoological Park**.

According to **Webster's Dictionary 1986- "Living things that are neither human nor domestic alead**". According to **Oxford's Dictionary- "The native flora and fauna of a particular area for protection of wildlife**". Central Board of Wild life was as organized, which is now known as IBWI.

There are two methods of conservation –

i) Ex-situ convervation
ii) In-situ conservation

1) Ex-situ conservation: In this type of conservation, plants and animal are conserved outside their natural habitat areas, for example- plants are grown in botanical garden and animals are protected inside the zoo etc.

Advantages of Ex-situ conservation

- Organisms are assured for food, shelter, water and security.
- In most of the endangered species, more offsprings are produced under ideal controlled condition.
- Genetic techniques (Biotechnology) could be used to improve, the species concerned.
- Insure the survival of endangered species
- By multiplying in zoo or garden, chances of survival of a species enhanced.

Disadvantages of ex-situ conservation:

- Wildlife is so-vast and varied, hence it is impossible to provide care and protection to a fraction of environmental collection.
- Maintenance and conservation in controlled condition (Ex-situ method) is very expensive.
- Genetic variations gradually become stagnant under human care. Following of the some examples of wildlife are conserved by ex-situ method e.g. **Hawai-goose or Nene**: In 1950, a single pair of Nene was located under protected area of Slimbridge England. Upto 1963, it's number was 230, while in 1978, it's number was around 1600 and it was again introduced in wild habitat.

  European Bison: In 1925, only few individuals about 40 were found in living state, they are protected by Ex-situ method and now a days its number is around 800.

  Pere Davids deer: First discovered in 1865. In 1900, it became totally extinct from nature and now a days more than 1000 species are available.

  Perz wask horse: This is rare South American animal, recently conserved by Ex-situ method.

  Some of the Indian examples are-

  Manipur deer: Only 17 deer were located in Calcutta and Delhi Zoo, now a days, it's number is 1050.

  Crocodile: It's number was also very less but now a days, they are more than 350 in different zoos.

  White-winged wood duck: It's number was also less than 200, but now a days, it's number is 350.

  Lion tailed macaque: It's native place is Western Ghat and its number was very less. Now a days it is conserved in Indian zoos.

2) In-situ Conservation: In this method, conservation is done inside the habitat area. For this, protected areas of different type are established. According to IUCN, PAS may be defined as, "An area of land and or sea specially delicated to the protection and maintenance of biological diversity and of natural and cultural resources managed by legal or other effective means".

According to IUCN, following of the important categories of PAS –

1) Strict nature reserves (Biosphere reserves (BR))
2) National Parks (NP)
3) Natural Monuments
4) Habitat and Species management
5) Protected land escapes
6) Managed resource protected area

Sanctuary (Sancts)

- First sanctuary in India was established in 1888 in Tamilnadu as **Vedathangal** bird's sanctuary.
- Sanctuary lack cruise area
- Established without Act of Parliament.

National Parks (NPs)

- Yellow stone NP situated in America is the first PAs in the world.
- National parks are divided into 3 zones-
    i) **Sanctum sanctorum**- Totally free from human interference
    ii) **Cruise Area**- Aisc no interference
    iii) **Buffer zone** In this tourism, forestry and other scientific activities takes place
- **NP established after Act of Parliament.**

Biosphere Reserves (BRs)

- **Biosphere Reserves established according to guidance of MAB 1972.**
- At present, there are 397 biosphere reserves in entire world, out of which 14 are in India. The first biosphere reserves in India is **Nilgiri Biosphere reserves** established in 1986 and latest is **Amarkantak** in 2005.

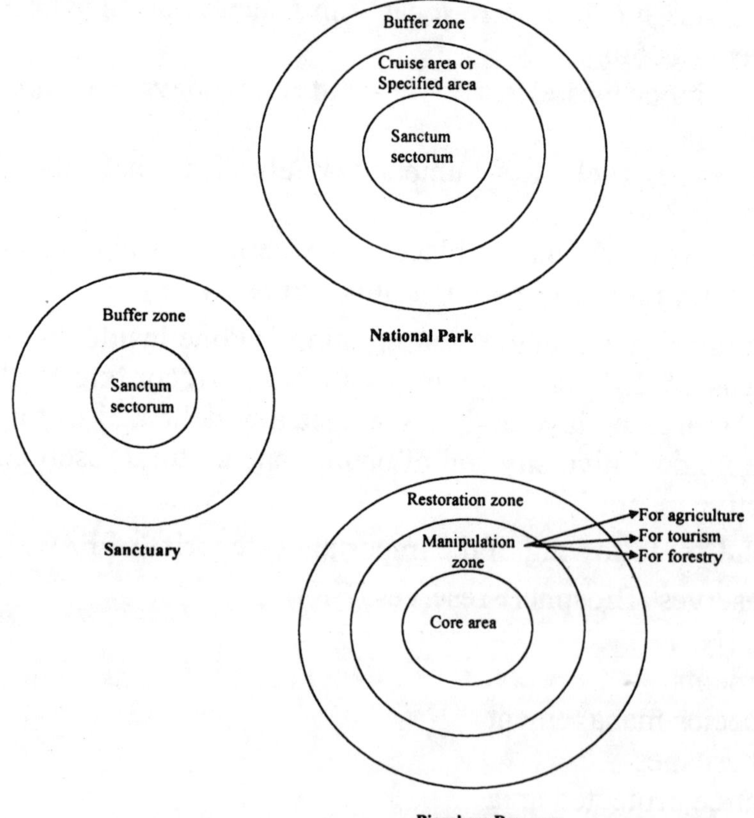

Important national parks and sanctuaries in India are-

Andaman and Nicobar Island-

1. Middle and Button Island National Park (Andaman)
2. Mount Harriet National Park
3. Marine National Park
4. North Button Island National Park (Andaman)
5. Saddle Peak National Park (Andaman)
6. South Button Island National Park (Andaman)
7. Narcondam Sanctuary
8. Barren Island Sanctuary (Bay of Bengal)
9. North Reef Island Sanctuary (Bay of Bengal)
10. South Sentinel Island Sanctuary (Bay of Bengal) + 90 sanctuaries

Andhra Pradesh-

1. Coringa Sanctuary (East Godawari)
2. Kawal Sanctuary (Adilabad)
3. Esturnagarm Sanctuary (Warangal)
4. Kolleru Sanctuary (West Godawari)
5. Kinnersani Sanctuary (Khammam)
6. Lanjamadugu Sanctuary (Adilabad and Karimnagar)
7. Majira Sanctuary (Medak)
8. Nagarjuna Sagar Srisailam Sanctuary (Guntoor Kurnool)
9. Pakhal Sanctuary (Warangal)
10. Neelapattu Sanctuary (Nellore)
11. Papikonda Sanctuary (East and West Godawari) + 5 sanctuaries

Arunchal Pradesh-

1. Moiling National Park
2. Itanagar Sanctuary
3. Namdapha National Park
4. Lali Sanctuary
5. Pakhui Sanctuary
6. Mehao Sanctuary

Assam-

1. Kaziranga National Park (Sibsagar and Nowgong)
2. Manas National Park (Barpeta)
3. Barandi Sanctuary
4. Garampani Sanctuary + 5 sanctuaries

Bihar-

1. Palamu National Park (Dalton Ganj)
2. Dalma Sanctuary

3. Bhimbandh Sanctuary
4. Nagi-Dam Sanctuary
5. Hazaribagh Sanctuary (Hazaribagh)
6. Nakti Dam Sanctuary
7. Rajaji Sanctuary + 9 sanctuaries

Chandigarh-

1. Sukhna Sanctuary

Delhi-

1. Indira Priyadarshani Sanctuary

Goa-

1. Bhagwan Mahavir National Park
2. Bondla Sanctuary
3. Bhagwan Mahavir Sanctuary
4. Cotigao Sanctuary

Gujrat-

1. Bansda National Park (Valsad)
2. Marine National Park (Jamnagar)
3. Gir National Park (Junagarh)
4. Velavadar National Park (Bhavnagar)
5. Dharamgadhra Sanctuary
6. Baroda Sanctuary
7. Nalsarovar Sanctuary (Ahemadabad)
8. Jessore Sanctuary
9. Shoolpaneshwar Sanctuary + 7 sanctuaries

Haryana-

1. Bhindawas Sanctuary
2. Sultan Sanctuary (Gurgaon)
3. Chotala Sanctuary + 4 sanctuaries

Himachal Pradesh-

1. Great Himalayan National Park
2. Pin Valley National Park
3. Chail Sanctuary (Chail)
4. Bandi Sanctuary
5. Lippa Asrang Sanctuary
6. Govind Sagar Sanctuary
7. Manali Sanctuary (Manali) + 24 sanctuaries

Jammu & Kashmir-

1. Dachigam National Park (Srinagar)
2. Hemis High Altitude National Park (Srinagar)

3. Kishtwar National Park (Srinagar)
4. Salim Ali National Park
5. Karakoram Sanctuary
6. Gulmarg Sanctuary
7. Nandini Sanctuary
8. Limber Sanctuary
9. Surinsar-Mansar Sanctuary + 8 sanctuaries

Karnataka-

1. Anshi National Park
2. Bandipur National Park (Mysore)
3. Kudremukh National Park
4. Bannerghatta National Park (Bangalore)
5. Cauvery Sanctuary
6. Nagarahole National Park (Coorg)
7. Puspagiri Sanctuary
8. Ghataprabha Sanctuary (Belgaon)
9. Ranganathittu Sanctuary (Mysore)
10. Sharavathi Valley Sanctuary
11. Shettihali Sanctuary + 12 sanctuaries

Kerala-

1. Eravikulam National Park (Idukki)
2. Silent Valley National Park (Palghat)
3. Periyar National Park (Idukki)
4. Nayyer Sanctuary
5. Idduki Sanctuary (Idduki)
6. Parambikulum Sanctuary
7. Peppara Sanctuary + 7 sanctuaries

Maharashtra-

1. Nawegaon National Park (Bandara)
2. Pench National Park (Nagpur)
3. Bar Sanctuary
4. Sanjay Gandhi National Park (Bombay)
5. Chandoli Sanctuary
6. Tadoda National Park (Chandrapura)
7. Chaprala Sanctuary
8. Great Indian Bustard Sanctuary
9. Nagzira Sanctuary
10. Koyna Sanctuary
11. Tansa Sanctuary + 18 sanctuaries

Manipur-

1. Keibul Lamjao National Park (Manipur Central)
2. Siroy National Park (Manipur East)

Meghalaya-

1. Balphakram National Park
2. Nokrek National Park
3. Siju Sanctuary
4. Baghmara Sanctuary
5. Nonglehyllem Sanctuary

Mizoram-

Dampa Sanctuary

Madhya Pradesh-

1. Fossil National Park
2. Banthavgarh National Park (Shahdol)
3. Indravati National Park (Bastar)
4. Kangerghati National Park
5. Kanha National Park (Mandla and Balaghat)
6. Panna National Park (Panna)
7. Madhav National Park (Shivpuri)
8. Pench National Park
9. Satpura National Park (Hoshangabad)
10. Sanjay National Park (Sidhi and Surguja)
11. Van Vihar National Park (Bhopal)
12. Bagdara Sanctuary
13. Gandhinagar Sanctuary
14. Bhairamgarh Sanctuary
15. Ghatigaon-Great Indian Bustard Sanctuary
16. Karera-Great Indian Bustard Sanctuary
17. Singhari Sanctuary
18. National Chambal Sanctuary
19. Tamor Pingla Sanctuary
20. Udanti Sanctuary + 22 sanctuaries

Nagaland-

1. Fakim Sanctuary
2. Intanki Sanctuary
3. Pulebatze Sanctuary

Orissa-

1. Simlipal National Park (Magurbhanj)
2. Lakhari Valley Sanctuary

3. Baisipalli (Mahanandi) Sanctuary
4. Balukhand Sanctuary
5. Debrigarh Sanctuary
6. Simlipal Sanctuary (Mayurbhanj) + 9 sanctuaries

Punjab-

1. Abohar Sanctuary
2. Bir-Mati-Bagh Sanctuary
3. Bir Bunnerheri Sanctuary
4. Bir Gurhiapura Sanctuary
5. Harike Lake Sanctuary

Rajasthan-

1. Desert National Park (Jaiselmer and Badimer)
2. Keoladaeo National Park (Bharatpur)
3. Ranthambore National Park (Savai Madhopur)
4. Sariska National Park (Alwar)
5. Mount Abu Sanctuary
6. Jawahar Sagar Sanctuary
7. National Ghariat Sanctuary
8. Sariksha Sanctuary (Alwar)
9. Ramgarh Sanctuary
10. Van-Vihar + 12 sanctuaries

Sikkim-

Khangchendzonga National Park (Gangtok) + 3 sanctuaries

Tamilnadu-

1. Guindy National Park (Madras)
2. Marine National Park
3. Kalakad Sanctuary
4. Anamali Sanctuary
5. Mudumalai Sanctuary (Nilgiris)
6. Pulcat Sanctuary + 6 sanctuaries

Tripura-

1. Gumti Sanctuary
2. Trishna Sanctuary
3. Rod Sanctuary
4. Sepahijala Sanctuary

Uttar Pradesh-

1. Jim Cordett National Park (National)
2. Dudhwa National Park (Lakhimpur Khiri)
3. Bihsar Sanctuary

4. Kedarnath Sanctuary
5. Nandadevi National Park (Chamoli)
6. Rajaji National Park
7. Valley of Flowers National Park (Chamoli)
8. Chandraprabha Sanctuary (Varanasi)
9. Hastinapur Sanctuary
10. National Chambal Sanctuary + 11 sanctuaries

West Bengal-

1. Neora Valley National Park
2. Gorumara Sanctuary
3. Singhlila National Park
4. Ballarpur Sanctuary
5. Sunderban National Park (24 Parganas)
6. Jalalapara Sanctuary (Jalpaiguri)
7. Narendrapur Sanctuary
8. Senctal Sanctuary + 12 sanctuaries

- Upto 2000, 87 National Parks and 497 Sanctuaries have been established in India.
- The maximum number of sanctuaries i.e. (94) and National park i.e. (11) are found in Andaman and Nicobar Islands and in Madhya Pradesh.

Hotspots

1. Hotspots are areas that are extremely **rich in species diversity**, have **high endemism** and are under **constant threat**.
2. The concept was developed by **Norman Myers** in 1988. Such spots are priority areas for *in situ* conservation. These areas are rich in diversity and high endemism.
3. India with 2.4% of land area accounts approximately 8% species of the world.
4. **25 terrestrial hotspots** have been identified globally, with an approximate area of 1.4%.
5. Among 25, **two hotspots (Western Ghats and Eastern Himalayas) are found in India**. Eastern Himalayas are active centre of evolution of many angiosperms and have many primitive angiosperms.
6. The 25 hotspots of world are as follows:
   1. Tropical Andes, 2. Mesoamerica, 3. Caribbean, 4. Brazil's Atlantic Forests, 5. Choco/Darien/Western Ecuator, 6. Brazil's Cerrado, 7. Central Chile, 8. California Floristic Province, 9. Madagascar, 10. Eastern Arc and Coastal Forests of Tanzania/Kenya, 11. West African Forests, 12. Cape Floristic Province, 13. Succulent Karoo, 14. Mediterranean Basin, 15. Caucasus, 16. Sundland, 17. Wallacea, 18. Philippines, 19. Indo-Burma, 20. South-Central

China, 21. Western Ghats/Sri Lanka, 22. South-West Australia, 23. New Caledonia, 24. New Zealand, 25. Polynesia, Micronesia.

7. The key criteria for determining a hotspot are:
   (i) Number of **endemic species**, i.e., the species which are found nowhere else.
   (ii) **Degree of threat**, which is measured in terms of habitat loss.
8. Tropical forests appear in 15 hotspots, Mediterranean-type zones in 5, and 9 hotspots are mainly or completely made up of islands. As many as 16 hotspots are in the tropics. About 20 per cent of the human population lives in the hotspot regions.

MASTER STROKES

Q. 1. Define the following terms.

(i) Point diversity
(ii) Alpha diversity
(iii) Gamma diversity
(iv) Epsilon diversity
(v) β-diversity
(vi) Delta diversity
(vii) Resource diversity
(viii) Habitat diversity
(ix) Differentiation habitat
(x) Organismic diversity

Ans. Types of Species Diversity:

(i) Point diversity: This is the diversity on the smallest scale, the diversity of a micro-habitat or sample taken from within a homogeneous habitat. It is but natural to expect difference in point diversity of the two sites of a habitat.

(ii) Alpha diversity: It is the second category of diversity. It is defined as within habitat diversity.

(iii) β diversity: This term was given by Whittaker (1960, 1977). It is a measure of how different or similar a range of habitats are in term of the variety of species found in them. It tells us how species diversity changes along gradient.

(iv) Gamma diversity: It is the third category of inventary diversity. It is defined as the diversity of a larger unit such as an island or landscape. It is considered to be the overall diversity of a group of areas of alpha diversity.

Gamma diversity = α diversity × β diversity

(v) Epsilon diversity: This is the fourth category of inventory diversity. The epsilon or regional diversity is defined as the total diversity of a group of areas of gamma diversity.

According to Magurran (1988), a single plant may be considered as a unit of α diversity, a leaf as an area of point diversity, a group of plant occurring together as an area of gamma diversity and the forest within which the plants are located as an area of epsilon diversity

(vi) Delta diversity: It is defined as the change in species composition and abundance between areas of gamma diversity, which occur within an area of epsilon diversity. It represents differentiation diversity over wide geographical areas.

(vii) Resource diversity: It means the diversity of resource that an organism utilizes. For example, some fish species in the hill streams have a wide trophic niche and depend on zooplankton, insects, algae and diatoms for their food.

(viii) Habitat diversity: It is the number of habitat types in a defined geographical area. This is an index which measure the structural complexity of a habitat. This structural complexity of environment, inturn, is responsible for the presence of a wide variety of spatial, trophic niches. This means that if any habitat supports more microhabitats, its biological diversity will be more as compared to a habitat which has less number of microhabitat.

(ix) Differentiation habitat: It is also called β diversity. It means degree of change in species composition beween sites or communities or along gradients.

(x) Organism diversity: In this, kingdom, phyla, families, genera, species, sub-species, population and individuals form a nested sequence, in which all elements at lower levels belong to one example of each of the elements at higher levels.

Q.2. What do you mean by ecosystem diversity?

Ans. Ecosystem diversity: It is variety of forms in the ecosystem due to diversity of niches, trophic levels, ecological processes like nutrient recycling, food webs, energy flow, role of dominant species, keystone species, various biotic interactions.

Diversity helps in producing more productive and stable ecosystems/ communities which can tolerate various stresses like prolonged drought.

ENTRANCE CORNER

1. Which of the following pairs of an animal and a plant represents endangered organisms in India? **(CBSE 2006)**
   (a) *Bentinckia nicobarica* and red panda
   (b) Tamarind and rhesus monkey
   (c) *Cinchona* and leopard
   (d) Banyan and black buck

2. Which of the following is considered a hotspot of biodiversity in India? **(CBSE 2006)**
   (a) Western Ghats (b) Indo-Gangetic Plain
   (c) Eastern Ghats (d) Aravalli Hills

3. Which one of the following is not included under *in situ* conservation? **(CBSE 2006)**
   (a) Sanctuary (b) Botanical garden
   (c) Biosphere reserve (d) National park

4. Which one of the following is the correctly matched pair of an endangered animal and a National Park? **(CBSE 2006)**
   (a) Lion - Corbett National Park
   (b) Rhinoceros - Kaziranga National Park
   (c) Wild ass - Dudhwa National Park
   (d) Great Indian Bustard - Keoladeo National Park

5. One of the most important functions of botanical gardens is that: **(CBSE 2005)**
   (a) they provide a beautiful area for recreation
   (b) one can observe tropical plants there
   (c) they allow *ex situ* conservation of germ plasm
   (d) they provide the natural habitat for wildlife

6. Biodiversity Act of India was passed by the Parliament in the year: **(CBSE 2005)**
   (a) 1996 (b) 1992
   (c) 2002 (d) 2000

7. Which one of the following pair is mismatched? **(CBSE 2005)**
   (a) Savanna - *Acacia* trees
   (b) Prairie - epiphytes
   (c) Tundra - permafrost
   (d) Coniferous forest - evergreen trees

8. What is most effective way to conserve plant diversity of an area? **(CBSE 2004)**
   (a) Tissue culture (b) Botanical garden
   (c) Biosphere reserves (d) Seed banks

9. Which group of vertebrates comprises highest number of endangered species? **(CBSE 2003)**
   (a) Mammals (b) Fishes
   (c) Reptiles (d) Birds

10. Which endangered animal is the source of the world's finest, lightest, warmest and most expensive wool-the shahtoosh? **(CBSE 2003)**
    (a) Nilgai (b) Cheetal
    (c) Kashmiri goat (d) Chiru

11. Plant genes of endangered species are stored in: **(CBSE 2000)**
    (a) gene library (b) gene bank
    (c) herbarium (d) none of these

12. Biosphere reserves are being threatened with: **(CBSE 2000)**
    (a) population growth (b) rains
    (c) pollution (d) all of these
13. Idri idri occurs in : **(CBSE 2000)**
    (a) India (b) Mauritius
    (c) Fiji (d) Madagascar
14. Best method to conserve genetic material of plants is : **(CBSE 1999)**
    (a) cold storage (b) tissue culture
    (c) seed storage (d) growing in natural habitat
15. Which communities are more vulnerable to invasion by outside plants and animals ? **(CBSE 1998)**
    (a) Tropical evergreen forests (b) Temperate forests
    (c) Mangroves (d) Oceanic island communities
16. MAB is : **(CBSE 1997)**
    (a) man and botany (b) man and biosphere
    (c) man and biotic community (d) man, antibiotic and bacteria
17. If long term basis ? **(CBSE 1996)**
    (a) Tribals living in these areas will starve to death \
    (b) Domestic animals in these and joining areas will die due to lack of fodder
    (c) Large biomes will become deserts
    (d) Crop breeding programes will suffer due to a reduced ability of variety of germplasm
18. Breeding place of flamingo (hansawar) in India : **(CBSE 1996)**
    (a) Chilka Lake (b) Sambar Lake
    (c) Rann of Kuchch (d) Ghana Vihar
19. Identify the correct match for tiger reserve and its state : **(CBSE 1995)**
    (a) Palamau - Orissa
    (b) Bandipur - Tamil Nadu
    (c) Manas - Assam
    (d) Corbett - Madhya Pradesh
20. Which one is matching sanctuary ? **(CBSE 1995)**
    (a)Gir - Lion
    (b) Kaziranga - Musk deer
    (c) Sunderbans - Rhino
    (d) N.E. Himalayan region - Sambar
21. Which Animal has become extinct from India ? **(CBSE 1994)**
    (a) Snow Leopard (b) Hippopotamus
    (c) Wolf (d) Cheetah

22. Ranthambore National Park is situated in : **(CBSE 1994)**
    (a) Maharashtra (b) Rajasthan
    (c) Gujarat (d) Uttar Pradesh
23. National Park associated with Rhinoceros is : **(CBSE 1994)**
    (a) Kaziranga (b) Ranthambore
    (c) Corbett (d) Valley of flowers
24. Black buck (Antilope cervicarpa) is an example of : **(AFMC 2006)**
    (a) endangered species (b) extinct species
    (c) vulnerable species (d) conservation dependent species
25. In which zone is limited human activity permitted ? **(AFMC 2002)**
    (a) Core zone (b) Buffer zone
    (c) Manipulation zone (d) Restoration zone
26. Keystone species deserve protection because these : **(AIIMS 2006)**
    (a) are capable of surviving in harsh environmental conditions
    (b) indicate presence of certain minerals in the soil
    (c) have become rare due to overexploitation
    (d) play an important role in supporting other species
27. In India, we find mangoes with different flavours, colours, fibre content, sugar content and even shelf-life. The large variation is on account of **(AIIMS 2006)**
    (a) species diversity (b) induced mutations
    (c) genetic diversity (d) hybridization
28. Biosphere Reserves differ from national parks and wildife Sanctuaries because in the former : **(AIIMS 2006)**
    (a) human beings are not allowed to enter
    (b) people are an integral part of the system
    (c) plants are paid greater attention than the animals
    (d) living organisms are brought from all over the world and preserved for posterity
29. One of the ex situ conservation methods for endangered species is : **(AIIMS 2005)**
    (a) wildife sanctuaries (b) biosphere reserves
    (c) cryoprservation (d) national parks
30. Genetic diversity in agricultural crops in threatened by : **(AIIMS 2005)**
    (a) introduction of high yielding varieties
    (b) intensive use of fertilizers
    (c) extensive intercropping
    (d) intensive use of biopesticides

31. If at high altitudes, birds become rare, the plants likely to disappear are : **(AIIMS 2004)**
(a) pine (b) orchids
(c) oak (d) rhododendrons

32. In case of extinction of Bengal tiger : **(AIIMS 2004)**
(a) wolves and hyenas shall become scarce
(b) wild areas will become safe
(c) gene pool will be lost forever
(d) population of deer and other herbivores will be stabilished

33. Which one of the following is a pair of endangered species ? **(AIIMS 2004)**
(a) Hornbill and Indian aconite (b) Indian peacock and carrot grass
(c) Garden lizard and mexican poppy (d) Rhesus monkey and sal tree

34. The map gives the former and present distribution of an animal. Which one it could be ? **(AIIMS 2003)** 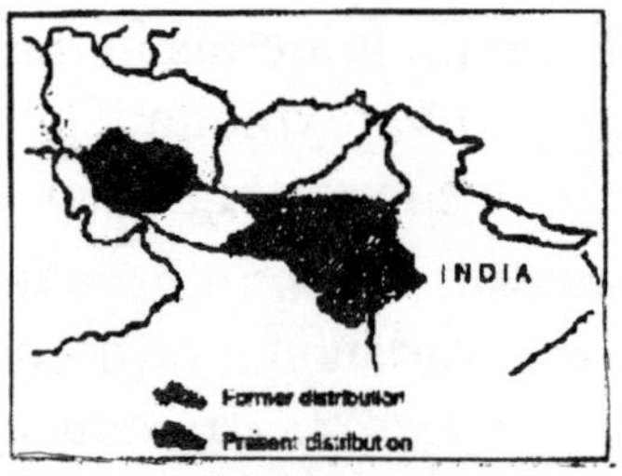
(a) Wild ass (b) Nilgai
(c) Black buck (d) Lion

35. Plant species on verge of extinction due to over-exploitation is : **(AIIMS 1998)**
(a) Centella (b) Podophyllum
(c) Gloriosa (d) All of these

36. Decrease in species diversity in tropical countries is mainly due to : **(AIIMS 1996)**
(a) urbanization (b) pollution
(c) deforestation (d) soil erosion

37. Bandipur (Karnataka) National Park is site of : **(AIIMS 1996)**
(a) deer project (b) peacock project
(c) elephant project (d) tiger project

38. IUCN, now called World Conservation Union (WCU) has its headquarter at : **(BHU 2006)**
(a) South Africa (b) America
(c) India (d) Switzerland

39. Maximum absorption of rainfall water is done by : **(BHU 2005)**
(a) tropical deciduous forest (b) tropical evergreen forest
(c) tropical savanna (d) scrub forest

40. Which of the following is not done in a wild life sanctuary ? **(BHU 2005)**
(a) Fauna is conserved (b) Flora is converved
(c) Soil and flora is utilized (d) Hunting is prohibited

41. The presence of diversity at the junction of diversity at the junction of territories of two different habitats is known as : **(BHU 2005)**
(a) bottle neck effect (b) edge effect
(c) junction effect (d) Pasteur effect

42. Which of the following species are restricted to a given area ? **(BHU 2004)**
(a) Sympatric species (b) Allopatric species
(c) Sibling species (d) Endemic species

43. Wildife conservation means the protection and preservation of : **(BHU 2004)**
(a) ferocious wild animals only
(b) wild plants only
(c) non-cultivated plants and non-domesticated animals
(d) all the above living in natural habitat

44. Species very near to extinction, if conservation measures are not promptly taken is : **(BHU 2003)**
(a) threatened species (b) rare species
(c) endangered species (d) vulnerable species

45. Which one is endangered member of flora ? **(BHU 2003)**
(a) Drosera indica (b) One horned rhino
(c) Flying squirrel (d) None of these

46. Kazirange National Park is located in the state of : **(BHU 2002)**
(a) Assam (b) West Bengal
(c) Kerala (d) Karnataka

47. In a National Park protection is provided to : **(BHU 2000)**
(a) entire ecosystem (b) flora and fauna
(c) fauna only (d) flora only

48. Tiger reserves in India for 'Project Tiger' are : **(BHU 1996)**
(a) more then 16 (b) 14-16
(c) 12-14 (d) 10

49. Which of the following causes a species to be different from other and maintain its own specificity ? **(Manipal 2003)**
(a) Micro-evolution (b) Genetic adaptability
(c) Macro-evolution (d) None of these

50. National 'Wildife Protection Act' was passed in : **(Manipal 2003)**
(a) 1962 (b) 1972
(c) 1982 (d) 1987

51. Related species living together comprise : **(Manipal 1999)**
(a) population (b) biotic community
(c) ecosystem (d) none of these

52. *Eichhornia crassipes* is a : **(Manipal 1997)**
(a) desert plant (b) parasite
(c) aquatic plant (d) terrestrial plant

53. A secies restricted to a given area is : **(CPMT 1999)**
(a) endemic species (b) allopatric species
(c) sympatric species (d) sibling species

54. Sal and teak is the dominant species in which of the following forests ? **(DPMT 2006)**
(a) Tropical dry deciduous forests (b) Temperate deciduous forest
(c) Temperate rainforest (d) None of the above

55. Biosphere reserves are different from national parks as : **(DPMT 2006)**
(a) plants and animals are protected in biosphere reserves
(b) humans are integral part of biosphere reserves
(c) humans are not involved in biosphere reserves
(d) none of the above

56. Hoolock gibbon (India's only ape) is found in : **(DPMT 2006)**
(a) Kaziranga Bird Sanctuary (b) Hazaribagh National Park
(c) Corbett National Park (d) Gir National Park

57. In situ conservation national genetic resources can be achieved by establishing : **(DPMT 2004)**
(a) National park (b) Wildife sanctuaries
(c) Bioshpere reserve (d) All of these

58. Rajaji National Park is situated in : **(DPMT 2003)**
(a) Tamil Nadu (b) Karnataka
(c) Uttaranchal (d) Rajasthan

59. The main reason for extinction of species is : **(DPMT 2003)**
(a) hunting (b) destruction of habitat
(c) pollution (d) none of these

60. Number of plant species reported in India are : **(DPMT 2003)**
(a) 45,000 (b) 40,000
(c) 80,000 (d) 58,000

61. Which one is endangered animal ? **(DPMT 2001)**
(a) Lion tailed macaque (b) Hanuman monkey
(c) Langoor (d) Antelope

62. What is the correct ascending order in respect of complexity of the following ?
1. Ecosphere 2. Species 3. Population
4. Community 5. Ecosystem **(DPMT 1998)**
(a) 1, 2, 3, 4, 5 (b) 3, 2, 4, 5, 1
(c) 1, 3, 2, 4, 5 (d) 2, 4, 3, 1, 5

63. Animal species protected in Kaziranga Nationa Park is : **(Karnatakd 2004)**
(a) Panthera leo (b) Rhinoceros unicornis
(c) Panthera tigris (d) Macaca mulatta

64. Khaziranga wildlife sanctuary is famous for : **(Karnatakd 2004)**
(a) tiger (b) musk deer
(c) elephant (d) rhino

65. The organization which has published 'Red Data Book' is : **(Karnatakd 2002)**
(a) International Union for Conservation of Nature and Natural Resource
(b) National Environmental Engineering Research Institute
(c) National wildlife Action Plant
(d) Convention on International Trade in Endangered Species of Wild Fauna and Flora

66. New approach to conservation is establishment of : **(Karnatakd 1999)**
(a) reserve forests (b) sanctuaries
(c) biosphere reserves (d) national park

67. Siberian Crane is a regular visitor of bird sanctuary : **(Karnatakd 1996)**
(a) Ranganathittoo (b) Bharatpur (Rajasthan)
(c) Vedanthgol (Tamil Nadu) (d) Lalbagh (Karnataka)

68. Conservation and maintenance of wildlife within the natural ecosystem is : **(Pb. PMT 2006)**
(a) In situ conservation (b) Ex situ conservation
(c) Botanical gardens (d) All of these

69. Which of the following is most dangerous to wild life ? **(Pb. PMT 2000)**
(a) Over exploitation (b) Man made forest
(c) Habitat destruction (d) Introduction of foreign species

70. Wildlife is conserved : **(Pb. PMT 2000)**
(a) in situ (b) ex situ
(c) both (a) and (b) (d) selective hunting of predators

71. Black buck is not allowed to be hunted by : **(Pb. PMT 1999)**
(a) Bishnois (b) Bhils
(c) Ahirs (d) Jats

72. Which of the following has high extinction prospects ? **(Pb. PMT 1998)**
(a) Himalayan bear and musk deer (b) Lion and leopard
(c) Tiger and bustard (d) Crocodile and elephant

73. Which of the following will help most in conservation of wildlife ? **(Pb. PMT 1998)**
(a) Making stringent laws (b) Making numerous zoos
(c) Making numerous sanctuaries (d) All of the above

74. The Environment Protection Act was passed in : **(HP PMT 2005)**
(a) 1968 (b) 1974
(c) 1981 (d) 1986

75. Which one is endangered species ? **(Haryana PMT 2000)**
(a) Cuscuta (b) Acacia nilotica
(c) Nepenthes (d) Both (b) and (c)

76. Dachigam sanctuary is associated with conservation of: **(Haryana PMT 2000)**
(a) hangul (b) rhino
(c) barking deer (d) both (b) and (c)

77. Animals and plants are best protected in : **(Haryana PMT 1994)**
(a) botanical gardens (b) zoos
(c) national (d) sanctuaries

78. National park and sanctuary has the common characteristic as : **(AMU 2006)**
(a) boundries are circumscribed by state legislation
(b) there is biotic interference
(c) tourism is permissible
(d) research and scientific management is possible

79. State bird of Rajasthan is : **(AMU 2005)**
(a) siberian crain (b) great Indian bustard
(c) flamingo (d) hornbill

80. Which one is not endangered ? **(AMU 2003)**
(a) Asiatic wild ass (b) Idri idri
(c) Lion tailed macaque (d) Addax antelopes

81. An endangered bird is : **(AMU 2001)**
(a) passenger pigeon (b) pink-headed duck
(c) great Indian bustard (d) vulture

82. One of the following is in endangered list : **(AMU 1998)**
(a) Pava (b) Felis
(c) Choriotis (d) Hemidactylus

83. One of the following is an endangered plant : **(AMU 1998)**
(a) Lycopersicon (b) Dalbergia
(c) Cedrus (d) Rauwolffia

84. 'Red Data Book' provides data on : **(AMU 1997)**
(a) biota of red sea (b) effect of red light or photosynthesis
(c) red pigmented plants (d) threatened species

85. Which one of these is an in situ method of conservation ? **(Kerala PMT 2005)**
(a) National Park (b) Botanical garden
(c) Tissue culture (d) Genetic engineering

86. The first biosphere reserve established in India for conserving the gene pool of flora and fauna and the life style of tribals is : **(Kerala PMT 2003)**
(a) Nilgiri biosphere reserve (b) Nandadevi biosphere reserve
(c) Uttaranchal biosphere reserve (d) great Nicobar biosphere reserve

87. The taxa believed likely to join the endangered category in near future is called : **(Kerala PMT 2003)**
(a) extinct (b) rare
(c) vulnerable (d) out of danger

88. World Wildlife week is observed during : **(Kerala PMT 2002)**
(a) first week of October (b) last week of October
(c) third week of October (d) first week of September

89. Red Data Book contains information about : **(Kerala PMT 2002)**
(a) red coloured insects (b) red eyed birds
(c) red coloured fishes (d) endangered plants and animals

90. WWF-N has logo for conservation of : **(Kerala PMT 2002)**
(a) tiger (b) giant panda
(c) red panda (d) polar bear

91. Animals like cockroach, lizard and mica share buildings of human dwellings. Such animals are : **(Kerala PMT 2001)**
(a) transgenic (b) transplants
(c) inquilines (d) cultigens

92. The management of biosphere in such a way that it may yield one greatest suitable benefit to present generation while maintaining its potential to meet the needs of future generation is : **(CET Chd. 2003)**
(a) Cynodon dactylon (b) Digitaria purpurea
(c) Cymbopogon citratus (d) Parthenium hysterophorus

93. Congress grass is : **(CET Chd. 1997)**
(a) *Cynodon dactylon* (b) *Digitaria purpurea*
(c) *Cymbopogon citratus* (d) *Parthenium hysterophorus*

94. Asiatic lion is protected in : **(MP PMT 2006)**
(a) Kaziranga National Park (b) Gir National Park
(c) Kanha National Park (d) Desert National Park

95. The most biodiversity rich zone in India: **(MP PMT 2005)**
(a) Gangetic planes (b) Trans Himalayas
(c) Western Ghats (d) Central India

96. From the point of view of *Rhinoceros* reserve which one of the following is correct? **(MP PMT 2004)**
(a) Corbett - Punjab
(b) Palamou - Orissa
(c) Kaziranga - Assam
(d) Nandan Kanan - Rajasthan

97. In India the *Rhinoceros* is the most important protected species in: **(MP PMT 2003)**
(a) Dachigam Naitonal Park (b) Kaziranga National Park
(c) Sunderbans National Park (d) Dudhwa National Park

98. Which step is required for better survival of human being ? **(MP PMT 2001)**
(a) Conservation of wilflife
(b) Afforestation
(c) Ban on mining
(d) Reduced utilization of resources

99. Dudhwa National Park is located in : **(MP PMT 1998)**
(a) Gujarat (b) Uttar Pradesh
(c) Rajasthan (d) Madhya Pradesh

100. The number of individuals of a species in a particular ecosystem at a given time remains constant due to: **(MP PMT 1998)**
(a) human beings (b) parasites
(c) predators (d) food availability

101. Kanha National Park is located in : **(MP PMT 1997)**
(a) Assam (b) Rajasthan
(c) Uttar Pradesh (d) Madhya Pradesh

102. Which one is a troublesome weed of agricultural fields? **(MP PMT 1996)**
(a) *Chenopodium album* (b) *Cyperus rotundus*
(c) *Parthenium hysterophorus* (d) *Eichhornia crassipes*

103. Jim Corbett National Park is known for: **(MGIMS Wardha 2001)**
(a) lions (b) tigers
(c) black buck (d) rhino

104. Widllife is : **(Orissa JEE 2005)**
(a) all biota excluding man, domestic animals and cultivated crops
(b) all vertebrates of reserve forests
(c) all animals of reserve forests
(d) all animals and plant of reserve forest

105. Species listed in Red Data Book are : **(Orissa JEE 2004)**
(a) vulnerable (b) threatened
(c) endangered (d) all of these

106. Similipal is : **(Orissa JEE 2004)**
(a) sanctury (b) biosphere reserve
(c) national park only (d) zoo

107. What is true of national park ? **(Orissa JEE 2004)**
(a) Tourism is allowed in buffer zone
(b) No human activity is allowed
(c) Cattle grazing is allowed in buffer zone
(d) Hunting is allowed in core zone

108. Asiatic lion is : **(Orissa JEE 2003)**
(a) extinct from wild (b) rare
(c) vulnerable (d) endangered

109. A threatened species is : **(Orissa JEE 2003)**
(a) only endangered species
(b) only vulnerable species
(c) endangered and rare species
(d) endangered, vulnerable and rare species

110. Ex situ conservation is carried out in : **(Orissa JEE 2003)**
 (a) sanctuary (b) national park
 (c) biosphere reserve (d) zoo

111. National wildlife Protection Act was formulated during : **(Orissa JEE 2003)**
 (a) 1972 (b) 1974
 (c) 1976 (d) 1978

112. Biosphere reserve project was started in India during : **(RPMT 1996)**
 (a) 1984 (b) 1985
 (c) 1986 (d) 1989

113. First national Park of India is : **(CMC 2002)**
 (a) Kanha National Park (b) Periyar National Park
 (c) Corbett National Park (d) Bandipur National

114. Mark the national tree : **(JIPMER 2002)**
 (a) Ficus religiosa (b) Mangifera indiaca
 (c) Ficus bengalensis (d) Azadirachta indica

115. Red Data Book is maintained by : **(BVP Pune 2001)**
 (a) IUCNNR (b) The Bombay Natural History Society
 (c) WPSI (d) IUCN

ANSWER SHEET

1. (a)	**2.** (b)	**3.** (b)	**4.** (b)	**5.** (c)	**6.** (c)	**7.** (b)	**8.** (c)	**9.** (b)	**10.** (d)
11. (b)	**12.** (c)	**13.** (d)	**14.** (b)	**15.** (d)	**16.** (b)	**17.** (c)	**18.** (a)	**19.** (a)	**20.** (a)
21. (d)	**22.** (b)	**23.** (a)	**24.** (c)	**25.** (c)	**26.** (a)	**27.** (c)	**28.** (v)	**29.** (c)	**30.** (a)
31. (d)	**32.** (c)	**33.** (a)	**34.** (a)	**35.** (b)	**36.** (c)	**37.** (d)	**38.** (s)	**39.** (b)	**40.** (c)
41. (b)	**42.** (d)	**43.** (d)	**44.** (c)	**45.** (c)	**46.** (a)	**47.** (b)	**48.** (a)	**49.** (a)	**50.** (b)
51. (b)	**52.** (c)	**53.** (a)	**54.** (a)	**55.** (b)	**56.** (b)	**57.** (d)	**58.** (c)	**59.** (b)	**60.** (a)
61. (a)	**62.** (d)	**63.** (v)	**64.** (d)	**65.** (a)	**66.** (c)	**67.** (b)	**68.** (a)	**69.** (c)	**70.** (c)
71. (a)	**72.** (a)	**73.** (c)	**74.** (d)	**75.** (c)	**76.** (a)	**77.** (c)	**78.** (c)	**79.** (b)	**80.** (d)
81. (c)	**82.** (c)	**83.** (d)	**84.** (d)	**85.** (a)	**86.** (a)	**87.** (c)	**88.** (a)	**89.** (d)	**90.** (c)
91. (c)	**92.** (b)	**93.** (a)	**94.** (b)	**95.** (c)	**96.** (c)	**97.** (b)	**98.** (d)	**99.** (b)	**100.**(c)
101.(d)	**102.**(c)	**103.**(b)	**104.**(b)	**105.**(d)	**106.**(b)	**107.**(b)	**108.**(d)	**109.**(d)	**110.**(d)
111.(a)	**112.**(c)	**113.**(c)	**114.**(c)	**115.**(d)					

Chapter 10
Environmental Pollution

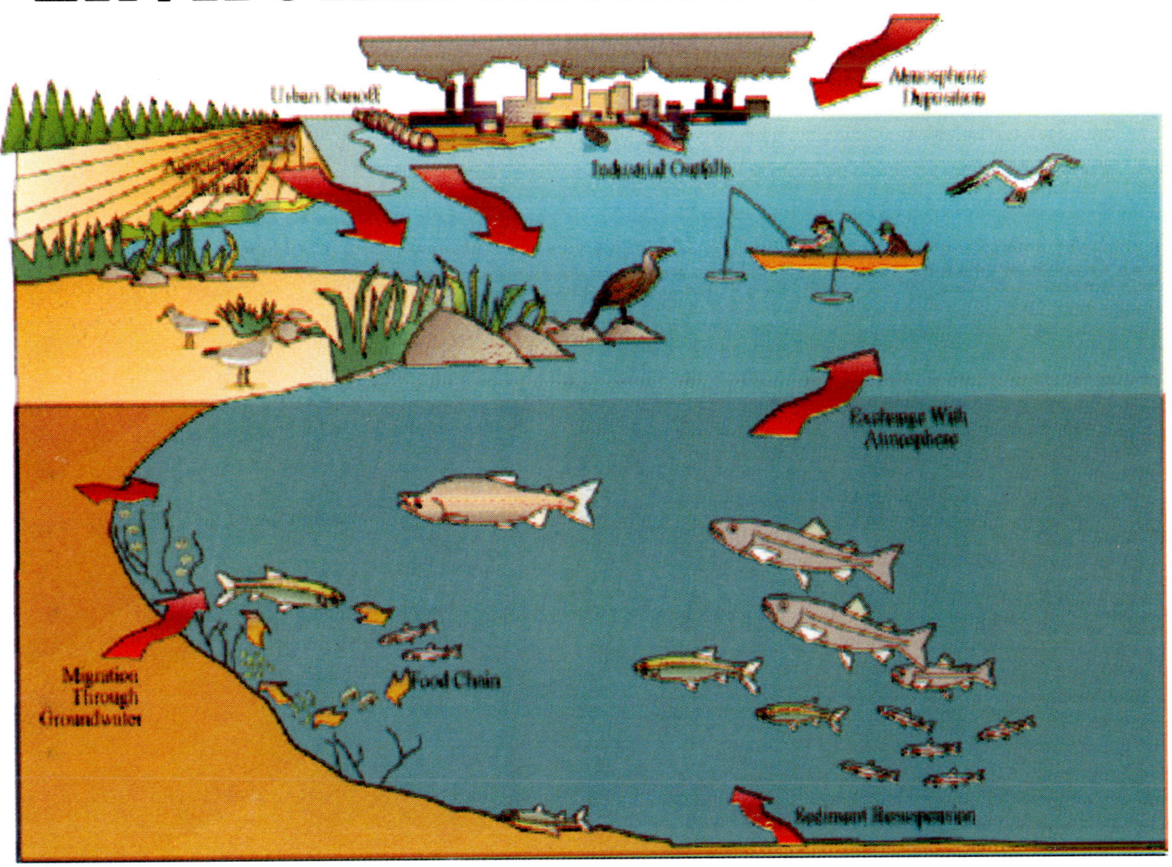

ENVIRONMENTAL POLLUTION

UNIVERSE

The vast space surrounding us is called **Universe**. The universe includes everything that exists- the most distant starts, planets, satellites, as well as our own earth and all the objects on it. The study of universe is called **astronomy** and experts as **astronomers**. The universe is made up of two types of bodies- first one are those which give off their own light (like sun and stars) and second one are those which do not give their own light (like planets and satellites). The position of sun is fixed. Those which revolve around the sun in a definite pathway (i.e. orbit) are called **planets** and those which revolve around the planets are called **satellites** (like Asteroids, Meteors, Meteorites, Comets, dust particles, gases). Our planet earth is one of the nine planets which move round the sun. The sun, planets and their satellites (or moons) form the solar system.

Solar system

Our solar system is just a tiny part of universe made up of billions of stars and vast areas of empty space. The sun is at the centre of the solar system and all these bodies are revolving around it. It holds them with a gravitational pull. In other words, the motion of all the members of solar system is governed by the gravitational force of the sun. The sun accounts for almost 99.9% of the matter of the whole system.

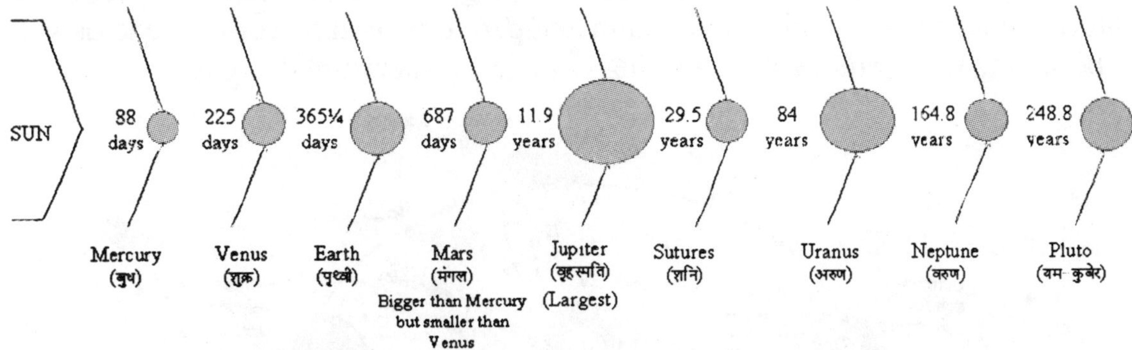

Fig. Planets of the solar system

Comparative position of sun and insolation

Heating due to solar energy or insolation depends on the inclination of the earth's axis and distance from the sun, while distance from the sun varies during the year, its effect is relatively small compared to that produced by the inclination of the earth's axis. Latitude plays most significant role in shaping weather and climate due to the angle of the sun-rays as well as the duration of sunlight.

The earth is inclined by 23½° and during half of the year the northern hemisphere is inclined toward the sun, while during the other half, it remains away from the sun. The amount of energy received per unit area of earth surface is determined by the angle at which sun-rays strike the earth. This angle varies place to place due to spherical shape of the earth and the inclination of its axis (as shown in figure).

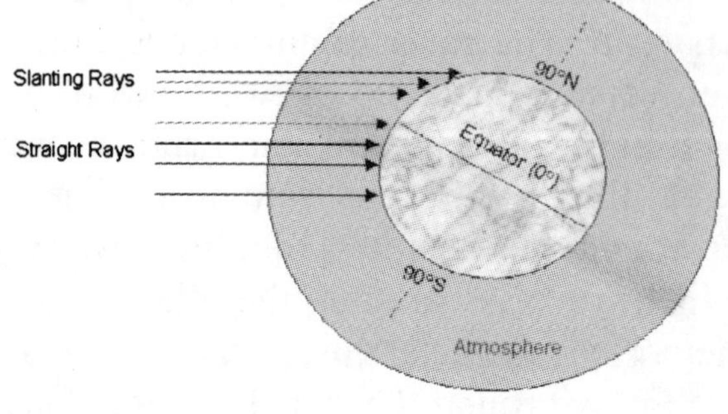

Fig. Rotation of earth on its axis and unequal insolation of solar energy

Distribution of insolation is different at different latitudes. Equatorial regions are characterized by 12 hour's day and 12 hour's night, this part receives maximum insolation round the year. Near polar regions, minimum insolation is received even during long light period of summer due to lesser angle of sun-rays with respect to the surface. It is also noteworthy that although equatorial region receives maximum insolation due to vertical rays but at these latitudes atmosphere remain cloudy, which reflects solar energy and consequently lesser energy is gained. This is the reason why equatorial line do not exhibit maximum temperature of the world. Conclusively it may be said that as a rule with increasing latitude, temperature decreases.

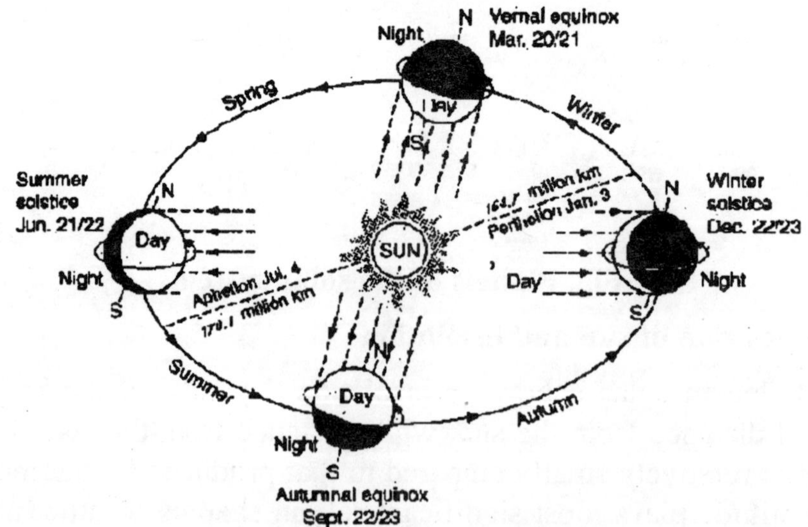

Fig. Earth's rotation around the sun and change of seasons

The earth rotates about its axis eastwards at a speed of **465** meters per second at the equator. The time interval of one complete round is **23** hours **56** minutes and **4.09** seconds. The rotating earth revolves round the sun in an elliptical orbit in which one complete revolution takes **365** days, **5** hours, **48** minutes and **45.51** seconds. This is a **tropical year**. The excess of nearly **6** hours (**1/4** day) is adjusted every four years as the extra day on **29th** February in the **leap year**. The plane of earth's orbit is **elliptic**. Earth's rotation bring seasonal changes as illustrated in the above figure. **September 22** and **March 23** are the two positions at the equator, when the days and nights are **equal** and thus they are called as **equinoxes**. The two extreme declination, **23.5°** north and south, are the **solstices**. When the northern hemisphere is inclined towards the sun during the northern solstice (**21 June**), it is summer there with long days and short nights. The opposite is the case when southern hemisphere is inclined towards the sun during the southern solstice (**22 December**). On 3rd January earth is nearest to the sun, it is called as **Perihelion**, while on **4th July** earth is at a maximum distance from the sun, it is called as **Aphelion**.

Longest Day (or Night) duration and latitude

Latitude	0°	17°	41°	49°	63°
Duration of day	12 hrs.	13 hrs.	15 hrs.	16 hrs.	20 hrs.
Latitude	66½°	67°21′	69°51′	78°11′	90°
Duration of day	24 hrs.	1 year	2 years	4 years	6 years

CONCEPT OF ENVIRONMENT

Literally, environment means "Surrounding an object". Term environment is large magnified and diversified word which indicates that, all natural resources are included in it. Usually environment of earth's planet exists in solar system because life is only possible on earth. It has optimum temperature and distance from sun is also optimum. "Earth is neither too hot as Venus and Mercury nor too cold as Jupiter". Environment consists of physical elements and biological elements. The physical elements are made up of air, water, snow, ice, sand, soil, rocks and mountain etc. whereas biological elements are plants, animals and microorganisms. Both physical and biological elements are dynamic in nature.

Components of environment: There are two segments of environment-

1) Abiotic components
2) Biotic components

1) Abiotic components: The non-living components of environment is called abiotic component. This includes air, water, soil, rocks and mountain etc. The abiotic component consists of following sub components-

1) Atmosphere
2) Lithosphere
3) Hydrosphere

1) Atmosphere: The gaseous segment of environment is called as atmosphere. It is component of air environment, which is just above the earth surface. Atmosphere consists of 5-layers-

i) Troposphere
ii) Stratosphere
iii) Mesosphere
iv) Ionosphere or Thermosphere
v) Exosphere.

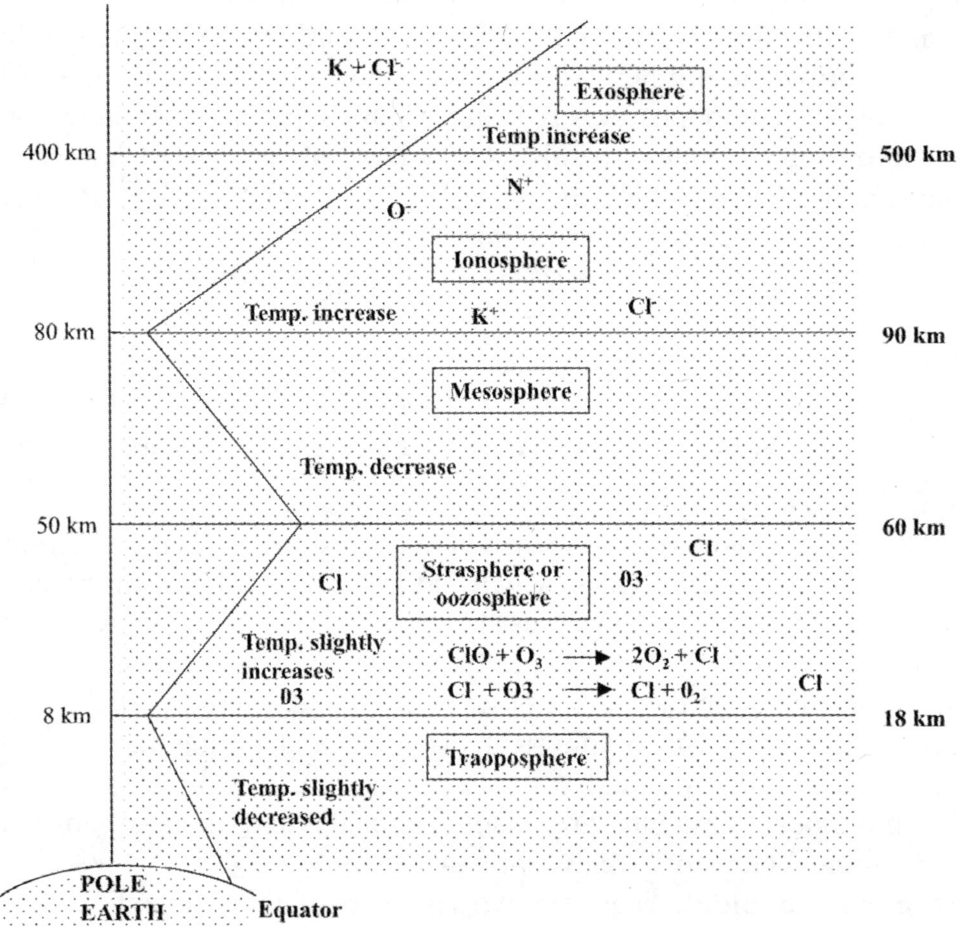

Fig. showing different layers of atmosphere

i) **Troposphere**: It is first layer of atmosphere. It extends 8 km from pole and 18 km from equator of earth. The upper limit of troposphere is known as **Tropopause**. Tropo means changes i.e. changes taking place in the climate. In troposphere, life saving gases such as CO_2, O_2, N_2, H_2O vapour, He, Ne, Ar, are present. Due to presence of life saving gases, life only exists in the troposphere. All kinds of plants, animals alongwith human-beings exist in the troposphere.

In troposphere, temperature slightly decreases with the increase in height. This decrease is about 0.65°C per 100 metres. This rate of decrease in temperature is called as **lapse rate**. Cloud formation, weather phenomenon and precipitation of rain takes place in troposphere. Different kinds of Aeroplanes and Aircrafts successfully fly in the troposphere of Atmosphere.

ii) **Stratosphere**: Just above the troposphere, stratosphere lies in atmosphere. It extends about 50 km from pole of earth and about 60 km from equator of earth. Temperature slightly increases with the increase in height in the stratosphere. The upper limit of stratosphere is known as **Stratopause**. Ozone gas is formed in the stratosphere after reaction with O_2 molecule and O atom with the help of photon of light

$$O_2 + O \xrightarrow{h\nu} O_3 \text{ (ozone)}$$

Due to presence of ozone gas, stratosphere is also called as **oozosphere**. Ozone gas of stratosphere act as a "**protective umbrella**", which absorbs the harmful UV-rays. UV-rays are harmful and injurious to health of human-beings and other biota. It can cause hazardous effect on the life of troposphere. In this way, stratosphere protects the life of troposphere.

iii) **Mesosphere**: Just above the stratosphere, mesosphere lies in the atmosphere. It extends 80 km from pole of earth and 90 km from equator of earth. The gases in free condition are absent in mesosphere, hence temperature rapidly decreases in the mesosphere with the increase in height. The upper limit of mesosphere is called mesopause. The lowest temperature occurs in the mesopause region of mesosphere.

iv) **Ionosphere**: Just above the mesosphere, ionosphere lies in the atmosphere. It extends 400 km from the pole of earth. Temperature rises in the ionosphere, with the increase in height. Due to high temperature, ionosphere is also called **Thermosphere**. The upper limit of ionosphere is called as **ionopause**. Different types of ions and particles are present in the ionosphere, which are responsible for the transmission of radiowaves of satellites. Artificial satellite is launched in the region of ionosphere. Microwaves, Laser rays and different kinds of websites communications are conducted in the ionosphere.

v) **Exosphere**: Just above the ionosphere, the region of atmosphere is known as exosphere. It extends about 1600 km to 1700 km from pole of earth. The temperature increases with the increase in height. In the upper region of exosphere, highest temperature is observed.

2. Lithosphere: The solid part of the environment is known as lithosphere. The centre of peripheral region of the earth along with soil and rocks are included in the lithosphere. Mountains are the modified form of the rocks which are also included in the lithosphere. Soil is formed from the rocks by the process of weathering and paedogenesis. In the region of upper crust of earth, each is differentiated into the following three regions (Basis of lithosphere):

(i) Core – contains Zn, Cd, Ni, Fe
(ii) Mantle – contains Quartz, Felspar, Mica, $CaCO_3$, $CaPO_4$ etc.
(iii) Crust

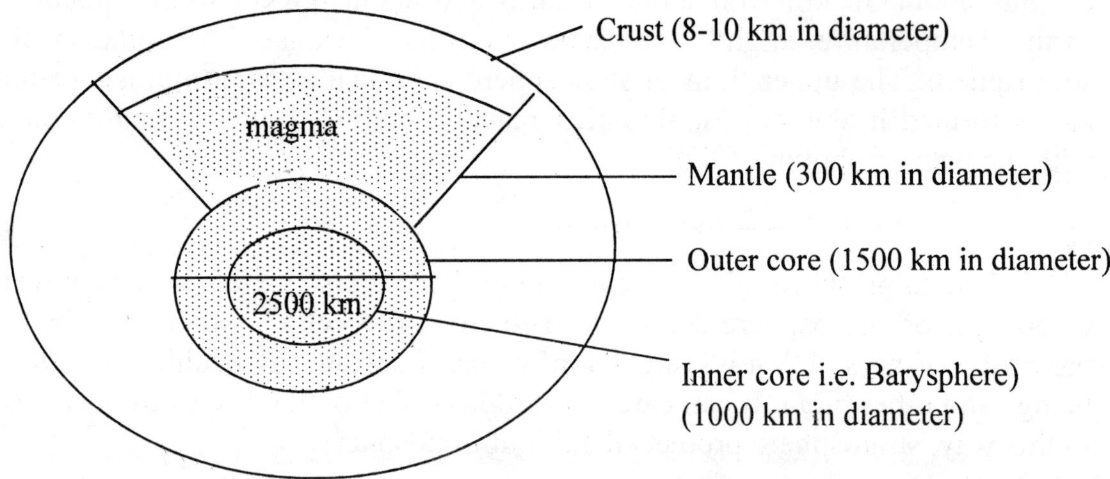

Fig. showing different regions of earth's crust

Due to heating and pressure, the weak region of crust breaks and **magma** of **mantle** is released outside in form of **volcanoes**. Magma of volcano is very hot, fluid like substance which is made up of different kinds of mineral salts. After cooling and solidification, magma of mantle is settled at a particular place in form of rocks. This rocks is known as **primary rock** or **igneous rock**. When primary rock breaks into the small segments and deposited on roof of water bodies in form of different layers are called **sedimentary rocks**. Such rocks are also known as **stratified rocks**. Igneous rocks and sedimentary rocks are **Metamorphic rocks** formed after metamorphosis.

There are two major and significant parts of crust of the earth-

a) Soil
b) Rocks

a) Soil: "Uppermost part of crust, which is fertile and in which plants successfully grow is called soil". Soil is the fertile part of the crust which is made up of minerals and humus and able for the succession of plant. The branch of biological science dealing with the study of soil is called **"PEDOLOGY"**. Soil formation is very complicated and time taking process. It is developed from the rocks by the process of weathering and paedogenesis. The formation of soil from the rocks is also known as "**Soil genesis**".

$$\text{Rocks} \xrightarrow{\text{Weathering}} \text{Small particles} \xrightarrow{\text{Pedogenesis}} \text{Soil}$$

i) **Mineral matters**: It is about 40% part of soil in which N, P, K, Ca, Mg, Fe, Zn and Ba etc. are found.

ii) **Organic matters**: It is about 10% of soil, after decomposition the organic matters changes into the humus. It responsible for the fertility of soil.

iii) **Soil water**: It is about 25% of soil. There are 4 types of soil water i.e. run away water, hygroscopic water, capillary water and gravitational water. Among them plants only absorb the capillary water.

iv) **Soil water**: It is also 25% of soil, responsible for aeration.

b) Rocks: Rocks are the significant part of Lithosphere, which consists of various kinds of minerals and mainly form by the magma of earth, which is erupted by means of volcano. Some rocks are formed after the modification of primary rocks. Rocks are of 3-types -

1) Igneous rocks
2) Sedimentary rocks
3) Metamorphic rocks

1) Igneous rocks: They are also called primary rocks and are formed by magma of erupted volcano after cooling and sedimentation. Igneous rocks usually occur at those places, where volcanoes are developed. They are non commercial, unspecific, unorganized rocks which consist of nickel, zinc, cobalt, chromium, iron, clay and sand etc. Volcano is erupted, when pressure increases in the mantle region of earth and site of volcano i.e. crust becomes weak. **Basalt** and **granite** are the classical examples of igneous rocks.

2) Sedimentary rocks: These are also called **Stratified rocks**, because they consist of different layers. The magma of erupted volcano, after cooling sediments on the floor of water bodies or roof of water bodies in form of different layers form the sedimentary rocks. Sedimentary rocks occur on lakes, rivers, ocean and springs etc. Water is released from the sedimentary rocks. It is origin of various rivers. Fossils are deposited in the sedimentary rocks. They usually consist of Ca, Zn, Co, Cr, Ba, Al, P, S and Si etc. **Lime stone** and **sand stone** are the classical examples of sedimentary rocks.

3) Metamorphic rocks: These are commercially valuable rocks, which are formed either from primary rocks or sedimentary rocks after metamorphosis. It means igneous rocks as well as sedimentary rocks both form the metamorphic rocks after erosion, pressure and temperature modifications. They consist of various kinds of minerals, which are present in igneous and sedimentary rocks. They are hard and solid, therefore having maximum durability. Various kinds of **marbles** and **quartzite** (Quartz or Mica) are the classical examples of metamorphic rocks. Lime stone of sedimentary rocks change into white marble or coloured marble after metamorphosis. Granite of igneous rocks changes into the black marble or green marble after metamorphosis. These marbles are also called **Granite marbles**. Sand stone of sedimentary rocks changes into **Quartzite** after metamorphosis.

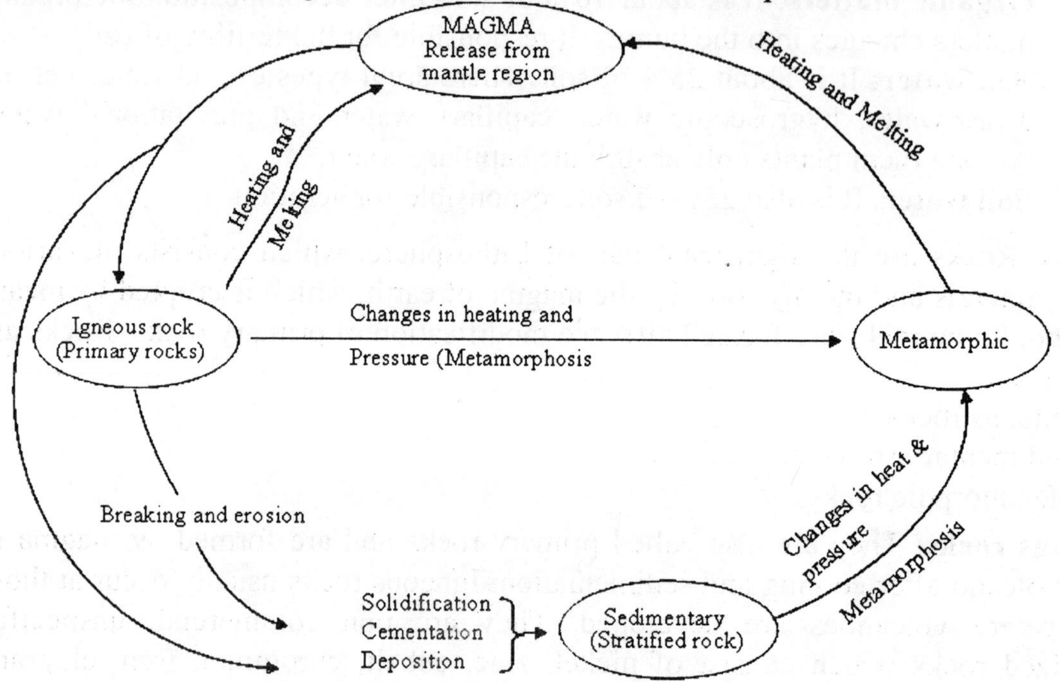

Fig. showing formation of different types of rocks

3) Hydrosphere: Water occurs in the environment in form of solid, liquid and gas. Presence of water in different forms, in different regions of environment is called "HYDROSPHERE". It is abiotic component of environment. Total water which is present in the hydrosphere is **about 2,86000 geogram** (1 geogram = 10^{20} gram). About 97% water of total water occur in form of vast ocean, 2% in form of ice sheath and 1% in form of fresh water, only 0.1% is the drinking water. Water occurs in three forms in the nature, on the basis of concentration of salts-

- **i)** **Fresh water**: In which concentration of salt is about 0.5%. It occur in rivers, lakes, ponds, pools etc.
- **ii)** **Brackish water**: In this water, concentration of salt is about 0.5-3.5%. It occurs in estuaries.
- **iii)** **Marine water**: The water with salt's concentration is 3-3.5% or above is called marine water. It occurs in ocean and sea.

Water is essential for all living things, either plants or animals. Water is colourless, tasteless, smellless and chemically a solvent. It is good solvent for different kinds of solutes. In living cells, water is about 90%. In case of plants, water is about 60-70%, but in xerophytes, water is about 90-95%. About 70% water is present in the human body. There are 4-vast oceans of the world having marine water. Among them, pacific ocean is the largest. These four oceans are divided into many seas. Arctic and Antarctic regions of earth and high altitude of rocks and mountains are covered with ice sheath or snow. The fresh water of earth is differentiated into 3-segments i.e. soil water, still water and running water. Soil water is of 4 types i.e. run away water, capillary water, hydroscopic water and gravitational water. Plants only absorb capillary water. Total amount of water present in soil is called **Holard**, water which is available for plants is called **Chesard** and which is not available is called **Echard**.

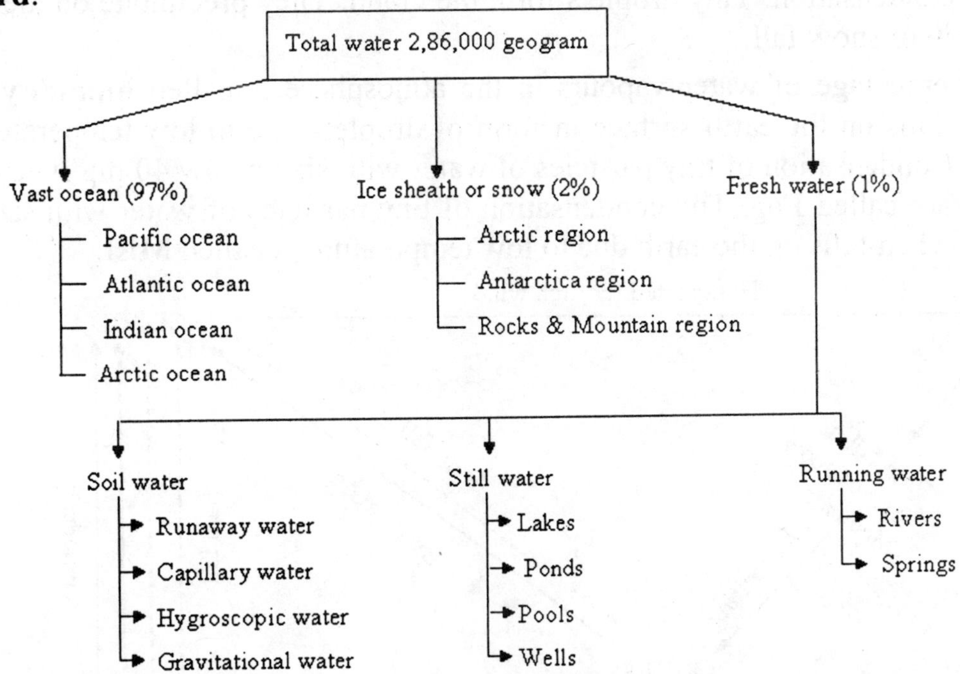

Fig. showing status of water in different forms and in different regions

Hydrological cycle

The circulation of water in various forms of atmosphere, biosphere and lithosphere is called hydrological cycle. In hydrological cycle, various physicochemical phenomenan take place, such as evaporation, condensation, sublimation and precipitation. A biochemical phenomenon i.e. transpiration is also involved in the hydrological cycle. All of the water of earth is affected by the moon and sun. In hydrological cycle, water is transferred from one system to other system and changes into one form to other form. When radiation of sun falls on the floor of water bodies such as ocean and river etc., water becomes hot and evaporates in form of vapours in the atmosphere. In atmosphere, it changes into tiny droplets after cooling and condensation. These cooling droplets form the cloud in atmosphere, they precipitate on earth, either as rainfall or snow fall. Like this, icesheath or snow of polar region or high altitude of mountain, change into water vapours due to radiations of sun, through process of sublimation. These water vapours change into tiny droplets by the process of cooling and condensation. Tiny droplets from the cloud precipitate on the earth in form of rain fall or snow fall.

Plants absorb the water from soil and lose the water in form of vapours through the process of transpiration. These water vapours also change into tiny droplets after cooling and condensation. Tiny droplets form the cloud. They precipitate on the earth in form of rain or snow fall.

The percentage of water vapours in the atmosphere is called **humidity**. The water which falls on the earth surface in form of droplets due to low temperature, is called **Dew**. Condensation of tiny particles of water with size below 40 mμ when falls on the earth are called **Fog**. The condensation of tiny particles of water with size 40-500 mμ size when falls on the earth due to low temperature is called **Mist**.

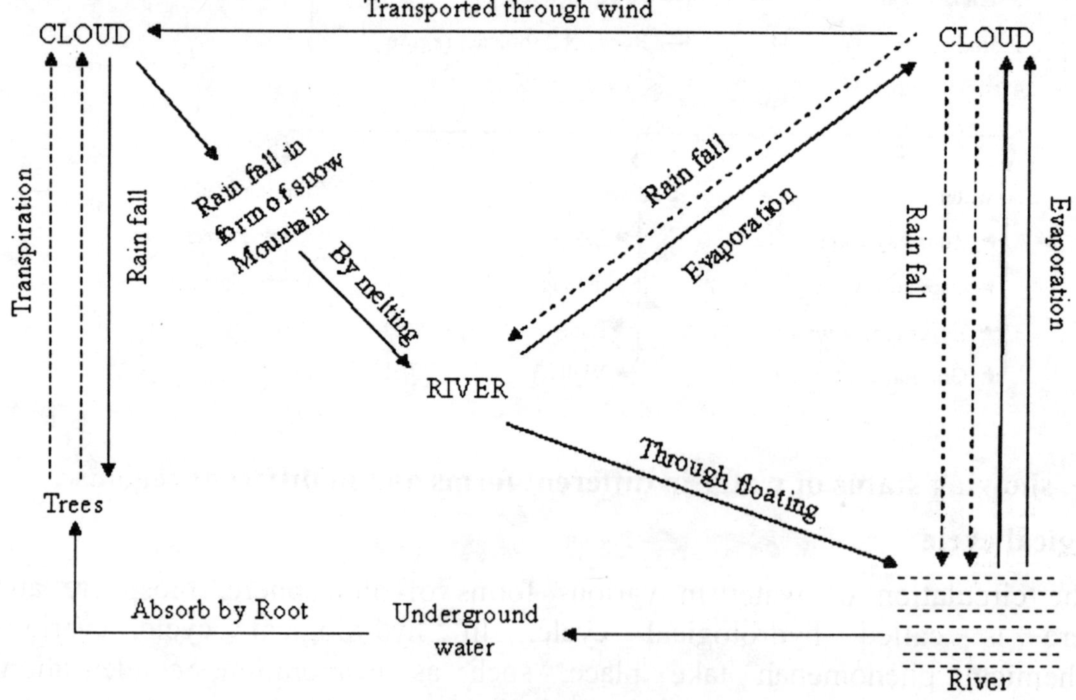

Fig. Hydrological cycle

2) BIOTIC COMPONENT

Living component of environment is called biotic component in which biosphere is included.

Biosphere: The region of Atmosphere, Hydrosphere and Lithosphere, where living organisms occur is called **Biosphere**. It is most significant component of the environment, because all biota create various types of biochemical and biophysical phenomenon and sustain the stability of environment. Various kinds of ecosystems are formed in the biosphere and we call it "**Ecosphere**". Various kinds of living things such as- plants, animals and microorganisms occur in the biosphere. The man dominated environment which is segment of biosphere is called "**Noosphere**". It is **Physio-socio** environment, where social and cultural aspects take place. Ecosystem is the sum of abiotic and biotic components. The biotic components consist of plants as producers, animal as consumers and microorganisms as decomposers. Ecosystems are of mainly of two types-

A) Natural ecosystem

1) Aquatic ecosystem – Which exists in waterbodies-

- a) **Fresh water ecosystem-** Exists in pond or river.
- b) **Marine ecosystem-** Exists in ocean or sea.

2) Terrestrial ecosystem – Exists in the terrestrial habit-

- a) **Grass land ecosystem**
- b) **Forest ecosystem**

B) Man made ecosystem or **Artificial ecosystem**

- i) **Cropland ecosystem**
- ii) **Garden ecosystem**
- iii) **An aquarium**

There are various types of interrelationships occur in the biosphere among plants and animals. They are of following three types-

1) Interaction between plants and animals

- a) **Insectivorous plants**
  - i) Nepanthes (Pitcher plant)
  - ii) Drosera (Sundew plant)
  - iii) Dionea (Venus fly trap)
  - iv) Utricularia (Bladder wort)
  - v) Addrovanda (Water flea plant)
- b) **Myrmecophily**: Interaction between plant and ant.

2) Interaction among plants

- i) Parasite
- ii) Epiphyte
- iii) Lianas

3) Interaction between microorganisms

i) Lichen

ii) Mycorrhiza

iii) Legume plant

POLLUTION

'Pollution is an undesirable change in the physical, chemical and biological characteristics of air, water and soil, which affects human life, lives of his related other useful living plants and animals, industrial progress and cultural assets etc."

Pollutant: The substance which cause pollution of environment is called pollutant. Pollutants are of two types-

a) On the basis of their forms, they exist in the environment after their use: On this basis, pollutants are of two types-

i) **Primary pollutants**: Those substances, emitted directly from an identifiable source. These pollutants exist as such after being added or released into the environment. For example- SO_2, NOx etc.

ii) **Secondary pollutants**: These are substances, derived from primary pollutants by chemical reactions. For example- primary pollutants, such as hydrocarbons and NOx, particularly in the environment, react in presence of sun light to form a group of nitrous compounds like PAN etc.

b) From ecosystem point of view: On this basis, they are of two types-

i) **Biodegradable pollutants**: Those substances, which are degraded by microorganisms and lose their toxicity, are known as biodegradable pollutants. For example- sewage, remain of plants and animal etc.

ii) **Non-Biodegradable pollutants**: Those substances, which are not decomposed by micro-organisms. For example- DDT, polythene, aluminium, Hg etc.

SOLID WASTE POLLUTION

Solid wastes are unwanted materials such as glasswares, crockeries, plastics, newspaper, paper bags, garbage, rubbish, automobile, spare parts etc., disposed off by man which can neither flow into stream nor escape immediately into the environment. These pile up at public places and cause obstruction in daily life.

The dumps create foul smell and are health hazards.

Types of solid wastes: The solid wastes are non-gaseous and non-liquid residues resulting from various human activities. The common type of solid wastes are as follows -

1) **Agricultural wastes**: These wastes include paddy and sugarcane husks and tobacco residues, forestry wastes, slaughter house waste, skeleton and hide of

dead animals, fruit and vegetable wastes and excretory wastes of domestic animals.

2) **Domestic and municipal wastes**: About 20% of India's total population lives in urban areas and generates about 15 millions tonnes of social wastes per annum. The average quantity of solid wastes generated **per person per day** in 8 major metropolitan cities is about 480 gm. The solid wastes generated in Indian cities, mainly include paper bags, glass and plastic containers, leather, rubber, sludge, garbage, waste from schools and other institutions, and wastes produced from cooking etc.

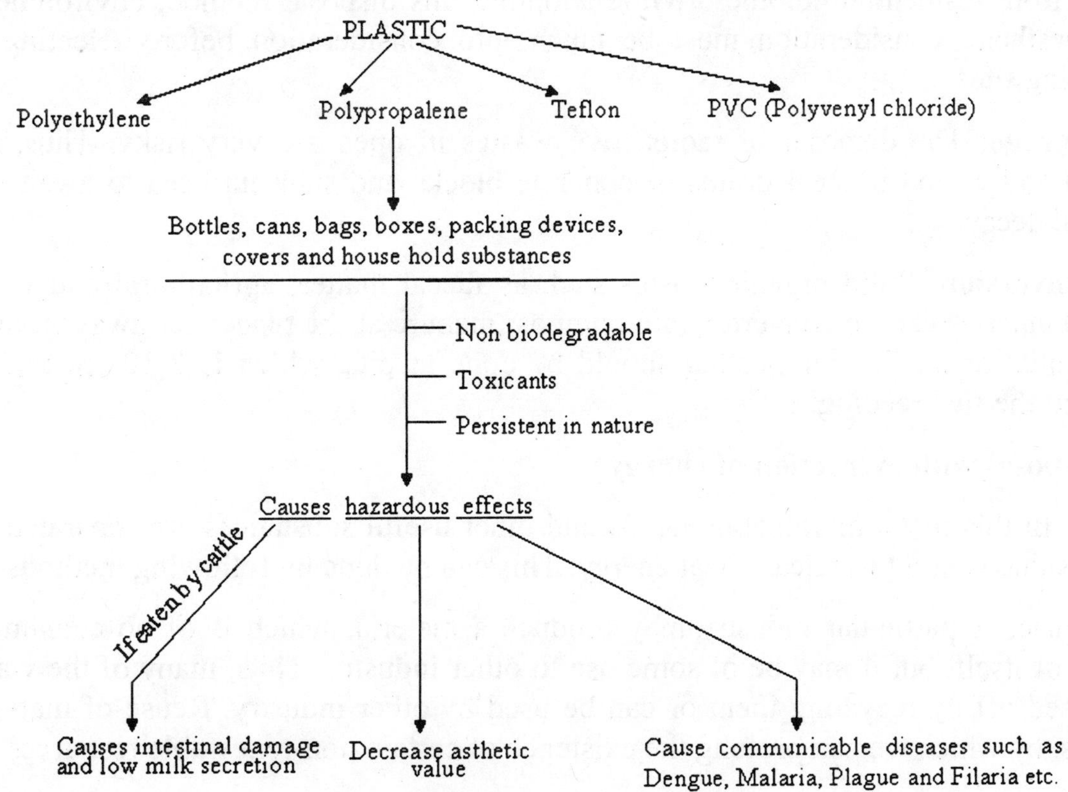

Fig. Impact of plastic on biota

3) **Industrial wastes**: 100% of the raw material does not convert into 100% of the product. Byproducts and wastes are integral parts of most of the manufacturing processes. Various types of solid wastes are generated by different industries. The industries, which are mostly generating the solid wastes are cigarette making industry, mining and metallurgy, fertilizers, pharmaceuticals, chemicals, paint and pesticides etc.

Disposal of solid wastes: Solid wastes may be disposed off by following two methods:

1) **Disposal without extraction of energy**
2) **Disposal with extraction of energy**

1) Disposal without extraction of energy

In this method, the solid wastes are dumped in places away from residential areas. This can be done by following three ways:

a) Land filling: Non combustible solid wastes, such as, tin, glass wares, ash etc. if not recoverable for usual purposes are disposed off by land fill method in low lying areas, away from residential colonies. While adopting this disposal method, environmental and aesthetic consideration must be taken into consideration before selecting the dumping sites.

b) Storage: The disposal of radioactive wastes in open are very risky. Thus, it is stored and cealed in steel drums or concrete blocks and sunk into sea to await their natural decay.

c) Conversion: Solid organic wastes such as faecal matter, agricultural and wastes from tanneries can be converted into compost manure at the places far away from the residential area. The composting should be done in pits, which is 8-10 cm thick to prevent the fly breeding.

2) Disposal with extraction of energy

In this method, valuable metals and other useful substances are separated and the residue is burnt to release heat energy. This can be done by following methods-

a) Reuse: A particular industry may produce a material which is of no economical value of itself, but it may be of some use to other industry. Thus, many of the wastes disposed off by recycling them or can be used by other industry. Reuse of materials means, by which we can prolong the existence of certain non-renewable resources.

Waste paper can be recycled to make paper, composition board and clip board. Waste paper can also be converted by hydrolysis into raw material for chemical industry. The remainder can be used to generate power in incinerators.

b) Incineration: The combustion of solid waste material in a scientifically designed incinerator (boiler) is known as incineration. Combustible solid wastes should be burnt in incinerators. This method is not much suitable to solve the problem in a real sense because in this, solid waste is being converted into gaseous wastes causing air pollution.

c) Pyrolysis: It is an alternative of incineration and is more suitable than incineration because much of the energy content of the waste material can be recovered by this method. In this process, waste material is burnt under controlled atmospheric oxygen and resultant gas and liquid are retained to be used as fuel. Pyrolysis of various carbonaceous wastes like firewood, coconut, palm waste, corn comb, cashew shell, rice husk, paddy straw etc. yield charcoal which is most suitable for industries along with products like tar, methyl alcohol, acetic acid and calcium acetate.

d) Chemical processing: More than 50% of agricultural and domestic waste material contains paper, paper products, grass, straw and cereals. They are mainly composed of cellulose (a polysaccharide) which yields glucose. Hydrogenation, esterification and organic fertilizer production are three important processing of cellulosic material. Hydrogenation of cellulose material is carried out in an autoclave, under controlled pressure, temperature and time in presence of a catalyst. Esterification of cellulosic material is carried out with caustic anhydride using H_2SO_4 as a catalyst. Cellulose acetate is formed, if acetic acid is used.

e) Synpyrol: It is a new technique, in fact it is a combination of two processes; synthesis and pyrolysis during which, the cellulose and water molecules react together to produce hydrolysed cellulose which thereafter breaks into fractions through an electrochemical process generating about 100 types of saturated and unsaturated hydrocarbons and other organics.

Power generation and production of cooking gas in cheaper rate is possible by using this unique technique.

Control of solid waste: To control the solid waste pollution, following methods have been suggested-

- Paper wastes packing boards can be used or recycled into fresh writing or printing papers.
- Agricultural wastes are put to various uses and then they can be converted into manure and slaughter house waste can be converted into animal and fish food.
- Among the main mining solid wastes are glass, iron, coal, dust, ash, bauxite mud etc., glass and iron are recycled easily, coat dust can be converted into combustible chips.
- Easily combustible material burns in incinerators for the use of power generation.
- Solid organic matter and faecal matter should be used as manure.

ATMOSPHERIC POLLUTION (AIR POLLUTION)

"Undesirable changes in atmosphere due to air pollutants cause atmospheric pollution". Atmosphere is non-living i.e. abiotic component of environment. Its gaseous components are-

$N_2 = 78.09\%$
$O_2 = 22\%$
$Ar = 0.1\%$
$CO_2 = 0.03\%$

He, Ne, Kr are in a trace amount, when undesirable gases such as- CO, CO_2, SO_2, NO, CFCs are mixed in atmosphere, they disturb the balance of chemical composition and causes pollution. There are following atmospheric pollutants which cause air pollution –

i) Carbon monoxide (CO)
ii) Carbon dioxide (CO_2)
iii) Sulphur dioxide (SO_2)
iv) Nitrogen oxides (NOx)
v) Chlorofluorocarbon (CFCs)
vi) Ozone (O_3)
vii) Hydrocarbons and lead (Pb)
viii) Peroxy acetyl nitrate (PAN)
ix) Photochemical smog
x) Dusts particle and aerosols

1) CO

It is toxic gas released from automobiles such as buses, trucks, four wheelers, three wheelers and two wheelers into the atmosphere. When its concentration become 1% in atmosphere, then causes atmospheric pollution. 1% or more CO inhaled in a proper amount, reach in lung capillaries, react with oxyhaemoglobin and impare the oxygen from haemoglobin and then after bind it with and form carboxyhaemoglobin.

$$\text{Oxyhaemoglobin} + CO \xrightarrow{\text{In lung capillary}} Hb + O_2$$

$$Hb + CO \rightarrow \text{Carboxyhaemoglobin}$$

Due to formation of carboxyhaemoglobin, oxygen deficiency takes place in blood capillaries and as a result **Hypoxia** condition occur in the body of human-beings. In this way, CO causes following hazardous effect on human-being:

- Headache
- Dizziness
- Convulsions
- Irritability

In "Traffic Jam" area, high amount of CO released from automobiles in atmosphere cause irritation and headache. Besides human-being, CO causes harmful effects on the plants also. The impact of CO on plants are as follows-

i) Rate of photosynthesis decreases

ii) Growth rate is affected

iii) Sensitive herbs unable to grow in polluted area.

2) CO_2

The percentage of CO_2 in atmosphere is 0.03% (300 ppm). When concentration of CO_2 increases upto 330 ppm to 350 ppm, it causes atmospheric pollution. CO_2 is released in atmosphere from burning wood, dung cake, dead animals and human-beings. Deforestation and Industrialization are two main cause by which concentration of CO_2 increases in the atmosphere. CO_2 has the covalent bond, which allows to pass all radiations of Sun, such as U.V. rays, VIBGYOR and I.R. rays. But this covalent bond re-radiate the I.R. radiation. Therefore, I.R. increases the temperature of atmosphere and causes global warming and this is called **"Green House Effect"**.

GREEN HOUSE EFFECT

Green house or Global warming is the burning problem of environmental pollution. **Arhinious** (Sweden) first of all, described the green house effect in 1956.

K. Thompson and **Chamberllin** of USA coined the term green house effect and discussed the phenomenon of green house effect. It may be defined as- "Earth retain heat due to air polluting gases such as CO_2", in other words "The temperature of earth increases due to re-radiation of IR-rays, is called green house effect". According to IUCN (International Union for Conservation of Nature and Natural Resources) "Temperature of earth is rising day by day, due to reradiation of IR-rays by covalent bonds of some atmospheric gases such as CO_2, CH_4, NOx etc."

We can say that the Green house effect i.e. Global warming is physiochemical phenomenon, due to activities of green house gases, which causes hazardous effects on Biota and Weather system.

In some cold countries, agriculture is performed in glass house, which is also called green house. The glass of green house allows to pass all radiations of sun but re-radiate, some radiations. So the temperature of green house increases and farmers grow summer crops in green house or glass house during the cold season. The green house gases act as a glass of green house and increase the temperature of earth.

Impact of green house effect

1. On climate
2. On biota (living organisms)

1. On climate: Green house effect is a major problem of the earth and its hazardous effects on the climate are as follows-

- Due to rise in temperature of earth, ice of Arctic zone and Antarctica melts and as a result, level of sea would increase. Thus, coastal cities would be submerged.
- Due to green house effect, summer will be longer and hotter but winter will be shorter and cooler.
- Rise in temperature of earth induces the lower limit of troposphere. As a result there is rainfall on earth.
- Rise in temperature interfere in the weather phenomenon of tropical and subtropical areas.

2. On Biota: Due to rise in temperature of the earth, new types of pathogenic microbes grow on earth, which may cause communicable diseases.

- High concentration of CO_2 increase the rate of photosynthesis upto certain level. But in a very high CO_2 concentration, the rate of photosynthesis falls, due to **Black man**'s limiting factor.
- At high temperature, the growth of some plants and animals is difficult.

Control measures

Green house effect is a major problem of the world, in which temperature increases day by day on earth. United Nations (UN), first time held an international conference as **Earth Summit** at **Reo-de-Zenerio (Brazil) in 1992** on **Global warming**. UNEP (United Nation Environmental Programme) suggests following practices for the control of global warming-

- To minimise the CO_2 concentration, burning of wood and dung cake should be stopped.
- Industries should use devices to minimise the release of NOx and CFCs etc. into the atmosphere.
- Eco-friendly trees should be grown in industrial areas and busy cities.
- Environmental education and awareness is also helpful in the control of global warming.

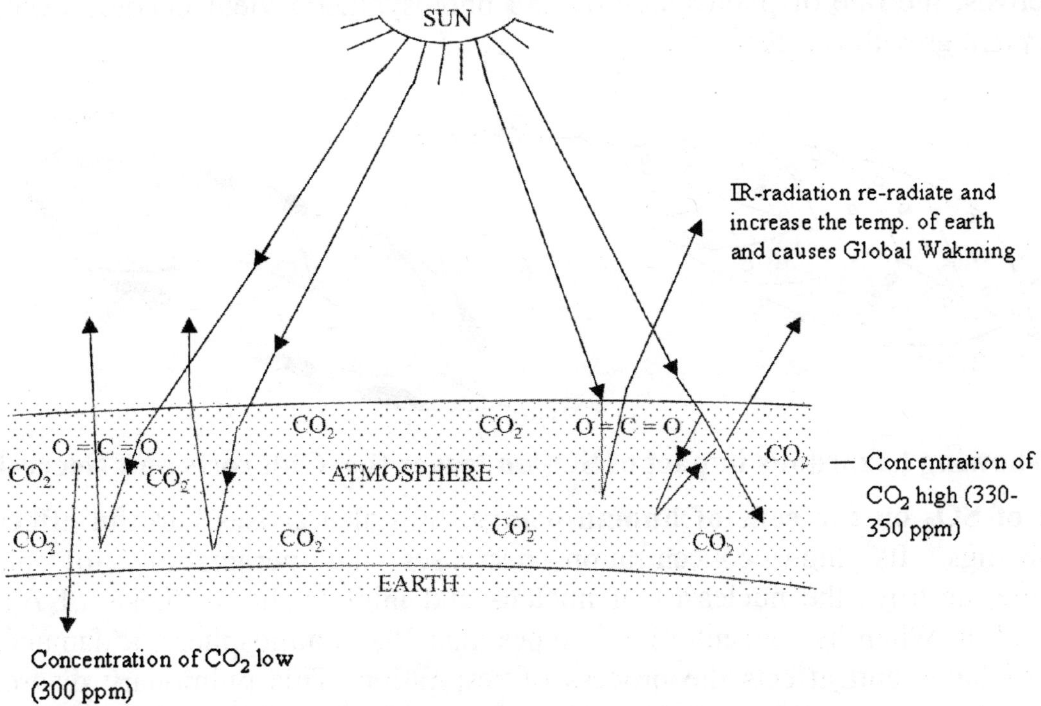

Fig. showing Green House Phenomenon

Global warming or green house effect is the burning problem of the earth, in which temperature of atmosphere increases due to high concentration of CO_2. The impact of green house effect on living things are as follows-

1. Increased temperature of the atmosphere, melt the ice of the polar region of earth therefore, sea level would increase and existence of coastal cities becomes affected.
2. Increased temperature, changes the climate and weather phenomenon, therefore, uncertain rainfall, coolness and hotness takes place.
3. In high temperature, new types of microorganisms exist in atmosphere which cause communicable diseases in human-beings.

3) SO_2

It emits from various industries, such as refineries, textile industry, bangle industries and motor car industries etc. It is a harmful gas, when its concentration goes upto 0.1% to 1%, it causes atmospheric pollution. SO_2 causes hazardous effect on plants, as well as on human-being. In case of plants, SO_2 causes **Chlorosis** and **Necrosis** in leaves. In chlorosis, chlorophyll becomes destroyed and leaves of plant show yellow colour, whereas in case of **necrosis**, margin tissues of leaves are destroyed and therefore affected area of leaves shows brown colour. Due to chlorosis and necrosis, the rate of photosynthesis and photosynthetic yield become decreased and the plant growth is affected.

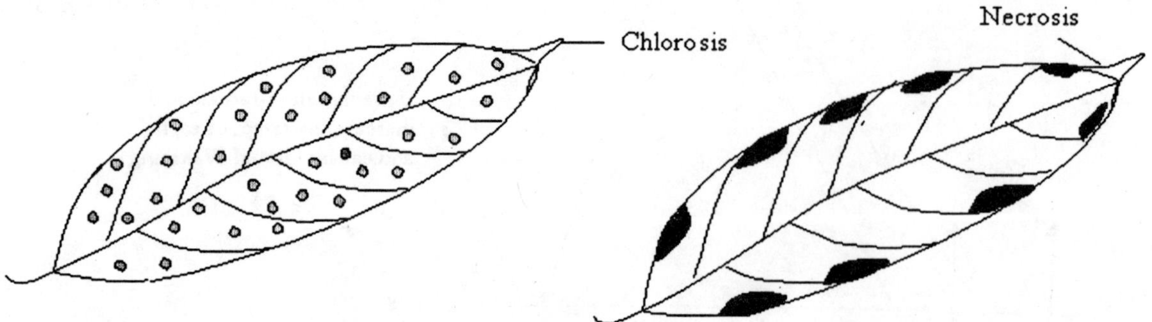

Fig. showing hazardous impact of SO_2 on leaves (i.e. Chlorosis and Necrosis)

Impact of SO_2 on the cells of human-beings: SO_2 also causes harmful effects on human-beings. Its high concentration decreases the permeability of plasma membrane, destroys the nucleases of nucleus and inhibits the synthesis of ATP in mitochondria. When its concentration is more than 1% in atmosphere, it damages the alveoli of lungs and affects the process of respiration. This pulmonary disorder is known as "**Pulmonary Edema**". SO_2 affects the metabolic activity in human-beings.

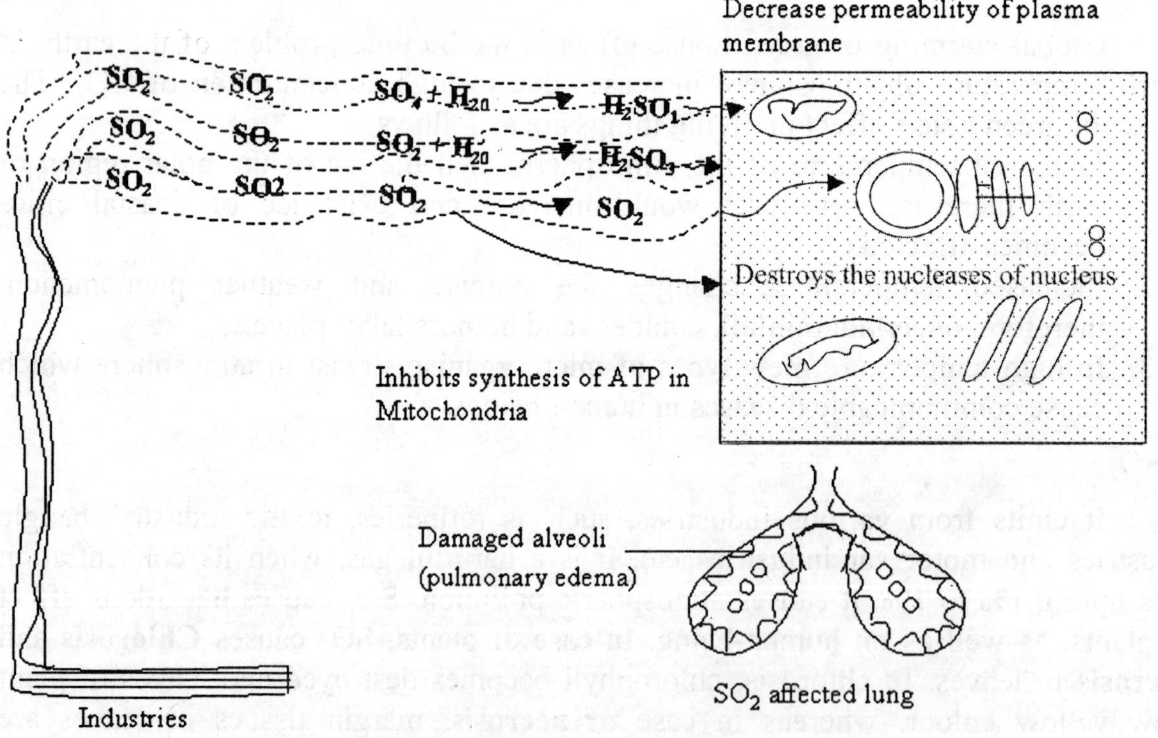

Fig. showing impact of SO_2 on the cells of human-beings

When H_2S gas reacts with water vapours of the atmosphere, then it forms the sulphuric acid (H_2SO_4) or sulphurous acid (H_2SO_3). They precipitate on the earth in form of **acid rain**. Acid rain causes acidity of soil, the acidic soil is not suitable for growing all types of crops. Acid rain is responsible for acidity in water bodies and causes "**Stone cancer**" on historical monuments e.g. **Taj Mahal.**

$$2SO_2 + 2H_2O \rightarrow H_2SO_4$$

$$SO_2 + H_2O \rightarrow H_2SO_3$$

Precipitation

Acid rain

4) Nitrogen oxides

Different types of nitrogen oxides NO, NO_2, N_2O, N_2O_5 are released in atmosphere from burning of fuels (Mainly petroleum) products. NO emits from industries such as oil refineries, fertilizer industries and motor vehicles such as 6 wheelers and 4 wheelers. Nitrogen oxides are toxic gases which are hazardous to the living organisms. When nitrogen oxides react with water vapours of atmosphere, they form the nitrous acid or nitric acid and they precipitate on the earth in form of acid rain.

$$NO + H_2O \rightarrow \underset{\text{(Nitrous acid)}}{HNO_2}$$

$$NO_2 + H_2O \rightarrow \underset{\text{(Nitric acid)}}{HNO_3}$$

Precipitation

Acid rain

When nitrogen oxides react with hydrocarbons in atmosphere, they form PAN. When nitrogen oxides react with hydrocarbon in presence of UV-rays of sunlight then they form the brown or grey fumes in atmosphere, known as **Photochemical smog**.

$$NOx + \text{Hydrocarbons} \longrightarrow PAN$$

$$NOx + \text{Hydrocarbons} \xrightarrow[\text{Sun light}]{\text{UV-rays}} \text{Photochemical smog}$$

NOx cause harmful effects on plants. When the concentration of NOx is 5-10 ppm, it causes **Necrosis** on young leaves, when the concentration of NOx is upto 10 ppm, it causes injury in reaction centre of chlorophyll and decrease the rate of photosynthesis. Flowering and fruiting of plants is affected in NOx polluted areas. Some sensitive plants, *Thuja, Araucaria, Ferns, Ixora* etc. with some vegetable crops do not properly grow in NOx polluted areas. Nitrogen oxide also causes harmful effects on human-beings. When its concentration is about 5-10% ppm in atmosphere then it causes **respiratory disorder**, in which **breathing** is affected. **Nostril** is blocked due to spasm and lungs become weak. When the concentration of NOx is about 10 ppm, it causes **pungent irritation** in eyes and stimulate tear's release.

According to recent concept, high concentration of NOx **dialate the blood capillaries of male and female sex organs** and causes slow blood circulation in sex organs and persons suffer from **sexual incompatibility** (sexual sterility). For removal of sexual incompatibility due to NOx, an important drug i.e. **VIYAGRA** discovered by **Ferid Murad Robert Frouchgott** and **Louis Ignora in 1996** is given. They received noble prize in 1998 for their work.

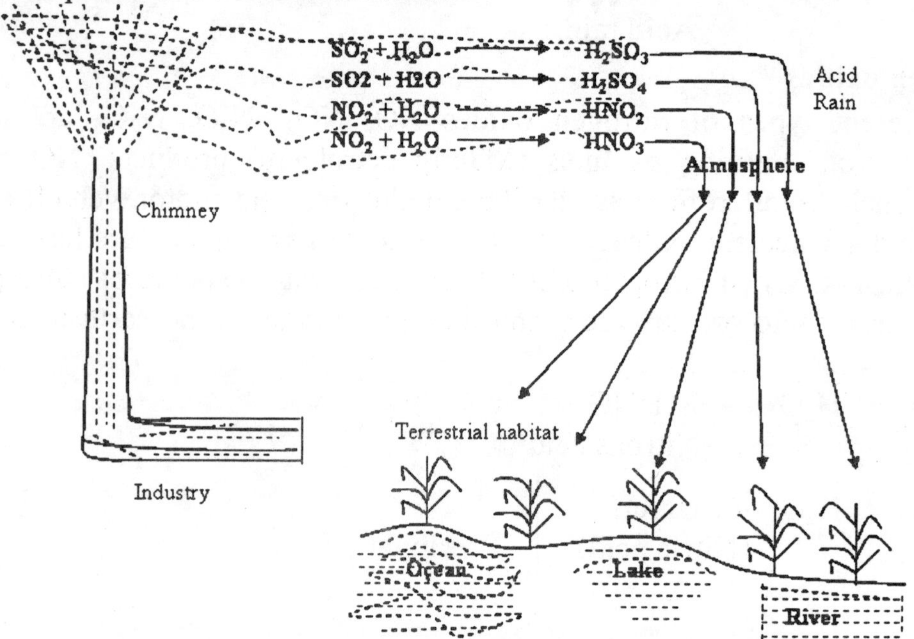

Fig. showing impact of acid rain on biota

5) Chloro-fluoro-carbons (CFCs)

It is also called **freon** gas, which emits from air conditioners and air propalators, cryogenic rockets, refrigerators and deep freezers. It is a harmful gas, which causes environmental pollution. In $Cl-\overset{F}{\underset{F}{C}}-Cl$ covalent bond is present, which reradiate the infrared radiation and causes global warming in troposphere. Like other green house gases, CFCs cause green house effect. In this, temperature rises in atmosphere.

CFCs move towards low temperature regions, therefore, when its concentration is high in troposphere, it moves into the regions of stratosphere. In stratosphere, it depletes ozone gases into oxygen and causes "**Ozone hole**". Through ozone hole, harmful UV-rays comes on the earth surface and cause hazardous effect on living organisms.

6) Ozone

Ozone of troposphere causes environmental pollution, whereas ozone of stratosphere protects the UV-rays and acts as a pollution controlling agent. When the concentration of O_3 is high in troposphere, then it causes adverse effect on plants and human-beings. It usually emits from industries. When its concentration is about 300-400 ppm, it affects the rate of photosynthesis in plants and decrease the growth and development. The more concentration of ozone i.e. above 400 ppm interfere the chlorophyll synthesis in plants and causes chlorosis. In case of human-beings, its high concentration is responsible for the breathing problems and bronchitis.

7) Hydrocarbons and Lead (Pb)

They are released from the motor vehicles into the atmosphere. Leaded petrol is used in motor vehicles in large scale. Tetraethyl lead (TEL) used in petrol as a "**Antiknocking Agent**". Hydrocarbons and lead in combination are released in atmosphere as a grey smoke from motor vehicles. Both lead and hydrocarbons directly inhibit the light reaction of photosynthesis in plants and cause respiratory disorders in human-beings. Unsaturated hydrocarbons react with nitrogen oxide to form PAN (Peroxyl acyl nitrate) which is toxic to living organisms.

Unsaturated hydrocarbons react with NOx in presence of UV-rays to form the "**photochemical smog**". It is also harmful for living organisms.

Lead is heavy metal which is emitted in the atmosphere through smoke of motor vehicles and is accumulated in hepatic cells of liver or other cells of the body of human-beings and inhibit the carbohydrate metabolism and fat metabolism; in this way it causes many disorders in human body.

8) Peroxy Acyl Nitrate (PAN)

When concentration of NOx, such as NO, N_2O, NO_2, N_2O_5 is high in the atmosphere, it reacts with unsaturated hydrocarbons and form the PAN. It is toxic pollutant which is usually formed in the busy cities, where motor vehicles are present in huge amount. Sometimes NOx also form the peroxy butyril nitrate (PBN) or peroxy propyronyl nitrate (PPN) or peroxy benzoyl nitrate (PBxN). They causes hazardous effect on living organisms.

NO + Unsaturated hydrocarbons $\rightarrow$ Peroxy Acyl nitrate (PAN)

NO_2 + Unsaturated hydrocarbons $\rightarrow$ Peroxy butiril nitrate (PBN)

N_2O + Unsaturated hydrocarbons $\rightarrow$ Peroxyproprionyl nitrate (PPN)

NO + Benzoic acid + Unsaturated hydrocarbons $\rightarrow$ Peroxy Benzoyl nitrate (PBxN)

Impact of PAN on living organism

- It interferes with light reaction of photosynthesis and affect the rate of photosynthesis.
- Plant growth and development are reduced.
- In case of human-beings, PAN causes respiratory problems and inhibits the breathing system.
- High concentration of PAN causes bronchitis in lungs.
- Eye irritation and nasal disorders are common features in PAN polluted areas.

9) Photochemical Smog

When nitrogen oxides react with unsaturated hydrocarbons in presence of U.V. rays of sunlight, it forms the photochemical smog. They are yellowish grey fumes, which are formed in day time. Photochemical smog causes air pollution in major metropolitan cities of world, such as – Los Angeles, New York, Tokyo, Mexico, New Delhi and Calcutta etc.

$$\left.\begin{array}{l} NO + \text{Unsaturated hydrocarbons} \xrightarrow{\text{UV-rays of sunlight}} \\ NO_2 + \text{Unsaturated hydrocarbons} \xrightarrow{\text{U.V.-rays}} \\ N_2O + \text{Unsaturated hydrocarbons} \xrightarrow{\text{U.V.-rays}} \end{array}\right\} \text{Photochemical smog}$$

Photochemical smog differs from smog in following respects-

- Photochemical smog is made up of yellowish grey fumes, formed in day time by NOx and unsaturated hydrocarbons, whereas smog is formed in any time either day or night by smoke of motor vehicles and dusts particles.
- Photochemical smog is only formed in the high temperature regions, whereas smog is formed in the moderate and high temperature region.

Impact of photochemical smog on plants: It causes following hazardous effects on plants-

1. In photochemical smog polluted area, some flowering herbs do not exist.
2. It causes senescence in plant.
3. It inhibits the light reaction of photosynthesis.
4. It reduces the proper growth and development of plants.

Impact of photochemical smog on human-beings: It causes following hazardous effects on human-beings-

1. Its high concentration may cause lung cancer.
2. It is 10 times more dangerous than PAN, hence affect seriously on alveoli of lungs and causes **Bronchitis**.
3. It usually causes eye irritation and headache.
4. It is responsible for breathing problem, coughing and sneezing.
5. It may cause mutation.

10) Dusts particles and aerosols

In city, town and industrial areas dust particles and aerosols cause air pollution. Among dust particles, soil particles, sand particles, pollen grains, carbon particles, silica particles combined with liquid droplets form the aerosols, which are hazardous for the living organisms. Soil particles, sand particles and pollen grains cause hypersensitivity, allergy, eye irritation and headache in human-beings. Soil particles and sand particles are deposited on leaves of plants and affect the rate of transpiration and rate of photosynthesis. In coal and cement industries, dust particles are abundant. Black coal particles from coal mines and silica particles from cement industry are released into the atmosphere and they cause **lung cancer** and **silicosis** respectively. In silicosis, particles of silica are deposited in the alveoli of lungs. The area around the brick industry is also polluted by carbon particles. These carbon particles cause side effects on human-beings, who are suffering from hypersensitivity, allergy, nasal and pulmonary disorders. These carbon particles are deposited on the leaves of the trees, particularly on mango trees and affect the growth, development, flowering and fruiting.

Others: Besides these harmful gases, some other toxic pollutants are released into the atmosphere due to some accident in the industry or any other reasons. Bhopal gas tragedy is a significant example of environmental pollution.

Bhopal gas tragedy

Venue : Bhopal (M.P.), India

Industrial Disaster Date	: 3 December 1984
Industry	: Union Carbide Ltd.
Industrial product	: Chemical insecticides (Mainly Methoxychlor)
Toxic gas released	: MIC (Methyl Iso Cyanate) and Phosgene gas
Persons died	: 2000-3000 (Approx.)
Affected persons	: 10,000 (Approx.)

OZONE DEPLETION

Ozone depletion is a great problem of the world, which is found in some parts of the earth, such as region of Arctica and Antarctica, Canada, USA, Mexico and some parts of Europe. Atmosphere is differentiated into 5-sublayers i.e. **Troposphere, Stratosphere, Mesosphere, Ionosphere** and **Exosphere**. In Stratosphere, ozone gases are found, hence this region is also called as **Oozosphere**.

Ozone gases of Stratosphere, absorb harmful U.V.-rays and protect the life of Troposphere. U.V.-rays are of 3 types-

i) **UV-A rays** : Its wavelength is about 320-380 nm.
ii) **UV-B rays** : Its wavelength is about 280-320 nm.
iii) **UV-C rays** : Its wavelength is about 220-280 nm.

Among them, U.V.-B rays are most dangerous. U.V.-rays cause hazardous effects on biota. At present time, ozone hole has formed in some parts of the earth, due to air pollutants, harmful. U.V.-rays come on the earth and cause adverse effects on life of troposphere.

How to check ozone depletion? : The following two gases are mainly responsible for the ozone depletion-

1. CFCs
2. NO_x

1) CFC gases: It is freon gas i.e. Dichloro-difluorocarbon. It is released from air conditioners (AC), Deep freezers, Regrigerators, Air propellators and Criogenic rocket. In the region of Tropopause or in Stratosphere, U.V.-rays break the chlorine from CFCs and as a result Cl_2 molecules become free in the atmosphere.

$$\text{CFCs} \xrightarrow{\text{U.V.–rays}} \text{CF} + Cl_2$$

Chlorine molecules react with ozone gas and form chlorine oxide (ClO). This chlorine oxide again reacts with ozone to form the oxygen (O_2) and chlorine molecule are released in the reaction. This chlorine molecule again reacts with ozone (O_3). In this way, one molecule of chlorine depletes about million ozone gases in the stratosphere, as a result "**OZONE HOLE**" is formed in the Stratosphere.

$$Cl + O_3 \rightarrow ClO + O_2$$
$$ClO + O_3 \rightarrow 2O_2 + Cl$$
$$Cl + O_3 \rightarrow ClO + O_2$$

Chlorine is a light gas which moves towards low temperature region on the earth, hence ozone hole is found in the Arctic and Antarctica regions, where CFCs are not released in troposphere.

2) NOx (Nitrogen oxide): Different types of NOx gases such as NO, NO_2, N_2O etc. are released from various industries and motor vehicles into the troposphere, out of which, NO is lightest gas which enters in the stratosphere. In stratosphere, NOx react with ozone and form NO_2 and release O_2 (oxygen). This NO_2 again reacts with O_3 and form oxygen and release NO. This NO again reacts with O_3 and form NO_2 and release O_2. In this way, NOx depletes about one million ozone gases and cause "**OZONE HOLE**".

$$NO + O_3 \rightarrow NO_2 + O_2$$
$$NO_2 + O_3 \rightarrow 2O_2 + NO$$
$$NO + O_3 \rightarrow NO_2 + O_2$$

Impact of ozone depletion

a) On climate: Ozone hole was first time observed in Antarctica region in 1975. Later on, it was also discovered in other parts of the earth. Through ozone hole, U.V.-rays of sun come on the earth and causes following adverse effects on climate-

- Due to U.V.-rays, temperature of the earth increases which change the nature of climate and affects weather system.
- Due to rise in temperature of the earth, new microorganisms appear and some microbes and insects become resistant affecting the ecosystem.

b) On plants: UV-rays come on earth through ozone hole and cause following adverse effects on plants-

- It destroys the guard cells of stomata in some plants and affect the transpiration.
- UV-B rays inhibit the synthesis of chlorophyll and destroy the chlorophyll, as a result, the rate of photosynthesis is affected.
- Recently, it is discovered that UV-B rays inhibit the synthesis of Auxin in apical regions of plant parts.

c) On human-beings: UV-rays come on the earth through ozone hole and cause following hazardous effects on human-beings-

- It causes retinal detachment in eye.
- It disturbs the metabolic activities in human-beings.
- UV-B rays causes skin cancer and mutation in human-beings.

Control measures

For the control of ozone depletion, it is necessary to minimize the use of CFCs and NOx. In 1992, UN held the International Conference in Reo-de-Zenero in **Brazil** known as "**EARTH SUMMIT**" to minimise the O_3 depletion. USA and Japan discovered the alternative of CFCs known as HFC (Hydrofluorocarbon), HFC-134 and HFC-22 used in A.C. refrigerators etc. It is free from chlorine molecule and do not cause ozone depletion.

Control of Air Pollution

Air pollution can be controlled by following methods:

1. By separation of pollutants
2. Avoidance of pollutants
3. Use of alternate sources of energy
4. Control through law

1) Separation of pollutants

i) High class chimney should be used in industries.

ii) The chimneys should be set up on high position.

iii) High class filters, which consist of silk fibres should be used in chimneys.

iv) A device known as electrostatic precipitator should be used to minimize air pollution.

v) Cyclone collector should be used in small and large scale industries to collect toxic pollutants.

vi) Silencer of motor vehicles should be modified, which emit low amount of pollutants into the atmosphere.

vii) Carburetor of motor vehicles should be modified, which is helpful to minimize pollution.

viii) Unleaded petrol should be used in motor vehicles.

ix) Instead of mobile oil, 2T-oil should be used in 2-stroke and 4-stroke motor vehicles.

2) Avoidance of pollutants

i) Purified fuel should be used in all motor vehicles.

ii) Engine of motor vehicle should be modified to emit minimum amount of air pollutants.

iii) Eco-friendly tree should be grown in cities and towns. They are useful to minimize air pollution. The common eco-friendly trees are as follows-

- *Azadirecta indica* (Neem tree)
  It absorbs CO_2 and CO
- *Emblica officinalis* (Amla)
  It absorbs SO_2
- *Terminalia arjuna* (Arjun tree)
  It absorbs H_2SO_4 and NO_2
- *Saraca indica* (Ashok tree)
  It absorbs CO and CO_2
- *Ficus religiosa* (Peepal tree)
  It absorbs CO_2
- *Syzizium cumini* (Jamun)
  It absorbs CO_2 and CO
- *Cassia fistula* (Amaltas)
  It absorbs CO_2 and hydrocarbons

Beside these, hedges of *Henna* and *Pitholobium* are also helpful in reducing the air pollution.

3) Use of alternative sources of energy

- Instead of petroleum fuels, some other alternative source of energy eg.: Gashole, CNG (compressed natural gas) and ethanol should be used.
- Instead of coal, petroleum fuel and electricity in industry, some alternative source of energy should be used e.g. Tide energy, Solar energy and Wind energy etc.

4) Control through law

- Air Pollution Act 1986 should be restrictly implemented by Government of India.
- Those motor vehicles, which emit high amount of air pollutants should be fined or banned.
- Those industries, which emit high amounts of air pollutants should be fined or banded.
- The best method to control the air pollution is, environmental education and environmental awareness.

WATER POLLUTION

"Undesirable changes in physical, chemical and biological characteristics of water in different water bodies, which cause hazardous effects on Biota are called water pollution". Water pollution is a big problem of the world because water bodies such as rivers, lakes and ponds etc. are polluted due to sewage water, agricultural chemicals, industrial wastes and various kinds of debris. In India, fresh water bodies as well as marine water bodies are affected by water pollution. Holy river "**Ganga**", "**Yamuna**", "**Narmada**", "**Kaveri**", "**Krishna**", and some famous lakes such as Ooty lake, Naina lake, Dul lake and Husain Sagar lake are polluted by various kinds of water pollutants.

There are following sources causing water pollution:

1) Sewage water
2) Industrial wastes (Industrial effluents)
3) Surface run off
4) Debris
5) Others

1) Sewage water: Wastes of houses in cities and towns are released into the water bodies through specific canal system. These wastes containing water, is called as sewage water. Sewage actually consists of human excreta, house hold detergents and plant debris. The colour of sewage water turns brick red due to decomposition of organic wastes by microorganisms. The turbidity of sewage water is very high due to pathogenic bacteria, fungi, protozoans and virulent viruses. The value of DO (Dissolved oxygen) becomes low. The parameter to measure the level of pollution is BOD. It is, Biochemical oxygen demand (old name – Biological oxygen demand) i.e. "**Amount of DO needed by micro-organisms in a given water sample, at 20°C for 5-days is called BOD**". In sewage water, **BOD value is very high but value of DO is low.**

When sewage water is mixed in river or lakes, it causes water pollution. Sewage mixed water is contaminated water for human-beings, animals as well as plants. Aquatic life becomes affected by various kinds of diseases caused by microorganisms in fishes, and hydrophytes. Sewage contaminated water is completely unfit for human-beings, because pathogenic microorganisms cause following communicable water born diseases:

(i) Cholera by *Vibrio cholerae*
(ii) Amoebiosis by *Entamoeba histolytica*
(iii) Diarrhoea by *Rheo* virus

(iv) Typhoid by *Salmonella typhae*
(v) Food poisoning by *Clostridium butyliticum*
(vi) Viral fever (influenza) by *Rhino* virus
(vii) Hepatitis–A, by H.A.V.
(viii) Dermatitis by *Trycophyton*
(ix) Ringworm by *Microsporum*
(x) Dandruff by *Microsporum & Trycophyton*

2) Industrial effluents: Industrial wastes such as effluents are released into the water bodies such as river, lakes etc. and cause water pollution. It consists of various type of chemical substances, such as heavy metals, sulphate, phosphate, chlorate, bromate, borate, nitrate of Na, K, Mg and Fe etc. Heavy metals are toxic metals, because they cause toxicity in aquatic life as well as cause hazardous effect on human-beings. Water bodies which are contaminated with industrial effluents have high COD value. COD is dissolved oxygen in sample water, used by oxidation of some chemical substances. Such water bodies have high turbidity.

The most dangerous substances of industrial effluents are heavy metals (toxic metals). Their toxic effects on biota are as follows-

i) Mercury (Hg): This heavy metal is released in water bodies from paint, shell, and some chemical industries. Mercury accumulate in fishes and some aquatic plants and causes toxicity in fishes and aquatic plants. Mercury contaminate water or mercury is accumulated in fishes which are taken up by human-beings which may suffer by **MINIMATA disease**. In this disease, memory loss and mental retardation takes place. It was first reported in Japan.

Minamata Incident:

Mercury is a well known toxic metal which came to limelight only after the incidence of **Minamata disease** in 1953-1960 in Japan. At Minamata bay in Japan more than 100 people lost their lives and many thousands were permanently paralysed from eating mercury-contaminated fish. In a particular village facing the bay 15% of the villager were either killed or permanently crippled. Genetic defects were observed in some 50 babies whose mothers had consumed the fish from the contaminated bay. The sea fish in the bay were found to contain 27-102 ppm of mercury in the form of methyl mercury. The source of mercury was the effluent discharged into the bay from an industrial plant, Minamata Chemical Company. The mercury was deposited on the bottom of the bay and remained there since 1950s.

In the history of water pollution, the Minamata incident was unique. The mystery of the existence of methyl mercury in sea fish was baffling at first but the extensive research revealed the way of mercury into the living beings and its subsequent biomagnification. This is the first known case where the natural bioaccumulation (in fish) of a toxic material (methyl mercury) killed hundred people and genetically a large population.

ii) Cadmium (Cd): This heavy metal is released from chemical, textile and motor vehicles industries. It also accumulates in fishes and aquatic plants and cause adverse effect on them. If cadmium contaminated water or fish is taken up by human-beings, Cd accumulate in their bones and cause porosity. As a result bone become fragile. This disease is called **Itai-Itai** (Ouch-ouch).

iii) **Arsenic (As)**: This heavy metal is released from paint, chemical, leather industries and refineries etc. They cause hazardous effect on aquatic life by disturbing their Krebs cycle. When arsenic contaminated water is taken up by human-beings, the respiration becomes affected and they come in passive form. High concentration of arsenic causes **Black-foot-disease** in human-beings, in which haemoglobin binding capacity with oxygen becomes very low in foot region.

iv) **Fluorine (F)**: This is released from cottage industry into the water bodies and cause toxic effects on aquatic life. Fluorine contaminated water is unfit for human-beings due to high concentration of fluorine. **Fluorosis** is a disease, in which tooth and bone decay takes place. Its very high concentration causes "**Knee Jerk**" in human-beings i.e. **knock-knee syndrome** (Fluoride).

v) **Chromium (Cr)**: It is released from different industries and causes adverse effect on aquatic life. Its high concentration affects the liver and kidney of human-beings by disturbing the metabolic activities.

vi) **Lead (Pb)**: It is released from the industries of pots and chemicals causing adverse effect on aquatic life. Its high concentration affects the fat metabolism and hormone secretion.

3) Surface run off: Various type of chemical fertilizers which contain nitrate, phosphate and chemical pesticides are used in modern agricultural bractices. These chemicals are mixed in water bodies after rainfall as surface run-off. Due to surface runoff, rivers, lakes and ponds become polluted. Such polluted water bodies having greater amount of nitrate and phosphate become "**eutrophicated**".

Eutrophication: Stagnant water bodies, which has high amount of nitrates and phosphates due to eutrophication, induce the growth of hydrophytes and phytoplanktons (green algae and BGA). In eutrophic water, some blue-green-algae such as *Microsystis, Anabaena* and *Oscillatoria* cover the entire surface of water bodies and cut off sun light reaching hydrophytes and other plankton in water bodies. Due to absence of light, no photosynthesis takes place by hydrophytes, as a result **Hypoxia** (deficiency of O_2) condition takes place. Due to deficient O_2 condition some animals die, therefore water becomes highly polluted in which D.O. becomes very-very low. The water becomes turbid, brown in colour and gives foul smell.

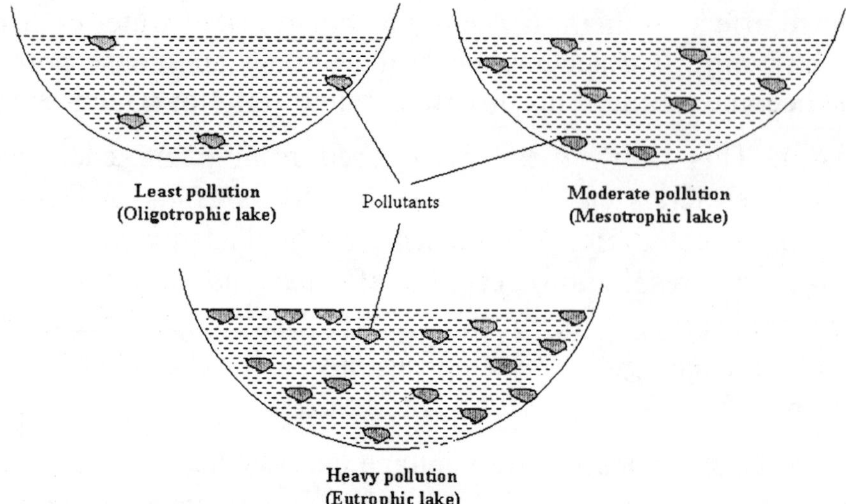

Fig. showing trends of eutrophication

Eutrophic water is unfit for drinking purposes, irrigation and even for recreation. Generally, fishes die in eutrophic water, which may lead to food problem in the affected area.

4) Debris: Various type of plant debris such as flower garlands, flower, leaves etc. are thrown in rivers or other water bodies due to religious thinking. Various type of animal debris such as, dead animals and dead human bodies are discharged into the water bodies, which cause water pollution. This animals and plants debris increase pollution level after biodegradation by microbes.

5) Others: Radioactive isotopes, toxic metals and other toxic chemicals also pollute water bodies, such water, is not suitable for drinking purposes. Water with high nitrate content is completely prohibited for human-beings because such water may cause "**Blue-Baby Syndrome**" in human-beings.

Ganga Action Plan: For the 1st time in 1973, it was noticed that the quality of river water of the Ganga is continuously degrading. Thus because of the sentimental attachment and also its multidimentional consumption, it was suggested to provide measures for the treatment of the river water. Government has decided all the basis of recommendation by Central Pollution Control Board (CPCB) and also the State Pollution Control Board (SPCB) that the river Ganga must be protected from pollution. Government has decided to control the pollution in various phase. In **Phase I of Ganga Action Plan**, crores of rupees were invested. By the time important sources of pollution of the river Ganga were located and sewage treatment plant was established in prominent cities. **Phase II of Ganga Action Plan** was also released in the year 2001-2002 but has not been available for the **remedial action**.

Narmada Bachao Andolan: This movement was initiated under the leadership of **Megha Patekar**. This is a movement against increased height of **Almati Dam** over the **Narmada** river. According to the protestants, by increasing the height a large area of fertile land will have to be **drowned** into the reservoir. As a result, the local inhabitants will have to leave the place. So that, they would be forced to suffer great economical loss. The chances of earthquake in that area will also increase. Therefore, it is a harmful act.

According to the supporters, by increasing height of the dam more electricity would be generated even during summer also, excess amount of water may be available for irrigation.

Control of water pollution: For control of water pollution, following strategies should be adopted-

1) Sewage treatment
2) Industrial effluent treatment
3) Using chemicals and living organism
4) By enforcing laws.

1) Sewage treatment: There are following three methods, involved in the sewage treatment:

i) **Primary treatment**: In this, mechanical process is involved, such as – filteration and sedimentation etc.

ii) **Secondary treatment**: In this, biological treatments are involved. Some aerobic and anaerobic microbes are used to decompose the organic matter of sewage water into simple inorganic form.

iii) **Tertiary treatment**: It involves chemical treatment by using chlorine, $KMnO_4$ etc. These chemicals help in purification of sewage water.

2) Treatment of industrial effluent: There are some physical and chemical processes involved in the treatment of industrial effluents, such as – filteration, centrifugation sedimentation, neutrilization, photocatalysis etc. First of all, industrial effluents are passed through a high class filter, which filters micromolecules. After this, effluent is transfered into sedimentation tank where, centrifugation process operate. In this way, heavy molecules in bottom sediments are filtered. After this, remaining effluent is treated by some chemicals for neutralization. In some cases **Titanium oxide** is mixed, which catalyse chemical compounds in presence of light.

3) Using chemicals and living organisms: In wells, water tanks of town and city, chlorinated tablets, bleeching powder and $KMnO_4$ are used for purification of water. In rivers and lakes etc., alum should be used for purification of water. **Tortoise** like aquatic animals are also significant for control of water pollution, because they phagocyte dead animals and human body. **In Ganga Action Plan 1985**, holy river Ganga was purified by using alum and tortoise.

4) By enforcing the law: There are following practices involved in control through laws-

- Water Pollution Act 1974 and Factory Act 1977 should be implemented strictly.
- Those industries releasing high amount of effluent should be fined or banned.
- Bathing with soap should be prohibited.

BOD and COD

1) BOD (Biological → Biochemical oxygen demand):

- It is biochemical oxygen demand, which is pollution indicator of sewage contaminated water.
- The old name of BOD was Biological oxygen demand, which indicates the degree of pollution in sewage contaminated water.
- BOD may be defined as, "The amount of oxygen needed for micro-organisms in given water sample at 20°C for 5-days".
- In BOD, micro-organisms usually require DO (Dissolved oxygen) for biodegradation of organic matter.
- Value of DO increases in unpolluted water and decrease in sewage contaminated water.
- Amount of oxygen which is dissolved in water is called DO. Approximate value of DO in unpolluted water is about 14 ppm. In sewage polluted water is 5-8 ppm. In strong sewage polluted water it is about 3-5 ppm.
- BOD is indicator for sewage water, its value increases in the polluted water by means of sewage and value decrease in non polluted water.
- According to WHO (World Health Organization), the BOD value of drinking water is about 1 ppm. Normally water with 1-2 ppm BOD is considered as drinking water.
- According to environmentalists, BOD of sewage contaminated water is about 20-25 ppm and in strong sewage polluted water, it is about 50 ppm.

2) COD (Chemical oxygen demand)

- It is chemical oxygen demand, which is pollution indicator for chemically polluted water.
- It is more significant than BOD value, because in industrial effluent polluted water bodies, it can be used as a measure of degree of pollution.
- COD may be defined as "Amount of oxygen required for biodegradation of organic matter in water sample by strong oxidizing agents, such as $K_2Cr_2O_7$ for hours".
- Value of COD can be measured in ppm.
- It is not significant for sewage polluted water and can be used only for chemically polluted water.
- Value of COD depends upon the nature of chemical substances and mode of decomposition of water bodies.
- COD value increases in industrial effluent polluted water, which is having high amount of heavy metals or toxic metals.

- COD value decrease in unpolluted water. The value of COD for drinking water is 1 ppm or below 1 ppm.
- According to environmentalists, COD value of the carpet or leather industry effluent contaminated water bodies is about 40-50 ppm.

Biological magnification: There is a wide range of pesticides and other chemicals used in modern agriculture to control diseases. Beside these agrochemicals some other chemicals also enter in our environment by other sources. These are most harmful in the sense that they are non-degradable or degrade very slowly for e.g. DDT, BHC and Aldrine etc. These are much hazardous if enter in our food chain. The concentration of such toxic substances increase successively in trophic levels in a food chain. For e.g. DDT enters in a pond ecosystem are taken as such by plants of the pond, then reaches to zooplanktons feeding on plants, then to minnows consuming the zooplanktons, then to fish which eats minnows and finally the body of birds or man (who eat fishes). It can be concluded that the DDT concentration continuously increase at successive trophic level in a food chain. This phenomenon is known as **Biological magnification** or **Biomagnification** or **Bioaccumulation**.

RADIOACTIVE POLLUTION

Pollution on land by means of radioactive elements, which emit α-rays, β-rays, γ-rays and cause hazardous effects on biota are called Radioactive pollution. According to recent concept, all types of radiation either non-ionic radiation or ionic radiations which cause harmful effects on human-beings, animals and plants are called radioactive pollution. It is only one type of pollution in which not only one generation but next generation is also affected due to harmful radiations. Some major mishap of world related to radioactive pollution are "**Atomic Bomb Explosion on Hiroshima-Nagasaki**" and "**Chernobyl Mishap**".

In (1944-45), during Second World War; USA bombarded the Atomic bomb on two cities of Japan i.e. Hiroshima and Nagasaki. About 10,000 people died and next generation was also affected due to gene mutation. In 1985-86, **Chernobyl Atomic Reactor Soviet Russia** was blasted and about 20,000 people died due to radioactive substances. Now a days various type of ionising and nonionising radiations emit from different sources and cause gene mutation not only in human-beings but also in animals and plants.

Sources of Radioactive Pollution

In nature, α, β and γ-rays emit from various type of radioactive elements, such as Cs-137, Sr-90, I-151, Pu-239, U-238 etc. X-rays emitted from X-ray diagnostic machine and X-ray crystallography, UV-rays emitted from ultrasound device and various type of scan devices. T.V.-sets and computer monitors also emit UV-rays.

Non-ionising and Ionising Radiation

Various types of UV-rays i.e. UV-A, UV-B and UV-C are nonionising radiations. The range of wavelength of UV-A rays is 320-380 nm, UV-B has 280-320 nm and UV-C has 220-280 nm. UV-A disturbs the metabolic activity, cause skin burn and retinal detachment, UV-B is most dangerous of all UV rays, which causes skin cancer in human-beings. Its continuous exposure is responsible for gene mutation. UV-B also affects the photphosphorylation in plants and decrease the growth rate. UV-C disturbs the metabolic activities in human-beings and may cause sun burn. In this way, non-ionising radiations i.e.-UV-rays cause hazardous effect on human-beings as well as plants.

Radioactive elements- Pu-239, Sr-90, Cs-137, Ra-226, Co-60, I-127 etc. emit ionising radiations, such as α-rays, β-rays and γ-rays. Some metallic ion emit cosmic rays and X-rays devices emit X-rays (Rontegon-rays). Ionising radiation causes harmful impact on biota. α-rays and β-rays change the metabolic activities and affect the metabolism in human-beings and animals. γ-rays are the electromagnetic rays having high penetration power and cause **blood cancer, bone cancer, skin cancer** and **gene mutation**.

Radioactive elements and their hazardous effects

i) **Cesium Cs-137**: Emit α, β and γ-rays – Causing blood cancer, bone cancer and gene mutation.

ii) **Strontium Sr-90**: Emit α and γ-rays – Causing blood cancer and gene mutation.

iii) **Iodine I-127 & I-131**: Emit α and β-rays – Disturb metabolic activities and affect the thyroid secretion.

iv) **Plutonium Pu-239**: Emit α and β-rays – Replace the Na^+ and K^+ and disturb the metabolism.

v) **Thorium Th-232**: Emit β and γ-rays causing gene mutation and metabolic disorders.

Except U-238, Xe-133, Ra-226, Co-60 and Kr-61 also disturb the metabolic activities and cause metabolic disorders, destroy co-enzymes and vitamins.

Control measures: Following practices can be adopted for the control of radioactive pollution:

- Atomic bomb explosion should be banned in whole world by United Nations (UN).
- Nuclear reactors and Atomic power stations should set up beyond urban areas.
- Eco-friendly trees like *Ficus religiosa, Saraca indica, Arjuna* sp. should be grown around the research centres.
- Use of Atomic power should be judicious.
- X-ray, ultrasound devices and scan devices should be used carefully.
- Computer monitor should be used carefully.
- Environmental awareness would be very effective.

NOISE POLLUTION

Unwanted sound intensity released in the atmosphere causes noise pollution or sound pollution. It is man made pollution. It is the gift of technology and modern civilization. Sound pollution is an important problem in the cities and towns. Major metropolitan cities like, Delhi, Mumbai, Kolkata and Chennai are affected by the noise pollution. Intensity of sound is measured in decibel (dB), but the frequency of sound is measured in **hertz**. Instensity of sound is mainly responsible for noise pollution.

In India, noise pollution causes hazardous effects on the human-beings. According to CPCB (Central Pollution Control Board), following three zones occur in India:

1. **Silent zone**- Average intensity of sound is 30-40 dB.
2. **Residential zone** - Average intensity of sound is 50-70 dB.
3. **Industrial zone** - Average intensity of sound is 70-90 dB.

Sources of Noise Pollution

More than 60-70 dB sound intensity cause noise pollution. Normal conversation of human-beings is about 20-30 dB or 30-40 dB. In cities and towns scooters, motorcycles, car, truck, tractors, buses, generators, loudspeaker, industrial machines, trains and aeroplanes cause noise pollution. According to CPCB, the sound intensity causing noise pollution for following substances, are as follows-

1) Two and three wheelers	-	70-80 dB
2) Car/Jeep	-	80-85 dB
3) Trucks and Buses	-	80-90 dB
4) Tractors	-	90-100 dB
5) Loudspeakers	-	90-110 dB
6) Generator sets	-	90-110 dB
7) Machines of industries	-	90-110 dB
8) Turbine machine	-	100-110 dB
9) Train	-	110-120 dB
10) Aeroplane	-	120-130 dB
11) Jet planes	-	above 130 dB.

Sonic Boom: A very high sound intensity released by Jet plane in tropopause region of troposphere is called Sonic Boom. It forms the cone like structure in atmosphere. Velocity of sonic boom is very high. It is measured in **Mach** (1 mach = 100 velocity). Sonic boom may cause rattling or breaking of glasses of doors and windows and may damage multistoried buildings.

Effects of Noise Pollution on Living beings: Noise pollution causes little impact on plants, its major effect is on human-beings. It causes deafness, anxiety, disturbance in convenience, increase in blood pressure, heart beat, loss of memory, headache and delivery of abnormal babies in human-beings.

MASTER STROKES

Q. What do you mean by Montreal protocol and Kyoto protocol?

Montreal protocol: In 1987, twenty seven industrialised countries signed the **Montreal protocol** for reduction in production and release of ozone layer depleting CFCs (chloro fluoro carbons) into the atmosphere. It was followed by increasingly stringent amendments in London in 1990 and in Copenhagen in 1992. To date, more than 175 countries have signed the Montreal protocol. First Earth summit held at Rio de Janeiro (Brazil) in 1992 for reducing green house gas emission.

Kyoto protocol: It was held in Kyoto (Japan) in December 1997 has specified the commitments of different countries to mitigate climate change. As per this protocol countries are committed to reduce their overall green-house gas emissions to a level at least 5 per cent below the 1990 level by the commitment period 2008-2012. Kyoto protocol came into force from 16 Feb. 2005 after the signature of Russia. 157 countries accepted this protocol but America and Australia did not sign till now.

ENTRANCE CORNER

1. The most harmful type of environmental pollutants are :
   (a) human organic wastes
   (b) wastes from fecal
   (c) non-biodegradable chemicals
   (d) natural nutrients present in excess

2. Acid precipitate kills fish by causing release of :
   (a) Al ions (b) CO
   (c) anticoagulants (d) mercury

3. Photochemical smog always contains :
   (a) Al ions (b) CH_4
   (c) O_3 (d) Phosphorus

4. Taj Mahal is threatened to the effect of :
   (a) Chlorine (b) SO_2
   (c) Oxygen (d) Hydrogen

5. The chemical that contribute to the destruction of ozone layer is:
   (a) Hg (b) CO
   (c) SO_2 (d) Chlorofluoro carbons

6. CO is a pollutant, because it :
   (a) inactivates nerves (b) combines with oxygen
   (c) combines with Hb (d) inhibits growth

7. Fly ash is environmental pollutant, produced by :
   (a) thermal power plant (b) oil refinery
   (c) fertilization plant (d) strip mining
8. Which type of pollution is caused by two wheelers :
   (a) air pollution (b) soil pollution
   (c) air and soil (d) water pollution
9. Most hazardous metal pollutant of automobile exhaust is :
   (a) Cd (b) Hg
   (c) copper (d) lead
10. Green-house gas is enhanced in the environment by the gas :
    (a) CO_2 (b) CO
    (c) fluorocarbons (d) NO_2
11. Acid rain occurs in areas :
    (a) where large plantation occurs
    (b) where small plantation occurs
    (c) there are big industries and the atmosphere is polluted with SO_2
    (d) there are no industries
12. Air pollution effects are usually found on :
    (a) leaves (b) flowers
    (c) stems (d) roots
13. Ozone day is observed on :
    (a) Jan. 30 (b) April 29
    (c) Sept. 16 (d) Dec. 25
14. The basic component of smog is :
    (a) O_3 (b) PAN
    (c) Both (a) and (b) (d) none of these
15. Acid rains occurs where atmosphere is heavily polluted with :
    (a) CO, CO_2 (b) smoke particle
    (c) O_3 (d) SO_2
16. CO is a major pollutant of :
    (a) H_2 (b) air
    (c) noise (d) soil
17. Increasing skin cancer and high mutation rates are generally the consequence of :
    (a) CO_2 (b) O_3 depletion
    (c) O_3 blame (d) acid rains
18. Acid rain is due to :
    (a) SO_2 pollution (b) CO pollution
    (c) pesticide pollution (d) dust particles

19. Catalytic converter is used in automobiles :
    (a) to remove lead (b) for converting NO_x
    (c) to remove H_2O (d) all of these
20. Which city has most arsenic pollution ?
    (a) Delhi (b) Kolkata
    (c) Ahmedabad (d) Maharashtra
21. The toxic effect of CO is due to its greater affinity for haemoglobin as compared to oxygen by :
    (a) 20 times (b) 1000 times
    (c) 2 times (d) 200 times
22. Photochemical smog is caused by :
    (a) CH_4 (b) SO_2
    (c) PAN (d) ethylene
23. Ozone layer is present in :
    (a) stratosphere (b) troposphere
    (c) ionosphere (d) mesosphere
24. The trace gas which is produced in rice paddy and is associated with global warming is :
    (a) CH_4 (b) chlorine
    (c) CO_2 (d) H_2S
25. Major aerosol pollutant present in the jet plane emission is :
    (a) SO_2 (b) Fluorocarbon
    (c) CCl_4 (d) CO
26. Main air pollutant is :
    (a) CO (b) CO_2
    (c) N_2 (d) sulphur
27. Which of the following is normally not an atmosphere classified pollutant ?
    (a) CO (b) CO_2
    (c) NO_2 (d) sulphur
28. The gas which was released in Bhopal gas tragedy was :
    (a) ethylene (b) methyl isocyanate
    (c) mustard gas (d) potassium isocyanate
29. Environment planning organization is :
    (a) CSIR (b) NEERI
    (c) ICAR (d) CEPHERI
30. NEERI is situated at :
    (a) Bhopal (b) New Delhi
    (c) Nagpur (d) Baroda

31. Natural sink of stratospheric ozone layer is :
    (a) SO_3
    (b) SO_2
    (c) ferons
    (d) sulphur flux of oceans

32. Mosses are indicator of :
    (a) air pollution
    (b) H_2O pollution
    (c) radiation pollution
    (d) soil pollution

33. The main indicators of SO_2 pollution are :
    (a) lichens
    (b) ferns
    (c) algae
    (d) bryophytes

34. Indicators of atmospheric pollution are :
    (a) liverworts
    (b) ferns
    (c) epiphytic lichens
    (d) horn worts

35. Pollutants of jet planes are :
    (a) fluorocarbons
    (b) CO
    (c) CO_2
    (d) NH_3

36. Pollutant causing leaf curling :
    (a) SO_2
    (b) CO
    (c) H_2S
    (d) O_3

37. Smog is :
    (a) smoke
    (b) moistened air gases
    (c) other name for dust storm
    (d) smoke

38. Main factor of water pollution is :
    (a) industrial waste
    (b) detergents
    (c) house hold water
    (d) pesticides

39. Decomposition of domestic wastes under natural process is included in :
    (a) industrial pollution
    (b) thermal pollution
    (c) non-biodegradable pollution
    (d) biodegradable pollution

40. Sewage water can be purified for recycling with the action of :
    (a) micro-organisms
    (b) penicillin
    (c) fishes
    (d) aquatic plants

41. If H_2O pollution continue at its present rate, it will eventually :
    (a) stop the water cycle
    (b) make O_2 molecule unavailable to water plants
    (c) make nitrate molecules unavailable to water plants
    (d) prevent precipitate

42. Often in water bodies subjected to sewage pollution, fishes die due to:
    (a) pathogens related to sewage
    (b) reduction of dissolved O_2 caused by microbial activity
    (c) clogging of their gills by solid substance
    (d) foul smell

43. Biodegradable pollutant is :
    (a) plastic (b) sewage
    (c) asbestos (d) all of these

44. If large quantities of domestic sewage are continuously emptied into a small stream, it leads to :
    (a) algal bloom
    (b) increase in temperature
    (c) eutrophication
    (d) depletion of O_2 content in stream water

45. The excess discharge of fertilizers into water bodies result in :
    (a) silt (b) death of hydrophytes
    (c) eutrophication (d) growth of fish

46. Shallow lake with rich organic products are called:
    (a) heterotrophic (b) eutrophic
    (c) mesotrophic (d) oligotrophic

47. Blue baby syndrome pollution of :
    (a) CO (b) fluorine
    (c) nitrate (d) mercury

48. Eutrophication caused reduction in :
    (a) dissolved O_2 (b) dissolved hydrogen
    (c) dissolved salt (d) all of these

49. Complete eutrophication of a lake renders it :
    (a) nutrient rich and productive (b) nutrient poor and unproductive
    (c) nutrient rich but unproductive (d) nutrient poor but productive

50. Population explosion of small organisms due to nutrient enrichment of a lake causes a visual aspect, called :
    (a) primary production (b) bloom
    (c) climax state (d) dominance

51. In polluted lake, index of pollution is :
    (a) frog (b) artemia
    (c) daphnia (d) none of these

52. BOD stands for :
(a) Biotic Oxygen Demand (b) Biochemical organic Decay
(c) Biological Organism Death (d) Biochemical Oxygen Demand

53. If BOD of a river water is found very high. This means water :
(a) clean (b) highly polluted
(c) having algae (d) contain dissolved minerals

54. By what method the quality of organic pollutant in water can be determined ?
(a) pH (b) Change of colur
(c) Measuring BOD (d) Transparency measurement

55. Huge amount of sewage is dumped into a river, the BOD will :
(a) increases (b) decreases
(c) unchanged (d) slightly decrease

56. BOD is connected with :
(a) microbes (b) microbes and organic matter
(c) organic matter (d) none of the above

57. Non-bio degradable pollutant is :
(a) plastic things (b) mud
(c) leaves (d) fruits

58. Use of DDT is undesirable because, it is :
(a) harmful (b) degradable
(c) accumulated in food chain (d) causes mutation

59. Increasing concentration of DDT in organisms of a food chain in higer trophic levels is known as :
(a) biological chain (b) biological value
(c) biological amplification (d) biotic potential

60. Minimata disease is produced due to water pollution of :
(a) silver nitrate (b) $HgCl_2$
(c) methyl isocyanate (d) none of these

61. The major drawback of DDT as a pesticide is that :
(a) it is not degraded in nature
(b) cost of production is high
(c) significantly less effective than other pesticides
(d) organism at once develop resistance to it

62. Itai-itai disease is caused by poison of :
(a) Hg (b) lead
(c) chromium (d) cadmium

63. Mimimata disease affects the :
    (a) circulatory system
    (b) skeletal system
    (c) nervous system
    (d) respiratory system
64. Thermal pollution of water bodies is due to :
    (a) discharge of waste from mining
    (b) discharge of agricultural run off
    (c) discharge of chemicals from industries
    (d) discharge of heat from power plants
65. BOD of drinking water is :
    (a) more than dirty water
    (b) less than dirty water
    (c) equal to dirty water
    (d) all of these
66. Water pollution causes :
    (a) increased photosynthesis
    (b) increased oxygenation
    (c) decreased turbidity
    (d) increased deoxygenating and turbidity
67. Type of pollution causes outbreak of javndice :
    (a) air
    (b) thermal
    (c) water
    (d) land
68. The first biocide used on a large scale was :
    (a) DDT
    (b) 2, 4-D
    (c) mercury
    (d) melathion
69. Noise pollution is measured in :
    (a) fathoms
    (b) decibels
    (c) tonne
    (d) kilogram
70. Most important cause of soil pollution is :
    (a) iron junks
    (b) glass junks
    (c) detergents
    (d) plastics
71. The most outstanding danger at present for survival of living beings on earth is :
    (a) glaciation
    (b) deforestation
    (c) radiation hazards
    (d) desertification
72. Chernobyl disaster was caused by a :
    (a) nuclear weapon accident
    (b) nuclear waster leak
    (c) nuclear reactor accident
    (d) nuclear test
73. Part of human body first to be affected by nuclear radiation :
    (a) brain
    (b) lungs
    (c) liver
    (d) bone marrow

74. The term 'Nuclear war' is associated with :
(a) soil pollution (b) water pollution
(c) radioactive pollution (d) all of the above

75. Most dangerous radioactive pollutant is :
(a) phosphorus-32 (b) strontium-90
(c) calcium-40 (d) sulphur-32

76. Chemical causes bone cancer and degeneration of tissue :
(a) I-131 (b) Ca-40
(c) I-127 (d) Str-90

77. Radioactive isotopes is used in the detection of thyroid cancer :
(a) I-131 (b) U-238
(c) C-14 (d) P-32

78. Algal bloom in a lake :
(a) lead to O_2 depletion (b) increase CO_2 level
(c) kills fishes and other organisms (d) all of the above

79. A pollutant is any substance, chemical or other factor that changes :
(a) natural geochemical cycle
(b) the natural balance of our environment
(c) the natural flora of our environment
(d) the natural wildlife of our region

80. Secondary pollutant is :
(a) SO_2 (b) CO
(c) O_3 (d) PAN

81. The chief source of pollutant H_2S is :
(a) decaying vegetation and animal matter
(b) automobiles
(c) oil refineries
(d) thermal power plants

82. The antiknock agent added to unleaded petrol is :
(a) tetraethyl lead (b) tetramethyl lead
(c) dibromo ethane (d) methyl tertiary butyl ether (MTBE)

83. Black foot/Flat foot disease in human is caused by :
(a) arsenic (b) SPM
(c) fluorine (d) Cd

84. Contamination of radioactive materials are dangerous because it causes :
(a) biological magnification (b) gene mutation
(c) photochemical smog (d) ozone destruction

85. The threshold of pain to human ears is :
    (a) 50-60 dB (b) 30-40 dB
    (c) 120-140 dB (d) 160-180 dB

86. The range of normal human hearing lies between :
    (a) 500 Hz to 10,000 Hz (b) 2,000 Hz to 2,00,000 Hz
    (c) 5000 Hz to 6000 Hz (d) 50 Hz to 15,000 Hz

87. Kyoto conference is connected with :
    (a) limiting production of CO_2
    (b) developing alternatives to ODS (Ozone Depleting Substance)
    (c) both (a) and (b)
    (d) reduction in use of energy

88. Radioactive element found in tobacco smoke :
    (a) I-131 (b) polonium-210
    (c) caesium-137 (d) strontium-90

89. Burning of plastics releases :
    (a) CFC (b) SO_2
    (c) benzopyrene (d) PCB

90. Component of living cell affected by pollutant SO_2 is :
    (a) nucleus (b) all cell membrane system
    (c) cell wall (d) none f these

91. Osteoporosis will be caused by pollutant :
    (a) chlorine (b) bromine
    (c) fluorine (d) none of these

92. The pollutant responsible for the luxuriant growth of algae which forms water bloom is :
    (a) phosphates (b) DDT
    (c) H_2S (d) sulphates

93. Leukemia is caused by :
    (a) Ca-40 (b) Sr-90
    (c) caesium (d) iodine

94. 3-4 benzopyrene causes :
    (a) leukemia (b) cytosilicosis
    (c) lung cancer (d) tuberculosis

95. Ozone hole over Antarctica was first detected by :
    (a) Molina and Rowland (b) Farman et. al.,
    (c) Augus (d) Molina and Molina

96. Lichens like Usnea is an indicator of :
    (a) SO_2 (b) CO
    (c) CO_2 (d) NO_2
97. Montreal protocol was signed in (Reduction in CFC) :
    (a) 1978 (b) 1987
    (c) 1991 (d) 1993
98. SPM causes :
    (a) respiratory disease (b) skin disease
    (c) bone deformities (d) all of these
99. Ozone hole is widest in :
    (a) equator (b) north pole
    (c) north temperate area (d) antarctica
100. In metro cities, the automobile causes air pollution :
    (a) 80% (b) 60%
    (c) 50% (d) 40%
101. Maximum air pollution is at :
    (a) Delhi (b) Bhopal
    (c) Kolkata (d) Bangalore
102. Pollutants directly emanating from human activities are :
    (a) qualitative pollutants (b) air pollutants
    (c) qualitative pollutants (d) water pollutants
103. Atmospheric pollutants formed by human activities constitute of the total :
    (a) 95% (b) 75%
    (c) 25% (d) 0.05%
104. Most atmospheric pollutant do not rise above :
    (a) 600 m (b) 1000 m
    (c) 2000 m (d) 200 m
105. Level of atmospheric CO_2 has increased from :
    (a) 280 ppm to 360 ppm (b) 250 ppm to 300 ppm
    (c) 300 ppm to 340 ppm (d) 330 ppm to 350 ppm
106. BOD is related to :
    (a) detergents (b) inorganic pollutants
    (c) organic pollutants (d) putrescibility
107. Climate of world is threatened by :
    (a) increasing concentration of atmospheric oxygen
    (b) decreasing amount of atmospheric oxygen
    (c) increasing amount of atmospheric CO_2
    (d) decreasing amount of atmospheric concentration

108. Green-house gases are :
   (a) absorbes of large-wave radiations from earth
   (b) transparent to both large and short
   (c) absorb of solar radiations for warming the atmosphere of earth
   (d) transparent to emissions from earth for passage into outerspace

109. Photochemical smog was first observed in :
   (a) Los Angeles (b) Tokyo
   (c) New York (d) Sydney

110. Classical smog was first observed in :
   (a) Tokyo (b) New York
   (c) Sydney (d) London

111. Pulmonary oedema is caused by :
   (a) SO_2 (b) carbon oxides
   (c) nitrogen oxides (d) hydrocarbons

112. Oil slick causes mass scale death of fish due to :
   (a) clogging of gills (b) non-availability of O_2 in water
   (c) non-availability of food (d) disruption of food chain

ANSWER SHEET

1. (c)	**2.** (a)	**3.** (c)	**4.** (b)	**5.** (d)	**6.** (c)	**7.** (a)	**8.** (a)	**9.** (d)	**10.** (a)
11. (c)	**12.** (a)	**13.** (c)	**14.** (c)	**15.** (d)	**16.** (b)	**17.** (b)	**18.** (a)	**19.** (b)	**20.** (c)
21. (d)	**22.** (c)	**23.** (a)	**24.** (a)	**25.** (b)	**26.** (a)	**27.** (b)	**28.** (b)	**29.** (b)	**30.** (c)
31. (c)	**32.** (a)	**33.** (a)	**34.** (c)	**35.** (a)	**36.** (a)	**37.** (d)	**38.** (a)	**39.** (d)	**40.** (a)
41. (b)	**42.** (b)	**43.** (b)	**44.** (d)	**45.** (c)	**46.** (b)	**47.** (c)	**48.** (a)	**49.** (c)	**50.** (b)
51. (c)	**52.** (d)	**53.** (b)	**54.** (c)	**55.** (a)	**56.** (b)	**57.** (a)	**58.** (c)	**59.** (c)	**60.** (b)
61. (a)	**62.** (d)	**63.** (c)	**64.** (d)	**65.** (b)	**66.** (d)	**67.** (c)	**68.** (a)	**69.** (b)	**70.** (d)
71. (c)	**72.** (c)	**73.** (d)	**74.** (c)	**75.** (b)	**76.** (d)	**77.** (a)	**78.** (d)	**79.** (b)	**80.** (b)
81. (a)	**82.** (a)	**83.** (a)	**84.** (b)	**85.** (c)	**86.** (d)	**87.** (a)	**88.** (b)	**89.** (d)	**90.** (b)
91. (c)	**92.** (a)	**93.** (b)	**94.** (c)	**95.** (b)	**96.** (a)	**97.** (b)	**98.** (a)	**99.** (d)	**100.**(a)
101.(c)	**102.**(c)	**103.**(d)	**104.**(a)	**105.**(a)	**106.**(c)	**107.**(c)	**108.**(a)	**109.**(a)	**110.**(d)
111.(c)	**112.**(a)								

INDEX

A

- Abrasion, 18
- Abundance, 136
- Abyssal region, 95
- Adaptation, 201
- Afforestation and reforestation, 229
- Agroforestry, 182
- Allelochemicals, 41
- Allelopathy, 41
- Alluvial soil, 33
- Alphadiversity, 265
- Amensalism, 41
- Anemophily, 18
- Antibiosis, 41
- Aphotic zone, 9
- Apostrophe, Epistrophe and Parastrophe arrangement of chloroplast, 10
- Approach effect, 38
- Aspectation, 141
- Autecology, 6

B

- β-diversity, 265
- Bathyscaphe, 96
- Benthos, 94
- Beufort Number, 45
- Bhopal gas tragedy, 305
- Biocoenosis, 65
- Biodiversity, 245
- Biogas, 185
- Biogenetic substance, 5
- Bio-geo-chemical cycle, 5, 83
- Biological magnification, 315
- Biological spectrum, 3
- Biomass, 187
- Biomass, 67
- Biome, 3
- Biosphere reserves, 258
- Biosystem, 3
- Biota, 3
- Black foot disease, 311
- BOD, 314
- Braun Blanquet, 143
- Brood parasite, 42
- Bruckner cycle, 42
- Bulk density, 24

C

- Calcifiation, 34
- Cannibalism, 42
- Canopy interseption, 17
- Canyon effect, 38
- Chapman mechanism, 44
- Chernobyl Mishap, 315
- Chesard, 289
- Chipko movement, 255
- Chlorellin, 41
- COD, 314
- Colluvial soil, 32
- Commensalism, 41
- Community, 4
- Competition, 39
- Components of soil, 24
- Compression wood, 19
- Continental shelf, 95
- Continental slope, 95
- Contour cropping, 228
- Cover, 137
- Cybernetics, 65

D

- Delta diversity, 265
- Density, 134
- Disphotic zone, 9
- Dunes, 33
- Dust storm, 17

E

- Ecads, 4
- Ecesis, 164
- Echard, 289
- Ecological boomerang, 232
- Ecological efficiency, 78
- Ecological Niche, 4
- Ecological Pyramid, 73
- Ecosphere, 291
- Ecosphere, 65
- Ecosystem, 65
- Ecotone, 7, 132, 133
- Ecotypes, 4
- Elfine scrub, 15

- El-Nino effect, 38
- Endangered, 251
- Environmental education, 252
- Environmental factors, 7
- Environmental lapse rate, 44
- Environmental legislation, 255
- Eolian soil, 33
- Epsilon diversity, 265
- Erosion of soil, 223
- Etiolated plant, 9
- Euphotic zone, 8, 9
- Eutrophication, 311
- Exfoliation, 29
- Exosphere, 284, 285
- Ex-situ and In-situ conservation, 256, 257
- Extinct, 251

F

- Fallowing, 228
- Fidelity, 142
- Field capacity, 27
- Flag tree, 18
- Fluorosis, 311
- Food chain, 69
- Food web, 71
- Forest rain, 17
- Formation of soil, 28
- Frequency, 134
- Fugitive species, 42

G

- Gamma diversity, 265
- GAP, 312
- Gasohol, 187
- General process of succession, 164
- Genetic pool, 4
- Gleization, 34
- Global warming, 297, 298
- Grazing and Browsing, 42
- Green data book, 250
- Green house effect, 297, 298
- Guilds, 69
- Gully erosion, 223

H

- Habitat diversity, 266
- Hadal region, 95
- Halophytes, 212
- Hanging water table, 27
- Hans Reiter, 3
- Hekisotherms, 15
- Heliophytes, 10, 11
- Holard, 289
- Homeostasis, 65
- Hot spots, 264
- Humus, 25
- Hydrogen, 186
- Hydrophytes, 201, 202
- Hydrosere, 167
- Hydroxylamine, 41

I

- Incomplete ecosystem, 66
- Invasion, 164
- Invasive or Alien species, 245
- Ionosphere, 284, 285, 305
- Itai-Itai, 311
- IVI, 143

J

- Juglone, 41

K

- Keystone species, 42

L

- La-Nina and El-Nino years, 43
- Laterization, 33
- Life form, 138
- Lindemann's 10% law, 77
- Litter, 25
- Lodging effect, 18
- Loess, 33

M

- Mesophytes, 201
- Mesosphere, 284, 285, 305
- Metamorphic rock, 287
- Microcosm, 65
- Milankovitch cycles, 42
- Mineralisation, 67
- Minimta disease, 310
- Moraines, 33
- Mulching, 228
- Mutualism, 40

N

- Narmada Bachao Andolan, 312
- National Parks, 258
- Nektons, 94
- Neustons, 94
- Neutralism, 39
- Noosphere, 291

O

- Oozosphere, 284, 285, 305
- Organismic diversity, 266
- Out of danger, 252
- Ozone depletion, 315, 316

P

- PAN, 313, 314
- Parasitism, 39
- Particle density, 24
- Pedology, 19
- Pedology, 289
- Periodicity, 141
- Periphyton, 94
- pf scale, 27
- Phenology, 141
- Photoblastic plant, 9
- Photochemical smog, 304
- Photomorphogenesis, 9
- Photoperiodism, 11
- Phytosociology, 131, 133
- Planktons, 94
- Plant indicators, 237
- Plant zero, 15
- Podozolisation, 33
- Point diversity, 265
- Pollutants, 292
- Population, 4
- Predation, 41
- Primary or igneous rock, 287
- Productivity of ecosystem, 79
- Protoco-operation, 40

R

- r and k selection, 171
- Rain shadow, 16
- Rainfed crops, 232
- Rare, 252
- Red data book, 250
- Renewable and Non renewable energy, 179, 183
- Resource diversity, 266
- Rill erosion, 223
- Riparian erosion, 223

S

- Sanctuary, 258
- Sciophytes, 10, 11
- Sedimentary rock, 287
- Sewage treatment, 313
- Sheet erosion, 223
- Shelter belts, 18
- Shifting cultivation, 224
- Shola forest, 17
- Silent valley, 188
- Slip erosion, 223
- Snow line, 15
- Sociability, 141
- Social forestry, 181
- Soil profile, 30
- Soil separates, 20
- Soil texture, 20
- Soil types of India, 34
- Solar system, 281
- Sonic Boom, 317
- Species diversity, 131
- Squall, 17
- Standing crop, 67
- Stenothermic sp. (Oligothermic sp., Polythermic sp.), 12, 13
- Stone cancer, 301
- Stratification, 131, 141
- Stratified rock, 287
- Stratosphere, 284, 285, 305
- Strip cropping, 229
- Structure of soil, 22
- Synecology, 6

T

- Tansley, 65
- Terrace-delta, 33
- Terracing, 230
- Thermal stratification, 13
- Thermoperiodism, 13
- Threatened, 252
- Tree line, 15
- Trophic levels, 68
- Troposphere, 284, 305
- Types of ecosystem, 90
- Types of succession, 163

U

- Universe, 281
- Urythermic sp., 12

V

- Van Allen radiation belt, 44
- Van mahotsava, 181
- Vant Hoffs Law, 11, 14, 15
- Vernalization, 13
- Vitality or Vigour, 142
- Vulnerable, 251

W

- Water holding capacity, 27
- Wind breaks, 18

X

- Xerophytes, 201, 204, 208
- Xerosere, 169

Z

- Zone of eluviation and zone of illuviation, 31